Human Rights in Language and STEM Education

Human Rights in Language and STEM Education

Science, Technology, Engineering and Mathematics

Edited by

Zehlia Babaci-Wilhite

University of California, Berkeley, USA

University of Oslo, Norway

SENSE PUBLISHERS

ROTTERDAM/BOSTON/TAIPEI

A C.I.P. record for this book is available from the Library of Congress.

ISBN: 978-94-6300-403-9 (paperback)
ISBN: 978-94-6300-404-6 (hardback)
ISBN: 978-94-6300-405-3 (e-book)

Published by: Sense Publishers,
P.O. Box 21858,
3001 AW Rotterdam,
The Netherlands
https://www.sensepublishers.com/

All chapters in this book have undergone peer review.

Printed on acid-free paper

This book is dedicated to everyone deprived of learning STEM subjects in ways that contribute positively to their personal and professional development as well as to their capacity to make a contribution to the development of their local, national and global communities.

TABLE OF CONTENTS

GREGORY JOHNSON

FOREWORD: PAULO FREIRE PEDAGOGIC LIBERATION THROUGH THE STEM

As/is
Becomes
discursive commentary/
Elemental
Transformation/
The
Word
Destination
Reflecting action/

A vessel
Of
Meaning/

Not
Separated
From reality
Or
Isolated
Independently
Not
(detached
From
The world/

Simultaneous consciousness
Is
A
Tapestry
Of
Information/

Knowledge
Exposed to the

True
Scientific
Source
Stream/
Posing questions/
(culturally
Responsive/
Problem
Solving/

Peeling back
Reflexive
Provocation/

Critical
STEM solutions/

A
Thought
For the educational
I
Equilibrium
That
Finds
It's
Own

Brave
Liberation

ACKNOWLEDGMENTS

My sincere thanks to the great reviewers and edits made by Professor Harold Wilhite from the University of Oslo (Norway), Professor Suzanne Majhanovich from the University of Western Ontario (Canada), Professor Tatyana Tsyrlina from the Seattle Pacific University, Seattle, WA (USA), Professor G. Ugo Nwokeji from the University of California Berkeley (USA), Dr. Shelley Terzian from the Dominican University in Chicago (USA), Dr. Kimmo Kosonen from Payap University (Thailand), Dr. Daniel Odinaka Abonyi from the Imo State University, Owerri (Nigeria), Pui Ki Patricia Kwok from the University of Hong Kong (Hong Kong) and PhD candidate Emily Linares from the University of California, Berkeley (USA).

A special thanks to Michel Lokhorst and Jolanda Karada for their great support as always.

Last but not least, to my grand-mother Fatima Babaci, to my aunt Hamama Babaci, to my sister Louisa Babaci, and to my children, Paul Kacem Babaci Wilhite, Alexandre Yanis Babaci Wilhite and Kahena Judith Babaci Wilhite, un grand merci pour votre généreux support tout au long de ma vie.

ZEHLIA BABACI-WILHITE

INTRODUCTION

The inspiration for this book emerged from my affiliation at the *Graduate School of Education and the Lawrence Hall of Science at the University of California, Berkeley, USA* where I was exposed to the importance of language in the teaching and learning of Science, Technology, Engineering and Mathematics (STEM) subjects. Professor David Pearson and Jacquie Barber, founders of the Seeds of Science and Roots of Reading curricula introduced me to insightful ways of learning and teaching science using a theory of inquiry that emphasizes the use of local context and thus fits well with learning science in the language that teachers and children master best. Professor Pearson and several other scholars from the Lawrence Hall of Science have made important contributions to this volume.

My interest in language and learning in Science and Mathematics began with my research in Zanzibar and Nigeria, where I studied language policies in African curriculum. A central finding from these studies was that the use of a local language was crucial to the learning of science. This was the subject of my previous book entitled "Local Languages as a Human Right in Education" (2015, Sense Publishers). I am currently continuing this line of research in Nigeria where I am affiliated to the *Imo State University (IMSU), Owerri*, as well as the *International Society of Comparative Education Science & Technology, Yenagoa, Bayelsa State, Nigeria.*

While teaching students from the Masters and Credential in Science and Mathematics Education (MACSME) program at the University of California, Berkeley, my understanding of the importance of using Cultural as and Artistic concepts to facilitate teaching and learning Mathematics grew, and my work benefited from using concepts from Ethnomathematics in the teaching of Science. Inspired by the extensive work of Professor Ubiratan D'Ambrosio who contributed an important chapter to this volume, my interest in Science and Mathematics in Africa expanded to include the use of digital Technologies for teaching Science and Mathematics.

I was fortunate to have the opportunity to audit Professor Jabari Mahiri's class on Digital Devices in Urban Education, which expanded my perspectives on the use of digital technologies in the classroom. Professor Mahiri has co-authored an interesting chapter in this volume with one of his PhD candidates, Jeremiah Sims.

Based on our collaborative work with another contributor to this volume, Professor Inga Bostad, the Director of the *Norwegian Center for Human Rights at the University of Oslo* where I am currently affiliated as a Visiting Scholar, Mahiri and I have developed a project focusing on Diversity, Technology and Human Rights in collaboration with the University of Tromsø. Dr. Lanette Jimerson who co-authored a chapter in this volume joined the project as well. This cross fertilization of perspectives on Science, Mathematics and Technology between the USA and Norway has linked STEM education to Language and Human Rights as well as to my work with the teaching of STEM subjects in Africa. The collection of chapters in this volume addresses various facets of the effort to improve the learning of Science and Mathematics.

This volume draws on a unique and diverse set of authors from the USA, Norway and Africa. There are excellent contributions by several scholars from Nigeria, Professors Macleans A. Geo-JaJa and Steve Azaiki as well as PhD candidate Basil Okonkwo, as well as from Malawi, Professor Mchombo and from Tanzania, Professor Ladislaus M. Semali. The contributions of the African authors demonstrates the linkages between local languages, human rights and STEM subjects in Education arguing that Africa has so far largely been excluded from the transformation that is going on in STEM pedagogy in the USA, where considerable emphasis has been placed on language and the development of English Language Learner (ELL).

My work in Africa started with another contributor, Professor Birgit Brock-Utne, the Norwegian leader of the Language of Instruction (LoI) in Tanzania and South Africa (LOITASA) project based at the University of Oslo, the University of Dar es Salaam (Tanzania) and the University of Cape Town (South Africa). The LOITASA project transformed to a new project called Transformation in Language and Education (TRANSLED) in collaboration with the University of Oslo, the State University of Zanzibar and the University of Dar es Salaam. As I have argued in previous publications, in its recent curriculum reform in Zanzibar, English was introduced as the primary LoI for Science and Mathematics subjects in higher primary grades without giving it the English Language Learner importance that it deserves. This change has thrown up new pedagogical challenges to the teaching of Science and Mathematics. My research about this curriculum change has revealed weaknesses in teacher preparation, teaching materials and pedagogy, and student comprehension in the transformation from learning in their own language to learning in English (see Babaci-Wilhite, 2013, 2014 and 2015).

Several of the contributions to this volume focus on the curriculum changes in language and STEM subjects within a global versus local curriculum. Other chapters explore the challenges of teaching and learning STEM subjects in local contexts in a range of countries around the world such as Suriname, addressed in the chapter of Emmanuelle Le Pichon and Ellen-Rose Kambel; the Philippines by Brad D. Washington; in Pakistan by Abbas Rashid, Irfan Muzaffar, Fatima Dar and Salaeya Butt; and by Lanette Jimerson and Page Hersey in the U.S.A.

I ended the previous volume I edited with a poem reminding us of Paulo Freire's legacy in Education Rights and his contribution to critical pedagogy. I decided to start this volume with Paulo Freire's legacy in Education Rights linked to the STEM introduced by a Poet-Artist, Gregory Johnson who articulated in his poem the importance of Art to creativity in the STEM subjects for a critical and innovative Education. Johnson's work in the Ethnomathematics and the Geometry of Art captured through his second poem concludes this volume.

ORGANIZATION OF THE BOOK

This volume comprises 15 chapters in addition to this introduction, the foreword and afterword through the 2 poems. The chapters are divided into four parts:

I – Human Rights in Language and Science Literacy
II – Equity and Critical Pedagogy in Technology and Human Rights Education
III – Language of Instruction in Science and Technology
IV – Human Rights in Mathematics and STEM Education

The contributions to *Part I* – entitled *Human Rights in Language and Science Literacy* address and analyze the knowledge as well as the methodology used to inquire science learning, the art of reading science and the practice of scientific argumentation. Reading should take place to support children to learn all subjects in order to acquire a language that will facilitate their learning process. The authors claim that its importance is a human right in education.

In the first chapter entitled "The Use of Local Languages for Effective Science Literacy as a Human Right", *Zehlia Babaci-Wilhite* addresses curriculum change in education and science literacy by exploring how to enable Science literacy through an inquiry-based approach in Africa. Further, it argues that learning in schooling should be based in local knowledge and that this ought to be defined as a right in education. The argument is framed with the theory that regards language as powerful tool for learning both in dominant and non-dominant cultures, and that science learning is improved when local knowledge and local languages are used. In other words, the theory of inquiry so important to science is best conducted in a local language. Using a local language in Science teaching will improve teaching and learning and will form a basis for new innovative learning. Implementing a new contextualized model of teaching and learning science will also support and enhance African languages and culture and should take its rightful place as a human right in education.

In their chapter entitled "Reading to Learn Science: A Right That Extends to Every Reader—Expert or Novice", *P. David Pearson* and *Alison Billman* argue that Literacy educators have traditionally assumed that "first you learn to read and then you read to learn." This view has been responsible for a host of practices that have led to a focus on basic skill literacy instruction in primary grades to the exclusion of reading to acquire the knowledge and inquiry skills available in disciplines like Science, literature, or history. In this chapter the authors explain and elaborate a model of integrated Science and literacy instruction that challenges many of these prevailing assumptions about early literacy learning. The key elements in the alternative model are (a) to employ both first hand (hands-on inquiry) and second hand (text-based inquiry) investigations on a topic from the outset of kindergarten, and to employ reading, writing, and language as tools (not goals), whose job is to aid in the acquisition of knowledge and inquiry in science. The authors close with a set of principles that teachers and schools can use to develop programs in which students read to learn while they are learning to read. The goal is to promote the human right to acquire knowledge about how the natural world works from the outset of schooling.

The third chapter entitled "Designing Standards-Driven Curriculum to Support Access to Science for All Students", by *Megan Goss, Helen Min* and *Carrie Strohl,* provides details on how one program created a CCSS- and NGSS-aligned Science curriculum for middle school using an argument-based, disciplinary literacy approach. This work started with the creation of Seeds of Science/Roots of Reading, as a joint project between the Lawrence Hall of Science and the Graduate School of Education at the University of California, Berkeley. The current curricular development followed a shift by researchers and practitioners to infuse more discipline-focused practices in content area classrooms. The examples included in this chapter focus on the practice of scientific argumentation because it is a central practice of Science, providing students authentic opportunities to engage in talk, reading, and writing around science ideas. Students actively collaborate in the construction of science knowledge as they develop the skills, strategies, and habits involved in scientific argumentation. This chapter supplies explicit curricular examples of argumentation from the units, to illustrate how students, over time and with practice, might build argumentation skills while also gaining Science content knowledge.

The contributions of *Part II* – entitled *Equity and Critical Pedagogy in Technology and Human Rights Education* take on the debate on technology in schools and the critical pedagogical approach to teaching and learning STEM subjects. Critical pedagogy will facilitate critical inquiry for a sustainable education drawing in the Universal Declaration of Human Rights (UDHR). The contributions address the broad implications that educational settings can have on STEM subjects through Technology and its connection to ethics and human rights. The contributions give their focus to African American learning challenges.

In the first chapter entitled "Engineering Equity: A Critical Pedagogical Approach to Language and Curriculum Change for African American Males in STEM" *Jabari Mahiri* and *Jeremiah Sims* demonstrate that pedagogical approaches to STEM education often fail particular student populations, and this is especially true of African American males. This study investigated a critical pedagogical approach to teaching and learning STEM subjects in an after school program that systematically connected the cultural backgrounds, experiences, and interests of middle school African American males to the development of identities with and competencies in STEM. A range of qualitative data was collected on the students' participation in the program for two to three Saturdays per month during a full academic year. Findings were that this approach significantly increased these students' competencies in STEM as well as their abilities to identify as emerging STEM practitioners in conjunction with greater understandings of how to use STEM in socially just and responsible ways. This work has implications for broadening access to STEM in school as well as after-school educational settings.

In their chapter entitled "Curriculum and Social Change in Education for a Sustainable Future? Ecophilosophy, Critical Inquiry and Moral Dilemmas", *Inga Bostad* and *Aled Dilwyn Fisher* contend that Education for Sustainable Development (ESD) is increasingly seen as depoliticised, leaving normative premises that promote neoliberalism and leave instrumentalist labour market demands unexamined. Likewise, environmental education's assumptions that education in, for and about 'nature' engender environmental consciousness require critical scrutiny regarding kinds of education in and about nature that are necessary, the role of teachers and, fundamentally, ideas about 'nature' promoted, particularly whether these encourage a romantic 'society'-'nature' dualism that ignores a socio-ecological approach based on the interconnectedness of humans, knowledge and physical systems. The authors explore how 'pluralistic' education can become relativistic or privilege normatively-loaded conceptions of 'pluralism,' while illustrating ESD's frequent instrumentalism. As an alternative, the authors explore socio-ecological pedagogy encouraging: critical reflection over values and ethics, including normative problems in scientific dissemination; a reconnection with the physical through ecology and the natural sciences; and seeing ecological issues as distinct from but, simultaneously, inextricably linked with socioeconomic relations and politics.

In their chapter entitled "Localizing Human Rights Education through Technology: Two Literacy Based Examples", *Lanette Jimerson* and *Page Hersey* argue that from the civil rights movement to 'black lives matter', California has been a geographic space of activism and struggles for social justice. These movements make visible how human rights activism is always local and global. Issues such as police brutality, worker's rights, and immigration impact students' lives and as such human rights education is fundamental to any education that seeks to develop a democratic citizenry. Furthermore, human rights education curricula that connect students to local, regional, and global struggles for justice through the use of technological tools can deepen students' ability to critically read the word and the world. In this chapter drawing on the UDHR, the authors present two critical literacy-based examples of localizing human rights research through technological tools. In the first classroom, ninth grade students investigate the promise and limitations of the pursuit of happiness in relation to UDHR Articles 23 and 25, writing from the philosopher John Locke's, and several canonical texts. Students engage technological tools to gather data, collect interviews, and communicate their lived experience and friends and family members' quest for human rights. In the second classroom, middle school special education students examine police brutality through blogging. Much like the research of Mahiri (2011), blogging provides the ability of youth to communicate and speak against experiences of societal violence both locally and globally. Implications of these two studies highlight the effectiveness of transformative classroom spaces wherein students engage technological tools to make personal and local connections to human rights.

In the contribution of *Part III* – entitled *Language of Instruction in Science and Technology* the authors address changes in Language of Instruction (LoI) in science and technology. A critic of the use of English as a dominant language is discussed within the context of Africa and Asia with the argument that the use of local languages in education is a necessity to facilitate the learning process. The authors conclude that English should be taught as a foreign language and indigenous languages should be used in science and technology. English as Medium of Instruction (MoI) has not been effective in the teaching of science and technology; on the contrary, using several examples, the authors demonstrates that it has had a detrimental effect.

In the first chapter entitled "English as the Language of Science and Technology" *Birgit Brock-Utne* starts with a story from the author's own research experience in Tanzania where she came across a student who hardly understood anything the teacher was saying yet felt that the LoI had to be English since "English is the language of Science and Technology." The author dwells on this unfounded belief in many so-called Anglophone countries in Africa that Mathematics and Science are best taught in English and not in an African language that the language pupils and teachers normally speak and command much better than English. Examples are given from Tanzania, South Africa, Botswana, Lesotho, Zambia and Ghana. The reintroduction of English from the 5th grade in primary school in Mathematics and Science in the Kiswahili speaking island of Zanzibar is discussed at some length. Examples from Africa are contrasted with examples from some Asian countries like Sri Lanka and Malaysia. The attitudes of parents which are built on a misconception are analyzed. The chapter ends on a more optimistic note claiming that misconceptions can be altered.

In the chapter entitled "Language, Scientific Knowledge and the 'Context of Learning' in Africa" *Sam Mchombo* takes on the debate on Nation building in post-independence Africa, which has revolved around the attainment of national unity and identity, as well as the adoption of programs of national development and continued modernization. Critical to the realization of those objectives are education and language. With regard to language some countries adopted the ideology of "one nation, one language" driven by "a deep-seated fear of linguistic diversity" (Kamwendo, 2013, p. 103). This was buttressed by strong faith in the power of a single indigenous language to facilitate the building of national unity and identity. With regard to education the policy reviews essayed to enhance school performance. In some cases, such as in Tanzania under the leadership of the first president, the late Mwalimu Julius K. Nyerere, there was a fundamental review of the content and overall goals of education in order to counter the basic Eurocentric system of colonial education that had been in place. Like other African languages elsewhere, most of which have yet to even get on the agenda of getting considered for use as LoIs, English is steadily replacing Kiswahili (cf. Bwenge, 2012; Neke, 2003). The salient reason is that English is the language of Science and Technology and of globalization. This chapter examines the connection between language and scientific knowledge with a focus on Mathematics (math) and Science education in Africa. The main thesis is that the appeal to Mathematics and Science education as grounds for the exclusion of African languages as LoIs in education in Africa is simply not defensible; that such exclusion derives from political and economic dependency of African states on European countries, and serves to solidify the "conceptual-cum-linguistic incarceration" of African education.

In the chapter entitled "Global Intersections of English Language Hegemony and Technological Innovation in the Republic of the Philippines" *Brad D. Washington* addresses the perceived necessity of the English language in promoting technological development via the vehicle of STEM education in the Republic of the Philippines. In a nation with over 200 heritage languages, there is an active debate between the preservation of multilingual cultures and histories with the acquisition of English as the prevailing lingua franca of the Philippines. Informed by Filipino communities, universities, and organizations, the chapter explores how conflict, government, and education intersect in documenting how Technology and the English language have been politically intertwined from the mid-20th century to present. In surveying reports by global non-governmental organizations, the chapter presents a framework to consider the inequitable development of education across regions and populations in the Republic of the Philippines in the 21st century. By documenting the voices of educators, researchers, and communities, this study will attempt to increase dialog on how the support of language rights can be a lens into human rights and regional development, especially as it pertains to educational and technological access.

In the chapter entitled "The Importance of Local Language to the Development of Technology" *Basil Okonkwo* argues that in today's world we notice the movement of consumerism and production. Production is a special culture of 'developed nations,' while consumerism is the exclusive culture of 'developing and underdeveloped nations'. It is not surprising that this situation has arisen in our contemporary society. Most countries are dependent on a few nations, which determine the economic strata of the world. Most economies are understood through the examination of the economies on which they depend. This is evidenced by the prices of international commodities such as crude oil, gold and other mineral resources. The cost of these could only be determined by demand of buyers. One begins to wonder why this situation has taken shape in the world. Some people will argue that most developing nations are being exploited and made to solely depend on other nations for aid, support and trade. But this excuse aside, we realize that developing nations are not unique in their cultural and geographical development. Indeed, most developing countries lost their identity to colonialism, but this excuse can no longer hold given that, in most colonized countries, the colonizers have been gone for more than forty years now. The question is 'why the dearth in cultural activity'? In this chapter the author explores the efficacy and indivisibility of language to the development of technology, tests the limitations to the import of language to the development of technology and projects a way forward by making recommendations to future use of local language.

In their chapter entitled "The Issue of English as a Medium of Instruction in Primary Schools in Pakistan: Learning English, Mathematics or Science?" *Abbas Rashid, Irfan Muzaffar, Fatima Dar* and *Salaeya Butt* argue that the Government of Pakistan (GOP) through the National Education Policy (NEP), 2009, made the instruction of English mandatory from class 1 and the MoI for Science and Mathematics from Grade level 4. This chapter deals with the issue of English as the MoI in Pakistan and the readiness of schools and all stakeholders to make this exigent transition. It examines the case of two provinces, Punjab and Sindh and discusses the effectiveness of the MoI policy in the case of Punjab and the readiness of the province of Sindh to undertake the task of teaching English as a subject in public schools from Grade 1 and the MoI in Grade 4. The findings reveal that the teachers' uncertain grasp of English leads to support neither the learning of English, Mathematics nor Science in the classroom—more likely to be achieved through the use of mother tongue, first or proximate language.

The volume ends with *Part IV* – entitled *Human Rights in Mathematics and STEM Education.* The contributions focus on the learning of Mathematics and its learning within the culture. Context matters, therefore through different studies from African and South America, the authors bring another perspective to the learning and teaching of STEM subjects as a human right. The authors argue that inconsistency in programs must be reversed to protect children's rights. Aid programs would be improved through the application of the right based approach to programs involving STEM subjects.

In the chapter entitled "Change in Space, Urban Culture and Ethnomathematics" *Ubiratan D'Ambrosio* points to a strand of Ethnomathematics focusing the State of the World and human occupation and urban culture. The effects of globalization and capitalist greed on the State of the World and the unplanned growth of cities demands a new approach to Mathematics. The author discusses how the Program Ethnomathematics may contribute to the emergence of new approaches to the teaching of Mathematics.

In their chapter entitled "Challenges of Mathematics Education in Multilingual Post-Colonial Context: The Case of Suriname" *Emmanuelle Le Pichon* and *Ellen-Rose Kambel* examine whether the language of assessment influences performance scores of young dual language learners in Mathematics and reading tests in order to determine the role of dual language support at primary school level in Suriname. The objectives of this chapter are to deepen insight into Suriname's linguistic landscape and to emphasize the need for plurilingual and intercultural education. They place the debate in the context of international legal obligations of Suriname which approved the UN Declaration on the Rights of Indigenous Peoples (2007). This declaration includes the right of indigenous peoples to provide education in their own language in accordance with their traditions. Given the current high academic dropout rate in Suriname, in particular in the rural areas, the results informed in this chapter are decidedly relevant. The authors conclude with a discussion of the potential didactic implications of these results for primary education in Suriname.

In his chapter entitled "Why Do Inconsistencies Persist in Children's Rights to 'Good' Education, Heritage Education and STEM Education?" *Ladislaus M. Semali* argues that champions of children's rights emphasize that children everywhere consist of a *vulnerable* age group and therefore, need protections. This chapter speaks for children's rights to "good" education, heritage education and STEM education, and argues that these rights are intertwined with children's health and overall wellbeing. Because young learners are vulnerable and need protections, inconsistencies in Children's Rights to "good" education, directly threaten children's wellbeing and future adult life. Inconsistencies exist when national and global policy statements that advocate for universal education for all children fall short of the political will, budgetary (fiscal) support, and lack system-wide educational planning and evaluation (assessment) of school infrastructure, curriculum in content areas, teaching staff and basic statistics to monitor progress. Consequently, it is imperative to respect and safeguard children's rights against these violations, abuse, or neglect.

In their chapter entitled "Human Rights in Development Aid for STEM Education in Nigerian Languages" *Zehlia Babaci-Wilhite, Macleans A. Geo-JaJa* and *Steve Azaiki* assess the weaknesses in the teaching and learning of STEM subjects, and propose new development approaches to STEM education. By defining access to effective STEM teaching and learning as a right in education, alternative development roadmaps can contribute to both sustainable development and to the satisfaction of the Paris Declaration of Aid Effectiveness. The premise of this chapter is that the right-capability based approach with a focus on human rights will secure best educational practice and insure that education aid systematically ameliorates poverty's multi-dimensionalities. Implicit to this approach is an acknowledgement that African knowledge and African languages are critical to the effectiveness of STEM education. This is demonstrated using Nigeria as a case study. In highlighting roadblocks to human rights of transformational change ownership, the authors recommend that local knowledge and linguistic rights in STEM should be the driving force in development aid and that this should be given the status of a human right in education for sustainable development in Africa.

PART I

HUMAN RIGHTS IN LANGUAGE AND SCIENCE LITERACY

ZEHLIA BABACI-WILHITE

1. THE USE OF LOCAL LANGUAGES FOR EFFECTIVE SCIENCE LITERACY AS A HUMAN RIGHT

INTRODUCTION

The chapter will give attention to several aspects aimed at improving the quality of learning science using local languages in education and local curriculum. The chapter draws on my research on the teaching of science subjects in Zanzibar, Tanzania (The United Republic of Tanzania)[1], South Africa, Nigeria, Norway and the United States, as well as on a review of research on critical issues related to the learning of science that arise from decontextualized teaching and learning in Africa. I will argue that the teaching and learning of science literacy in Africa will be improved through applying a robust theory of inquiry to the learning process.

The teaching of science in Africa suffers from a pedagogy grounded in its colonial history and continued in much of post-WWII development, the stripping of local cultural and natural contexts from science teaching as well as the absence of inquiry-based learning. African countries continue to absorb the standards of the world without the inclusion of local culture in education (Geo-JaJa & Azaiki, 2010; Babaci-Wilhite, 2013a; Okonkwo, 2014). Colonial education in Africa is not transmitting the values and the knowledge of African society from one generation to the next; it has involved a deliberate attempt to change those values and to replace traditional knowledge by the knowledge from a different society. To motivate the active mind, one has to take into consideration the variations in different societies, differences in knowledge and different ways of teaching to achieve quality education. If education is conceived of as imparting knowledge about the world, then schooling should be regarded as only one aspect of education, since it does not cover all forms of knowledge.

Science teaching and learning can benefit from recapturing local language and a pedagogy that emphasizes critical thinking in the search for evidence which provides teachers and students with the capacity to understand and deploy new pedagogical tools for teaching and learning science subjects. Recent studies have shown that students exposed to such models made significantly greater gains based on measures of science understanding, science vocabulary, and science writing (Pearson et al., 2010).

The focus of UNESCO´s Education for All (EFA) initiative has been on basic literacy, stopping short of emphasizing the importance of science literacy in the development of a nation. According to EFA´s global monitoring report, literacy

Z. Babaci-Wilhite (Ed.), Human Rights in Language and STEM Education, 3–15.

skill is more important than ever in today´s knowledge societies, benefiting individuals, communities and nations as a whole. Literacy has been defined as a basic human right, yet in January 2014 the cultural agency of the United Nations issued a report that documented falling education standards around the world. The report pointed out that a quarter of a billion children worldwide are failing to learn basic science principles, creating an education crisis that costs governments $129 billion annually.

Language is central to achieving science literacy and UNESCO (2003) supports mother tongue instruction as a means of improving educational quality by building upon the knowledge and experience of both learners and teachers. Furthermore the support of bilingual and/or multilingual education at all levels of education as a means of promoting equality is a key element of linguistically diverse societies such as most nations in Africa (Skutnabb-Kangas, 2000; UNESCO, 2005; Benson & Kosonen, 2014). Nonetheless, non-localized languages are still chosen as Languages of instruction (LoI) in most African nations' from primary schools through university (Babaci-Wilhite, 2014a, 2014b). Based on my own research and that of many scholars (Pearson et al., 2010; Brock-Utne, 2011; Cervetti et al., 2012), I will argue first that achieving science literacy is much more than rote learning, but rather involves comprehension, the capacity to inquire and to read strategically, and second that use of a local language is critical to access concepts and facilitate deep learning (Cervetti et al., 2006).

The approach I will put forward incorporates the importance of local context in the learning of science, emphasizing the development of local capacity on local terms. This draws heavily on the work of David Pearson (2007) on the important role of language and literacy in the learning of science and in the capacity to think critically and flexibly across many domains of knowledge. In this context Paulo Freire (1970) and his insight that inclusive education, accomplished through the integration of formal and informal knowledge dignifies learners rather than oppressing them. Drawing on the theory of education for self-reliance, developed by Julius Nyerere (1968) and applied in Africa mainly Tanzania, I will argue that education for self-reliance has relevance today in Africa´s efforts to achieve equal access and fairness in education and in the society at large. Finally, I will argue that deep science comprehension is accessible only through the use of a language that students understand best and that science learning in a local language is essential to the achievement of human rights in education (Spreen & Vally, 2006; Babaci-Wilhite, 2012). LoI plays a critical role in cognitive learning and in the development of critical thinking and new knowledge. Drawing on my research in Tanzania, Nigeria and the United States, I will argue that acknowledging local knowledge in educating for science literacy will be a bridge to improve teaching and learning as long as it is contextualized in local languages. This will make a positive contribution to the achievement of quality education in science subjects in Africa.

HOW DO WE ACQUIRE AND IMPROVE SCIENCE LITERACY?

Improving science learning can be addressed by improving literacy, which will facilitate inquiry. Jacqueline Barber (2005) argues that "Inquiry is curiosity-driven… It involves reading books to find out what others have learned… Inquiry requires the use of critical and logical thinking… Good readers inquire information gathered from text." Therefore, Pearson and Barber´s approach to improve literacy through inquiry science engages students in "real-world" interaction to achieve better results for science as well as literacy. This approach resolves the problems of teaching and learning science associated with poorly trained teachers and inadequate teaching aids and facilities. A major challenge in science education is how to support teachers in understanding and enacting inquiry-based instruction. Pearson and Barber in the Seeds of Science/Roots of Reading (S/R) model[2] require teachers to be clear on long-term goals, identify measurable indicators of success, and be accomplished in the practices known to meet the linguistic needs, such as using graphic representations of abstract concepts (Pearson et al., 2013). This fundamental concept builds a curriculum that gives emphasis to literacy through "texts, routines for reading, word-level skills, vocabulary, and comprehension instruction." These work in the service of acquiring the knowledge, skills, and dispositions of inquiry-based science (Cervetti et al., 2012).

Berit Haug and Marianne Ødegaard (2014) described the inquiry-based approach as having "the potential for students to learn how to do science, learn about science, and learn science by doing science." In their study of science teaching in Norway, Haug and Ødegaard (2014) focus specifically on the aspect of "learning science by doing science," which is how to teach for conceptual understanding by emphasizing word knowledge development in an inquiry-based setting. Haug (2014) argues that a call for research on how knowledge is constructed when engaging students in hands-on activities has come up time and again over the past several decades.

Access to literacy must involve making sense of the physical world through first and second hand experiences while addressing foundational dimensions of literacy. Pearson and Barber assembled science and literacy experts to study, enact in the form of curriculum, and test the limits and potential of the science-literacy interface (Pearson et al., 2010). This model of science inquiry involves students in searching for evidence to support their ideas and investigations. Students also engage in critical and logical thinking to learn how to make and revise explanations based on the evidence found. This curriculum addresses the ways that reading, writing, and discourse can be used as tools to support inquiry-based science learning and in which benefits accrue to reading, writing and discourse when they are embedded in inquiry-based science.

The Models designed on the inquiry based-approach has the potential to bring the needed outcomes, results and accomplishments that will improve the process of teaching and learning science. Furthermore Pearson et al. (2013) link their emphasis on existing background knowledge to Goldenberg (2008) who claimed that:

> Beginning reading instruction is guided by neither a theory nor a goal of knowledge development. In fact, just the opposite: children are presented with texts—mostly narrative—chosen to reflect their existing background knowledge, the assumption being that they can use that knowledge to comprehend familiar content.

Goldenberg's point reflects the current situation in Tanzania, Nigeria and other African countries. According to Haug (2014) findings from several studies highlight the importance of the communication and discussion phases of inquiry to foster conceptual understanding in students and stress the importance of the communication phase for learning to take place. They state that to develop understanding of scientific concepts, students must explain and justify their conclusions instead of just presenting their findings as collected evidence. Similarly, Bigozzi et al. (2002) considered the ability to justify data, the most evident feature that distinguishes deep and lasting learning from learning that is purely oral and superficial.

As argued above, engagement with local language and local knowledge is necessary to facilitate the teaching and learning process. Furthermore each outcome in the pathway of change is tied to an intervention, revealing the complex web of activity that is required to bring about change. These principles of learning address the connections between early, intermediate and long-term outcomes and the expectations about how and why the proposed interventions will bring them about (Cervetti et al., 2007). This inquiry based-approach aims for deep conceptual understanding, implementation of a program of planning and evaluation, and a shared cross-disciplinary understanding of the long-term goals and on how they will be reached, as well as what will be used to measure progress along the way.

Improving science literacy in Africa today is mainly done by bringing universal principles to African students through a foreign language and using non-local contextualization and non-local examples (Babaci-Wilhite, 2013a). A model with emphasis on local contextualization, reading comprehension and vocabulary development using a LoI has a great potential to improve science learning. The materials of this model consist of a number of units covering several topics within the different sciences: life science, physical science and earth science. Each unit has a detailed step-by-step teacher's guide describing when to introduce, and how to combine, the different modes of learning (do-talk-read-write). Also included are in-depth science background, instructional suggestions, and statements of what students should master at specific points in the unit, for example, knowing how the targeted scientific concepts interrelate to make meaning.

A multi-modal approach used for example in the S/R model, which provides students with opportunity to access every essential concept to be learned in a unit through a range of different learning modalities. By doing experiences, it engages students in discussing the essential concepts learned and it makes it easy to understand them by reading and enabling students to write them. For instance, for students in coastal areas, they might read about shorelines, then investigate sand, gather evidence from sand and write a text about its properties, which lead to an understanding of the original source of the sand. They discuss their work at regular intervals (in their local language), eventually forming expert groups focusing on particular sand samples. They read about shorelines, then do, then talk, then read again, then do more, then talk, then write, then talk again, then finish what they have written (Barber, 2005). These multiple modalities provide opportunities for students to apply, deepen, and extend their knowledge of the learned concepts.

This effective research-based curriculum offers students an explicit focus on disciplinary literacy and in the specialized knowledge and skills involved in reading, writing, and talking about science. This approach promotes substantive science knowledge through inquiry skills involving students in deep forays into learning about the natural world by searching for evidence through firsthand experiences (hands on activities) and secondhand experiences (text) in order to construct more accurate, nuanced, and complete understandings of the natural world. Furthermore, students engage in written and oral discourse with the goal of communicating and negotiating evidence-based explanations, then evaluate and revise explanations based on that evidence. This is in opposition to the usual approach that simply adds literacy tasks onto a science curriculum, without connecting those additional tasks directly to the advancement of the understanding from the initial investigation and does not provide explicit instruction on how to read and write science text (Pearson et al., 2010). Recent studies have shown that students exposed to the program made significantly greater gains on measures of science understanding, science vocabulary, and science writing (Cervetti et al., 2012).

A curriculum model that links firsthand experiences, discussions, and writing to the ideas and language in informational texts to foster development of core science knowledge and literacy skills is crucial to improving science literacy in Africa, where the contexts of everyday life contrast sharply with the North American and European contexts, where science theory and science curricula are most advanced (Afflerback et al., 2008).

GLOBAL TO LOCAL IMPLEMENTATION OF SCIENCE LITERACY

Around the world, inquiry-based science is recognized in policy documents and curriculum materials as essential to the learning of science; however, research in classrooms indicates that teachers have not fully applied inquiry-based science (Haug, 2014, p. 80). By observing in classroom instructional practices in different phases of science inquiry and the interactions that occurred between teachers and students, Haug (ibid) illuminated different teaching approaches and how they influenced students' conceptual understanding. According to the studies, it is clear that conceptual learning occurs when students are required to apply key concepts in their talk throughout all phases of inquiry, with the students' use of language. In contrast, conceptual understanding is not promoted when teachers do the talking, rephrase students' responses into the correct answer, or fail to address students' everyday perceptions of scientific phenomena. The frameworks applied for word knowledge and link-making are effective in terms of student conceptual learning only if the students are the ones doing the talking and the ones actively engaged in making the links (Haug & Ødegaard, 2014).

Furthermore Haug (2014) argues that when asking Norwegian students in Norwegian schools to apply their new knowledge and think about why it is important to consider when designing new things, nobody responded. She found that students needed clarification and further explanations to develop a higher level of conceptual knowledge. The example shows that knowing the definitions and being able to use science concepts properly in short answers represents only the first step toward conceptual understanding (Bravo et al., 2008). Since reinforcement of new knowledge and development of conceptual understanding typically take place when students discuss their empirical findings and link them to established science, teachers must plan carefully to include enough time for discussion and communication during inquiry-based science instruction.

Pearson et al. (2010) point out the connection between word knowledge and conceptual knowledge by emphasizing that when science words are taught as concepts, applied in a context and in relation to other science words and concepts, word knowledge is consistent with conceptual knowledge. The use of local language and context is critical to learning to use the language of science (Wellington & Osborne, 2001; Barber, 2005). Thus, it is important to emphasize students' development of word knowledge using local language. This application of contextualized science learning through the use of language is consistent with the theories and policies of African education put forward by Nyerere. He insisted on the need for rethinking the relationship between general education and formal education, asserting that the basic system of education, which the Tanzanians took over at the time of independence, should be questioned and the use of an African language should be used in school. His strongly held view was that education must acknowledge local culture, which includes language, social settings and non-material dimensions of life and should be an integral part of daily life, not separated from it. Education should address both the needs of the local people and the country in which they live.

According to Freire (1970), much of the knowledge that forms the basis for schooling has its origins from another place and another time: "Knowledge emerges only through invention and re-invention" (Freire, 1993, p. 53). The students who catch on to this form of learning will be successful in school, but might actually have less knowledge in the broad sense of the word than those who do not attend school. However education is most often equated with schooling (Babaci-Wilhite, 2010). The language issue in Tanzania is deeply related to how one conceptualizes education and the debate around whether or not education should encompass the local cultural context. I agree with Freire (1993) and others scholars such as Hassana Alidou (2003), Martha Qorro (2004), Zubeida Desai (2004), Kwesi-Kwaa Prah (2005), Birgit Brock-Utne (2011) and Jerome Okonkwo (2014) who argue that using a local language as medium of education fosters the broader view of learning which softens the barriers between real-life and classroom experience (Babaci-Wilhite, 2014c). This is consistent with the thinking behind Pearson's approach to literacy that knowledge and wider vocabulary are a consequence as well as a cause of reading comprehension. Therefore the model developed by Pearson and Barber is based on the evidence that language matters and reading comprehension cannot be achieve in a language, which is not mastered by the teacher and the students. There is evidence from several countries around the world that learning through the local languages is the best way to learn science. Most of the countries of Europe and North America teach science in their local language, yet, when it comes to the African contexts, local languages continue to compete with colonial languages in education. It is time for African countries to rethink science literacy using local languages since language is not everything in education, but without language, everything is nothing in education (Wolff, 2006). Thus it can be said that the use of a local language is essential to inquiry-based learning.

LOCAL LANGUAGES AND KNOWEDGE AS A RIGHT TO ACCESS SCIENCE LITERACY

Languages with a colonial legacy, such as English, French, Portuguese and Spanish continue to be used as official languages in many developing countries today. English is a particularly powerful globalizing language that carries with it a cultural context foreign to the local contexts for education (Bamgbose, 2003). Today, English is used as an official or semi-official language in over 60 countries in the world and has a prominent place in a further 20, making it the most learned language in the world (Majhanovich, 2013, 2014). Many language policy makers have adopted this view both in wealthy nations like the United States of America and the United Kingdom, where large amounts of 'foreign aid' money is spent on promoting English, particularly in sub-Saharan Africa where English is now often the sole official LoI at all levels of education in many African countries (Mazrui, 2003).

This language imperialism will need to be arrested if quality learning and science literacy is to be secured. As discussed above, language is the key when it comes to science inquiry. Students gain a better understanding of science concepts when taught through local languages rather than through a foreign language (Brock-Utne, 2012; Babaci-Wilhite, 2015). To develop conceptual knowledge students need help in linking scientific concepts to their everyday experiences, to assimilate new and unfamiliar science words and concepts, and learn how to use concepts in context (Bravo et al., 2008).

In Tanzania, South Africa, Nigeria and elsewhere in Africa, reforms and policies connecting local culture and knowledge to education have been neglected (Babaci-Wilhite, 2015). According to Joel Samoff (2007, p. 60) "effective education reform requires agendas and initiatives with strong local roots" (see also Geo-JaJa, 2013). In other words, local knowledge (also referred to as indigenous knowledge) should be included in the curriculum (Odora, 2002; Babaci-Wilhite & Geo-JaJa, 2011;

Semali et al., 2012). This knowledge should be conveyed in local languages, which is critical to the preservation and development of local knowledge. The choice of the LoI is extremely important not only because of the implications for quality learning, but also because of the intimate ties between language, culture and identity (Brock-Utne, 2012; Babaci-Wilhite, 2013b; Okonkwo, 2014). Local knowledge and languages are being severely strained through globalization, which is a shorthand way of describing the spread and connectedness of production, communication and technologies across the world. We cannot think simplistically about the unidirectional nature of global flows of products and culture, but when it comes to languages there are no doubt that the flow is unidirectional from the Global North to the South. This process has its roots in earlier stages of imperialism and continues to the present (Ngugi wa Thiong'o, 1994; Geo-JaJa & Yang, 2003).

In Africa today, knowledge is equated with what the learner is taught in schools. The knowledge that forms the basis for school curricula is decontextualized. The educated person is one who has mastered sets of facts, propositions, models and cognitive skills that are fundamentally separate from the context in which they were learned. Knowledge is also typically viewed as relatively stable. In mainstream approaches to education, including Africa, schooling often involves the transmission of isolated, portable bodies of knowledge. Schools make sense as institutions only because stable knowledge and reasoning procedures can allegedly transfer and have value in other contexts where students will use the knowledge they learned in school. Because the context is not integral to the knowledge or skill, the isolated "bodies of knowledge" often hold little meaning for anyone other than the members of the community who generated that knowledge. The problems students solve in school are thus problems of the disciplinary communities from which the knowledge originated. This often makes schooled knowledge and skills less useful outside of schools. Moreover, given the decontextualized, insular nature of the knowledge being passed on, there is generally little opportunity for students to question the claims on which the knowledge is based.

Given that most science education in Africa today is neither context sensitive, conducted in language students use to relate to the natural environments nor grounded in an inquiry based curriculum, I argue that the way that science is taught is violating children's rights in education. Article 26 of the United Nations Declaration of Human Rights (1948) states that everyone has the right to education, however it says little about the nature, kind and quality of that education. Rights in education imply that rights are not ensured unless the education offered is of high quality. The rights-based approach is based on the premise that the use of a local curriculum should be regarded as a right in education (Tomasevski, 2006; Babaci-Wilhite & Geo-JaJa, 2014).

Education is an important contributor to human rights effectiveness as it increases human capabilities, functions and opportunities in societies. This further leads to the linkage between human rights and development and enables policy makers and developers to incorporate the "Common Understanding" of a human rights-based approach, assuring these principles: indivisibility, equality, participation and inclusion (UNDP, 2006, pp. 17–18). "Human rights in education" is a powerful notion as it is intimately connected to the social, occupational, political, cultural, religious and artistic life of the people (Babaci-Wilhite et al., 2012; Bostad, 2013). UNESCO's convention on the Protection and Promotion of the Diversity of Cultural Expressions emphasizes the importance of linguistic diversity (2005).

The globalization trends for language in education, in which local curricula are de-contextualized, are in contravention of the tenets of the rights-based approach to language in education. In an increasingly interdependent world it is important to facilitate the mastery of foreign languages as well as mastery of subject matter. The policies for developing countries should also be context sensitive and in addition permit developing countries to remain partners in the global society. Therefore education should be regarded as a set of processes intended to enhance glo-local learning. This is a radical departure from most mainstream educational research and practice, which is designed to enhance global rather than local learning. An innovation to teaching and learning science which couples global developments in science theory with local knowledge, local languages and local teaching can be the bridge to science literacy and to securing African students rights in education

CONCLUSION

In this chapter, I have reviewed the theory of inquiry relevant to understanding the importance of the use of local languages in the achievement of science literacy. Science cannot be taught without contextualized inquiry. As explained in this chapter when science content is addressed through a combination of inquiry and literacy activities, students learn how to read, write, and talk science simultaneously. These literacy activities support the acquisition of science concepts and inquiry skills. Furthermore the recent studies discussed in this chapter emphasize the connection between word knowledge and conceptual understanding. Therefore the synergy between science and literacy rests upon the understanding that an active level of word knowledge in science (understanding of words as they are situated within a network of other words and ideas) can be described as conceptual knowledge. An inquiry-based model embraces and builds on this science/literacy integration, and especially the connection between word knowledge and conceptual knowledge. As I have argued, conceptual understanding is not promoted when teachers do the talking, rephrase students' responses into the correct answer, or fail to address students' everyday perceptions of scientific phenomena. The frameworks applied for word knowledge and link making are effective in terms of student conceptual learning only if the students are doing the talking and are actively engaged in making the links. Therefore in order to enable inquiry, African languages need to be valued and preserved. This will facilitate the learning process and support students in their preparation to engage with the world in their own terms.

There is ample evidence that embracing local context through the use of local language in an inquiry-based approach leads to improved literacy, scientific knowledge, and personal efficacy for students and greater professional efficacy for teachers. I recommend mapping African teachers' ideas and thoughts regarding science inquiry and their view on their own level of content knowledge. This research would examine the whole inquiry cycle in different stages and how this could be planned for and utilized in teaching. This would allow teachers to engage students in discussions that build on evidence collected through investigation and to be more aware of what to look for in student responses and how to act upon these to promote conceptual understanding. This implies more research from African classrooms on how to conduct science inquiry in ways that enhance student learning. This model, emphasizing the local context and using a local language, will contribute to rights in education and to children's engagement in science subjects for quality education.

NOTES

1 Tanganyika gained its independence from Britain; and Zanzibar became independent in 1963. On 26 April 1964, Tanganyika joined with Zanzibar to form a new state, now named the United Republic of Tanzania.

2 The S/R model was co-founded by Professor David Pearson and Jacqueline Barber at the Lawrence Hall of Science (UC-Berkeley) California. This model is a primary curriculum that supports teaching in which students learn science concepts in depth simultaneously as they are taught how to read, write, and discuss through inquiry-based science.

REFERENCES

Afflerbach, P., Pearson, P. D., & Paris, S. G. (2008). Clarifying differences between reading skills and reading strategies. *Research Based Classroom Practice, 61*(5), 364–373.

Alidou, H. (2003). Medium of instruction in post-colonial Africa. In J. W. Tollefson & A. Tsui (Eds.), *Medium of instruction policies: Which agenda? Whose agenda?* (pp. 195–214). Mahwah, NJ: Lawrence Erlbaum Publishers.

Babaci-Wilhite, Z. (2010). Why is the choice of the language of instruction in which students learn best seldom made in Tanzania? In B. Brock-Utne, Z. Desai, & M. Qorro (Eds.), *Educational challenges in multilingual societies: LOITASA phase two research* (pp. 281–305). Cape Town, South Africa: African Minds.

Babaci-Wilhite, Z. (2012). A right based approach to Zanzibar's language-in-education policy. Special issue on right based approach and globalization in education. *World Study of Education, 13*(2), 17–33.

Babaci-Wilhite, Z. (2013a). The new education curriculum in Zanzibar: The rationale behind it. In Z. Desai, M. Qorro, & B. Brock-Utne (Eds.), *The role of language in teaching and learning science and mathematics: LOITASA phase two research* (pp. 127–151). Cape Town, South Africa: African Minds.

Babaci-Wilhite, Z. (2013b). An analysis of debates on the use of a global or local language in education: Tanzania and Malaysia. In D. Brook Napier & S. Majhanovich (Eds.), *Education, dominance and identity* (pp. 121–133). Rotterdam, The Netherlands: Sense Publishers.

Babaci-Wilhite, Z. (Ed.). (2014a). *Giving space to African voices: Right in local languages and local curriculum* (Vol. 33). Rotterdam, The Netherlands: Sense Publishers.

Babaci-Wilhite, Z. (2014b). E-Learning science literacy as a right in education in Africa. *International Society of Comparative Education, Science and Technology of Nigeria, 1*, 115–130.

Babaci-Wilhite, Z. (2014c). Local language as a human right in education: Comparative cases from Africa. In A. Pitman & M. A. Pereyra (Series Ed.), *Comparative and international education: A diversity of voices* (Vol. 36). Rotterdam, The Netherlands: Sense Publishers.
Babaci-Wilhite, Z. (2015). Zanzibar's curriculum reform: Implications for children's educational rights. *Prospects, UNESCO's Quarterly Review of Comparative Education, 45*(2), 181–195. doi:10.1007/s11125-015-9341-6
Babaci-Wilhite, Z., & Geo-JaJa, M. A. (2011). A critique and rethink of modern education in Africa's development in the 21st century. Papers in Education and Development (PED). *Journal of the School of Education, University of Dar-es-Salaam, 30*, 133–154.
Babaci-Wilhite, Z., & Geo-JaJa, M. A. (2014). Localization of instruction as a right in education: Tanzania and Nigeria language-in education's policies. In Z. Babaci-Wilhite (Ed.), *Giving space to African voices: Right in local languages and local curriculum* (pp. 3–21). Rotterdam, The Netherlands: Sense Publishers.
Babaci-Wilhite, Z., Geo-JaJa, M. A., & Lou, S. (2012). Education and language: A human right for sustainable development in Africa. *International Review of Education, 5*, 619–647.
Bamgbose, A. (2003). A recurring decimal: English in language policy and planning. *World Englishes, 22*(4), 413–431.
Barber, J. (2005). *The seeds of science/roots of reading inquiry framework.* Retrieved from http://scienceandliteracy.org/sites/scienceandliteracy.org/files/biblio/barber_inquirycycle_pdf_54088.pdf
Benson, C., & Kosonen, K. (Eds.). (2013). *Language issues in comparative education: Inclusive teaching and learning in non-dominant languages and cultures*. Rotterdam, The Netherlands: Sense Publishers.
Bigozzi, L., Biggeri, A., & Boschi, F. (2002). Children "scientists" know the reasons why and they are "poets" too. Non-randomized controlled trial to evaluate the effectiveness of a strategy aimed at improving the learning of scientific concepts. *European Journal of Psychology of Education, XVII*(4), 343–362.
Bostad, I. (2013). Right to education- for all? The quest for a new humanism in globalization. *World Studies in Education, 14*(1), 7–16.
Bravo, M. A., Cervetti, G. N., Hiebert, E. H., & Pearson, D. P. (2008). From passive to active control of science vocabulary. In *The 56th yearbook of the National Reading Conference* (pp. 122–135). Chicago, IL: National Reading Conference.
Brock-Utne, B. (2011). *Language and inequality: Global challenges to education.* BAICE Presidential Address at the 11th UKFIET International Conference, Oxford, England.
Brock-Utne, B. (2012). Language policy and science: Could some African countries learn from Asian countries? *International Review of Education, 58*(4), 481–503. doi:10.1007/s11159-012-9308-2
Cervetti, G. N., Barber, J., Dorph, R., Pearson, P. D., & Goldschmidt, P. G. (2012). The impact of an integrated approach to science and literacy in elementary school classrooms. *Journal of Research in Science Teaching, 49*(5), 631–659.
Cervetti, G. N., Pearson, P. D., Barber, J., Hiebert, E. H., & Bravo, M. A. (2007). *Integrating literacy and science: The research we have, the research we need.* New York, NY: Guilford.
Cervetti, G. N., Pearson, P. D., Bravo, M. A., & Barber, J. (2006). Reading and writing in the service of inquiry-based science. In R. Douglas, M. Klentschy, & K. Worth (Eds.), *Linking science and literacy in the K–8 classroom.* Arlington, VA: National Science Teachers Association.
Desai, Z. (2004). Starting a research project: Some lessons to be learnt. In B. Brock-Utne, Z. Desai, & M. Qorro (Eds.), *Researching the language of instruction in Tanzania and South Africa* (pp. 117–131). Cape Town, South Africa: African Minds.
Freire, P. (1970, Reprinted in 1993). *Pedagogy of the oppressed.* London, England: Penguin Books, Ltd.
Geo-JaJa, M. A. (2013). Education localization for optimizing globalization's opportunities. In S. Majhanovich & M. A. Geo-JaJa (Eds.), *Economics, aid and education: Implications for development.* Rotterdam, The Netherlands: Sense Publishers.
Geo-JaJa, M. A., & Azaiki, S. (2010). Development and education challenges in the Niger Delta in Nigeria. In M. A. Geo-JaJa & S. Majhanovich (Eds.), *Education, language, and economics: Growing national and global dilemmas* (pp. 53–69). Rotterdam, The Netherlands: Sense Publishers.

Geo-JaJa, M. A., & Yang, X. (2003). Rethinking globalization in Africa. *Chimera, 1*(1), 19–28.

Goldenberg, C. (2008). Teaching English language learners, What the research does and does not-say. *American Educator, 32*(1), 8–22.

Haug, B. S. (2014). Inquiry-based science: Turning teachable moments into learnable moments. *Journal of Science Teacher Education, 25*, 79–96. doi:10.1007/s10972-013-9375-7

Haug, B. S., & Ødegaard, M. (2014). From words to concepts: Focusing on word knowledge when teaching for conceptual understanding within an inquiry-based science setting. *Research Science Education, 44*(5), 777–800. doi:10.1007/s11165-014-9402-5

Majhanovich, S. (2013). English as a tool of neo-colonialism and globalization in Asian contexts. In Y. Hébert & A. Abdi (Eds.), *Cultural perspectives on international education* (pp. 249–261). Rotterdam, The Netherlands: Sense Publishers.

Majhanovich, S. (2014). Neo-liberalism, globalization, language policy and practice issues in the Asia-Pacific Region. *Asian-Pacific Journal Education, 34*(2), 168–463.

Makalela, L. (2005). We speak eleven tongues. Reconstructing multilingualism in South Africa. In B. Brock-Utne & R. K. Hopson (Eds.), *Languages of instruction for Africa emancipation: Focus on postcolonial contexts and considerations* (pp. 147–175). Cape Town, South Africa: CASAS and Dar es Salaam, The United Republic of Tanzania: Mkuki na Nyota.

Mazrui, A. (2003). *English in Africa: After the Cold War.* Sydney, Australia: Multilingual Matters Ltd.

Ngugi wa Thiong'o, J. (1994). *Decolonising the mind: The politics of language in African Literature.* London, England: James Curry Ltd.

Nyerere, J. K. (1967, Reprinted in 1968). Education for self-reliance. In *Essays on socialism* (pp. 44–75). Dar es Salaam, The United Republic of Tanzania: Oxford University Press.

Odora, C. H. (Ed.). (2002). Stories of the hunt – Who is writing them? In *Indigenous knowledge and the integration of knowledge systems. Towards a philosophy of articulation* (pp. 237–257). Claremont, South Africa: New Africa Education.

Okonkwo, J. I. (2014). Appropriate language in education: The strategy for national development in Nigeria. In Z. Babaci-Wilhite (Ed.), *Giving space to African voices: Right in local languages and local curriculum* (pp. 131–146). Rotterdam, The Netherlands: Sense Publishers.

Pearson, P. D. (2007). An endangered species act for literacy education. *Journal of Literacy Research, 39*(2), 145–162.

Pearson, P. D., & Hiebert, E. H. (2013). Understanding the common core state standards. In L. Morrow, T. Shanahan, & K. K. Wixson (Eds.), *Teaching with the common core standards for English language arts: What educators need to know* (Book 1: Grades PreK-2; Book 2: Grades 3–5) (pp. 1–21). New York, NY: Guilford Press.

Pearson, P. D., Moje, E., & Greenleaf, C. (2010). Literacy and science: Each in the service of the other. *Science, 328*, 459–463. doi:10.1126/science.1182595

Prah, K. K. (2005). Language of instruction for education, development and African emancipation. In B. Brock-Utne & R. K. Hopson (Eds.), *Languages of instruction for African emancipation: Focus on postcolonial contexts and considerations* (pp. 23–87). Cape Town, South Africa: CASAS.

Qorro, M. A. S. (2004). Popularising Kiswahili as the language of instruction through the media in Tanzania. In B. Brock-Utne, Z. Desai, & M. Qorro (Eds.), *Researching the language of instruction in Tanzania and South Africa (LOITASA)* (pp. 93–116). Zanzibar, The United Republic of Tanzania: African Minds.

Samoff, J. (2003, re-edited 2007). Institutionalizing international influence. In R. F. Arnove & C. A. Torres (Eds.), *Comparative education: The dialectic of the global and the local* (pp. 52–91). Lanham, MD: Rowman & Littlefield Publishers.

Semali, L., & Khanjan, M. (2012). Science education in Tanzania: Challenges and policy responses. *International Journal of Educational Research, 53*, 225–239.

Skutnabb-Kangas, T. (2000). *Linguistic genocide in education or worldwide diversity and human rights?* Mahwah, NJ: Lawrence Erlbaum Associates.

Spreen, C. A., & Vally, S. (2006). Education rights, education policies and inequality in South Africa. *International Journal of Educational Development, 26*(4), 352–362.

Tomasevski, K. (2006). *Human rights obligation in education: The 4-A scheme*. Enfield, NH: Enfield Publishing & Distribution Company.
UNDP. (2006). *Human development report*. New York, NY: United Nations.
UNESCO. (2003). *Education in a multilingual world* (UNESCO Education position paper). Paris, France: UNESCO.
UNESCO. (2005). *Convention on the protection and promotion of the diversity of cultural expressions*. Paris, France: UNESCO.
United Nations. (1948). *The universal declaration on human rights* (Adopted by the General Assembly on 10 December 1948). Geneva, Switzerland: United Nations.
Wellington, J., & Osborne, J. (2001). *Language and literacy in science education*. Philadelphia, PA: Open University Press.
Wolff, H. (2006, March 27–31). *Optimizing learning and education in Africa – the language factor. A stock-taking research on mother tongue and bilingual education in Sub-Saharan Africa* (pp. 26–55). Association for the Development of Education in Africa (ADEA): UNESCO Institute for Education-Deutsche Gesellschaft für Technische Zusammenarbeit. ADEA 2006 Biennial Meeting, Libreville, Gabon.

Zehlia Babaci-Wilhite
Graduate School of Education
University of California, Berkeley
USA

Norwegian Center for Human Rights
University of Oslo
Norway

P. DAVID PEARSON AND ALISON K. BILLMAN

2. READING TO LEARN SCIENCE

A Right that Extends to Every Reader—Expert or Novice

INTRODUCTION

A long-standing mythical distinction in reading pedagogy is captured by this oft-repeated advice:

> *First you learn to read, and then you read to learn.*

Just such an assumption is engraved in many of the stage-models of reading (e.g., Chall, 1983, 1996); it is embodied in our commercial reading programs (Pearson Scott Foresman, 2015; Macmillan, 2015), and it is a key plank in many reform movements, such as No Child Left Behind (NCLB, 2002). The two of us question the validity of the distinction and have been committed for several years to conducting research to eradicate this phrase from the professional discourse around early reading pedagogy. Why would we take issue with a position that has become almost an article of faith in the circle of reading pedagogy? Because we think it implies that learning something of value as one reads is not a necessary part of early reading instruction. A second adage, wholly consistent with the first, reinforces just such a perspective:

> The content of early reading materials is unimportant because in those early stages kids are so consumed by the magic of turning letters into sounds and print into speech that the content of their texts doesn't really matter.

This is the view that has been responsible for a host of practices, mostly misguided, for a very long time. First is the long tradition of decodable texts that are specially constructed to allow students to practice the decoding skills they have learned up to a particular point in the curriculum—gems in the tradition of, "Dan can fan Nan. Nan can fan Dan. Dan, fan Nan. Nan, fan Dan. Fan Dan in the van." Second is the equally long-standing practice, coming from a different early reading philosophy, of using only narratives in the primary grades, especially K-1, documented so dramatically in Duke's classic study (Duke, 1998, 2000). Third is a belief, accompanied by a set of practices, that privileges decoding and fluency over text comprehension as dominant activities in the early stages of learning to read—master the basics, this philosophy suggests, and then you can focus on the good stuff, like comprehension, critical reading, and responding to text.

Z. Babaci-Wilhite (Ed.), Human Rights in Language and STEM Education, 17–34.

The two of us have been engaged in a counter-curricular effort—namely a quest to learn how to engage beginning readers in *reading to learn* AT THE SAME TIME as they *learn to read*. We have been figuring out how to make this wedding work for about six years. We spent the first three (2010–2012) in a thought experiment about what it would look like if we taught kids to read with books filled with information about the natural world rather than stories about everyday or imaginary events. We have spent the last three years (2013–2015) engaged in an ongoing design experiment that grew out of the thought experiment—making it happen by building and testing an integrated science-literacy program in real classrooms with real first graders[1] and teachers.

Over the course of those six years, we have built a conceptual model that challenges many of the assumptions that we, as literacy educators, have traditionally made about the best way of teaching beginning reading, most notably that ubiquitous assumption that "first you learn to read and then you read to learn." From the process of building and testing that conceptual model, we have discovered a set of principles that we believe would serve as better (compared to the current conventional wisdom) guideposts for developing programs for teaching beginning reading.

In this chapter, we share key mileposts in our journey, as a way of convincing you that it is a journey worth taking. But what, one might ask, does a chapter on building and validating a model for integrated science-literacy curriculum have to do with the primary theme of this book—promoting human rights in underdeveloped parts of the world? On the face of it, perhaps little. After all, receiving or learning from an integrated curriculum can hardly compare to basic human rights such as food, shelter, medical care, freedom, justice, and universal education. But underneath the surface, access to this sort of curriculum, with its emphasis on USING the tools of reading, writing, and language to comprehend, critique, and construct scientific explanations of the natural world, is pretty important to one of the most basic of human rights—the right to learn, to gain access to the knowledge and inquiry skills needed to participate in the free and democratic processes of political and social self-determination. We hope to persuade you that this right should be included along with those that are more transparently regarded as basic human rights in the global world that in which all of us must live and thrive.

WHAT IS OUR STORY?

The story behind our work and the insights that have emerged from it begin in the early 2000s, before Alison arrived at Lawrence Hall of Science (LHS), with an earlier group of scholars equally as committed as the two of us to the idea that science and literacy should be integrated and synergistic rather than isolated and disconnected from one another. The first foray into the work was a 5-year effort to build and test an integrated curriculum dubbed *Seeds of Science/Roots of Reading* (S*eeds and Roots* for short) for the elementary grades 2–5. By beginning the work with grade 2, the first round of *Seeds and Roots* development finessed the very issues that the two of us have used to frame this chapter, namely the question of whether students can read to learn while learning to read. Focusing on grades 2–5 as a first approximation of an integrated approach was probably a wise choice because the team could assume, if starting with grade 2, that even the most novice of readers in the program, would bring a modicum of sub-lexical skills (within word skills such as alphabet knowledge, phonemic awareness, decoding and word recognition) to the learning experience. This also meant that texts could be used as a medium for second-hand investigations of key science concepts in order to reinforce what kids were learning in their first-hand (hands-on) investigations.

But, as we built the original curriculum up (grades 2–3, then 3–4, then 4–5) we knew that we would eventually build all the way up to grade 8 and all the way down to kindergarten. And that is where Alison (and by implication the two of us) came into the picture in a more central way—precisely because we were tasked with the challenge of finding a funding source that would take the risk of supporting us to take on the challenge of creating an integrated approach that would work in K-1.

KEY STEPS ALONG THE WAY

Building Our Own Understanding

Our first "step" involved building our own understanding of the challenges that come with designing an integrated approach to use in first grade classrooms—classrooms filled with beginning readers. We chose to set ourselves a high bar by targeting issues of language and knowledge development for first graders who were also learning English. While we hypothesized that contextualizing literacy instruction would benefit all students, it would likely be especially supportive of English Learners (EL), who, in comparison to their English-only peers, bear the additional task of learning a new language while learning how to read and how to engage in science practices. We suspected that a content-rich integrated literacy and science curriculum could provide an authentic language-rich context that would support first grade ELs in building content knowledge as well as oral language, vocabulary, comprehension, decoding and word reading strategies—the tools known to contribute to reading comprehension development (Storch & Whitehurst, 2002; García, 1998; Lesaux & Geva, 2006).

Accepting the Risk

In our second step we needed to acknowledge that we were exploring new territory, potentially rife with false hopes—potentially full of risk. After all, there is a long history of research that informs the instruction that is common in today's first grade classrooms. However, there is little research that explores the design or impact of integrated curricula with this population of learners, namely young ELs learning to read. The work would be challenging and would require a keen focus on defining and keeping in mind the legitimate goals of the project as well as systematic and principled reflection along the way. In short, we knew that it was "risky business," but with that risk, we also knew, came the prospects of important outcomes.

Building a Framework

Our third step was to construct a defensible theoretical framework—a conceptual network that would champion the very sorts of synergies between learning to read and learning to think and act like a scientist that would have to be part and parcel of this approach (and had been the basis of our grade 2–5 effort). Grounded in the research of our colleagues and forebears we developed a theory of change (Figure 2.1), to frame our thought experiments and ultimately the design of integrated literacy and science curriculum. As a starting point we hypothesized an integrated literacy and science curriculum for grade 1 ELs (*Input*) drawn on the *Seeds of Seeds and Roots* model of literacy-science with the ultimate goal of improving EL students' reading comprehension of informational text that might occur outside *Seeds and Roots* (the *Distal Outcome*).

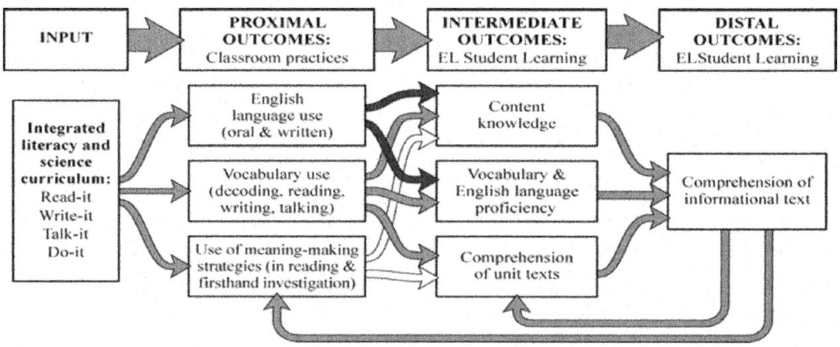

Figure 2.1. Theory of change

The *Seeds and Roots* model of literacy-science integration supports teachers in implementing classroom practices (*Proximal* Outcomes) that provide children with learning experiences that capitalize on the overlap between literacy and science in order to address learning goals in both domains. The model integrates, within the context of coherent bodies of conceptual knowledge, systematic opportunities for *reading* (encounters with text—focused on science content and reading processes), *writing* (documenting investigations through journals and class charts), *talking* (discussions to make sense of investigations and texts), and *doing* science (engaging in firsthand experiences). The model also provides explicit instruction in the meaning-making strategies that are shared by scientific inquiry and reading comprehension—strategies such as posing questions, predicting, and synthesizing. As we theorized the model for grade 1, we knew the design would involve adjusting the focus on literacy practices and learning goals to include those relevant to the development of reading comprehension skills in beginning readers—decoding, oral language, vocabulary, and word reading strategies—along with the shared meaning making strategies/comprehension strategies themselves.

We further hypothesized that changes in classroom practices—how teaching and learning happen in the classroom—would promote increases in EL learning (the *Intermediate Outcomes*) in (1) science knowledge, (2) vocabulary knowledge and English language proficiency, and (3) comprehension of unit texts. And finally, the intermediate learning outcomes are those known to support the *Distal Outcome*, reading comprehension (Droop & Verhoeven, 2003; Kintsch, 1988; Neuman, 2001; Storch & Whitehurst, 2002)—in particular of informational text. The increase in comprehension skill should continue to promote increases in subject matter knowledge and vocabulary, propelling the entire system into a positive loop, or virtuous cycle, in which knowledge begets comprehension begets knowledge, and so forth.

Convincing Others

Our fourth step was convincing key constituencies—our *Seeds and Roots* colleagues and external funders. The internal persuasion was needed not to convince our colleagues that this effort was worthwhile but that (a) it ought to be a high priority and (b) that we could actually involve print (reading and writing) in a major way in grade 1 and, ultimately, kindergarten as well.

The external persuasion focused on funders, both public and private, who would support the time intensive design-based research (DBR) methodology we would need to employ in order to test, revise, and refine the pedagogical models and units we created over time. We needed to convince them that DBR, with its inherent iterative and provisional spirit, was not only worth doing but also it was the *best* approach given the goals of the project, the context of the project, and the very limited research-base around the question of integrated curriculum for beginning readers.

The good news is that in the Spring of 2013, we learned that we had been funded for a July 1, 2013 start-up. That is, of course, when we faced the daunting reality of turning our ever so pleasant thought experiment (the program in our minds) into a palpable, living curriculum that would allow grade 1 students to *read to learn, learn to read*, and become budding scientists, all at the same time. And that has been our consuming passion for the last two years, as we set this chapter to pen and ink.

Instantiating the Model in Two Units

Our fifth step was to build it! Along with defining the model came designing and creating the materials needed to further test the curriculum model. We developed two full first grade units. One unit targets ELs entering first grade and the second unit targets ELs at mid-first grade. Using the Next Generation Science Standards (NGSS, 2013) as a guide we chose to focus the units on two different science domains, Life Science and Physical Science. Selecting both a Life and Physical Science topic pushed us to ensure that the model could be successfully applied in a range of science domains, each with its own core concepts and particular practices.

Each unit includes five types of materials: Teacher guides, student books, a student investigation notebook, an assessment system, and a materials kit to support hands-on learning, as detailed in Table 2.1.

Table 2.1. Materials for the Grade 1 Units

Teacher's guide	A comprehensive guide provides step-by-step instructions and options for implementing and modifying the unit lessons to meet the unique needs of students. The guide is educative; that is, it is designed to provide just in time support to help teachers successfully implement the unit. Additional teacher notes provide information to educate teachers about the subject matter and the pedagogical approach, especially for ELs.
Student books	Each unit includes five, four-color, custom-written student books that serve as a central feature to support students' knowledge development.
Student notebook	The student investigation notebook provides a place to track observations, data, important vocabulary and other written work that reflects the thinking and writing practices of science.
Assessment system	The assessment system includes formative and summative assessment tools that provide teachers with options for engaging in formative conversations with students and tracking progress toward unit learning goals. Each unit includes formative and work sampling measures for (a) science knowledge, (b) reading comprehension, and (c) vocabulary acquisition and use.
Materials kit	Importantly, each kit includes all of the materials necessary to implement the first-hand activities in the unit.

Testing the Model in Classrooms

The sixth and final step in this journey is designing and evaluating the model in first grade classrooms. Beginning in Fall of 2013, we started working with classroom teachers to implement a three phase design experiment. The basic commitment of this design based research and development effort was to test the efficacy of the intervention in the crucible of the classroom at every step along the way. In phase one, a team of three designer/researchers and two classroom teachers designed and taught sets of lessons to discover and validate the influence of variables such as text difficulty, text features, instructional supports, and the relationship between first-hand investigations, second-hand investigations, writing and talking.

Armed with some insights about what variables to emphasize in our pedagogy, we moved to phase two in which we worked with teachers to design, refine, and implement two units, a light and sound unit and an animal defenses unit. During this stage, we made constant, almost instantaneous modifications to lessons as we received feedback from our teachers and, of course, their students about what was and was not working as planned.

Phase 3, currently in progress as we write this chapter, involves what we call a Curriculum Implementation Trial (CIT). The CIT allows us to directly compare the nature, quantity, and quality of learning in the integrated science and literacy intervention (i.e., our units) with a text-only version of the same curriculum and a business as usual (separate science and literacy curricula) treatment.

The eventual hope for the outcome of the project, of course, is that the final model of integration will ultimately lead to more effective programs for promoting both science and literacy learning in our schools. Whatever else this approach has taught us, it has taught us that holding ourselves to such a high standard—being accountable to classroom tryouts as a basis for evaluating the efficacy of pedagogical approaches—is worth the effort. The end result is a more effective and engaging curriculum for budding readers and scientists.

WHAT HAVE WE LEARNED SO FAR?

As we have taken this journey we have learned more than we expected, even hoped, about teaching and learning reading, writing, and language in the context of acquiring knowledge of science concepts and inquiry skills in first grade classrooms.

Insight 1: First Grade Students Can Handle the Integrated Approach

The first and most important insight from our work at this point (Spring of 2015) is that the integrated approach is just as effective and motivating for grade 1 students as it is for grade 2–5 students. We know that grade 1 students are hard-wired to learn and naturally curious about the world around them, but even with these two factors in their favor, it takes plenty of energy and grit to persist when ideas are new and complicated. The pedagogical trick is optimizing the content and design of lessons so young students can operate in their Goldilocks zone—enough novelty and wonder to make the instructional experience interesting and motivationally challenging (same old, same old can be boring), but enough familiar information to allow for those all important connections between what's new and what's already known. We believe several key design decisions were responsible for our success with young ELs.

Finding the Sweet Spot Between the Novel and the Familiar

Designing inquiries that bridged students' experiences with the world to the science concepts they were about to learn triggered just the right amount of wonder, interest, and motivation to engage in the hard work of first hand investigations. For example, a book such as *Can you See in the Dark?* posed an intriguing question that set up a reason to conduct a first-hand investigation, such as trying out materials and light sources to understand how much light can be blocked by materials with different properties.

Engaging All the Experiential/Language Modes

A key finding in working with grades 2–5 was that concepts worth learning were worth teaching in four experiential/language modes (reading, writing, talking, and doing). This same combination of multimodal experiences worked equally well grade 1. This multimodal approach supports learning key concepts, the words we use to name those concepts, and the ways of talking about those concepts. They also work for learning and practicing strategies for making sense such as predicting, asking questions, or visualizing. In short the combination of modes promotes knowledge development and inquiry skills, and promote it to the point where students can use what they have learned to create their own projects and products. For example, in our physical science unit: Students experience a combination of investigating with materials and light, reading books like *Can You See in the Dark* and *What Makes a Shadow Like This?* and talking about shadows to build an understanding of what it means to block light. Armed with this information, students take on the role of light engineers and apply what they know about light to create a background scene—the illusion of buildings, trees, mountains, or other aspects of the scene for a puppet show—using materials to create the patterns of light effects they want.

Insight 2: Vocabulary, Vocabulary, Vocabulary

A second major insight focuses on the role that words play in the integrated curriculum model—a role that cannot be underestimated. But we also discovered a subtle, almost counterintuitive insight about vocabulary: *words are not the point of good vocabulary instruction, knowledge is*. The resulting essential guiding principle is that words are concepts and need to be treated as such.

We began our work in grades 2–5 with a commitment to de-emphasize the isolated vocabulary instruction long associated with text-centric science curricula (endless lists of words and definitions that drive serious inquiry science scholars to distraction). We achieved that goal, not by de-emphasizing the role that words play in learning to do science; but instead by avoiding isolated vocabulary instruction. This means that opportunities to learn and use words are embedded in a host of contexts and activities, including the single most important context—while children are engaged in first-hand investigations. In this respect, we make sure that students have a chance to encounter each word in each context appearing in Figure 2.2, which illustrates how students encounter the concept of habitat in a soil science unit. Notice that the settings from *relationships* to *synthesis* emphasize the conceptual (i.e., the knowledge) bases of word learning which is exactly the point of effective vocabulary instruction.

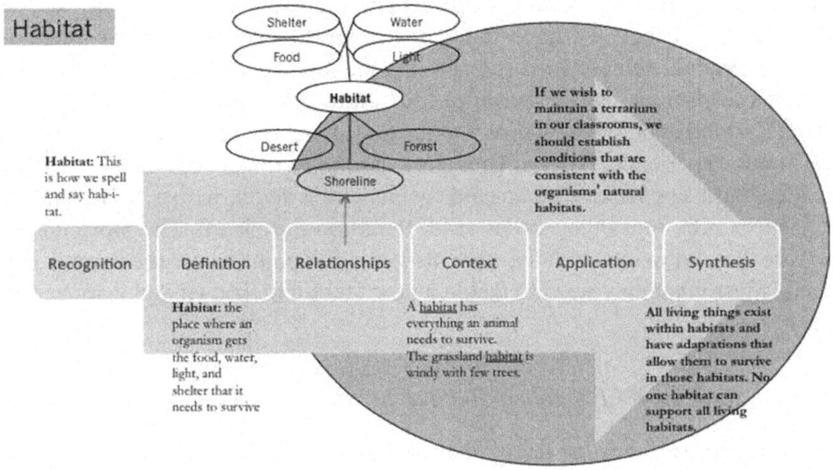

Figure 2.2. Visual depiction of the various contexts in which a student should meet a word in order to gain control of its meaning. As a student moves from left to right, she moves from decidedly passive to increasingly active control of the concept underlying the word. (Cervetti et al., 2006, reprinted with permission)

Multiple Encounters in Multiple Contexts

If there is one stable finding from decades of work on vocabulary acquisition, it is that systematic and varied encounters with new concepts is the surest route to developing rich and stable concepts (Elleman et al., 2009; Stahl & Fairbanks, 1986; Van Daalen-Kateijius et al., 2001). Besides encountering each word in each context, students should also meet these words (really the concepts they represent) in each mode of use—reading, writing, talking, and doing. And finally, in the design of our units, this breadth of contact occurs across texts and first-hand activities as well. As students in our grade 1 unit on light and sound read each new book (5 in total) and conducted each new investigation (16 in total), we made sure that they encountered and had a chance to use a select set of key words again and again, in some cases (for example, light, shadow, block, surface), more than 20 encounters per word.

Selectivity in Choosing Key Words

If we are to ensure that students meet key words in 6 contexts across 4 modes for each of 5 books and 5 or more investigations, then we must limit the number we choose for this full treatment in each unit. Our guiding principle is to choose words that are essential for articulating key ideas that represent the conceptual territory of the unit. Our commitment is to design instruction to ensure that each word is taught thoroughly and encountered in each language/experiential mode.

The Role of Text in Acquiring Vocabulary

Texts are key tools in providing scaffolds and repeated encounters with key words and concepts. In fact, we have found that texts can be designed to intentionally provide a gradual release of responsibility for reading to learn about the topic of study. It's a virtuous cycle: the more kids know (including the words), the more they understand about those words and the concepts they represent. This happens, at least in our experience, only when we are careful to make sure that kids encounter key ideas and vocabulary again and again within texts, across texts, and across experiences. The more they understand, the more they learn. The more they learn, the more they know. Eventually, students get to the point where the knowledge they gained is just what they need to tackle a new text on their own and learn something from it.

Insight 3: Students WERE Able to Read to Learn as They Learned to Read

Perhaps the most important lesson that we have learned is that the approach works: Students can read to learn while they are learning to read! In our first teacher taught classroom implementation trial of a unit based on the revised model of integration (more later), several indicators of growth from the beginning to the end of the trial suggest that students were learning what we had hoped they would learn and could comprehend texts about what they had learned.

The Classroom Implementation Trial

Our classroom implementation trial (CIT) involved a total of six classrooms. Two teachers taught the integrated science-literacy curriculum. A single control classroom received a business as usual control treatment and three classrooms taught using the 5 books from the unit without the first-hand science materials. So, the books only group got the same content as the full treatment group, but in the more traditional text-based format so roundly criticized by science educators over the past 35 years. Thus it was a rigorous test of the intervention.

Some preliminary findings. The results we have available when this chapter was "put to bed" are limited to the experimental group; we have yet to conduct the comparative analyses. Nonetheless, the results give us a good reading on the sorts of growth readers at the end of first grade experience in the program. First, students made statistically and educationally significant growth on a comprehension measure (a MAZE, a type of multiple-choice cloze test, which requires students to select the best choice to fill in a missing word—every seventh word is missing—on a topically related passage. This finding is all the more impressive when one considers that the MAZE texts, despite considerable effort on our part to keep the text complexity (and readability scores) in check, scaled at about a 2.5 level on the Spache Readability Formula.

Second, students demonstrated substantial growth in their capacity to describe challenging science concepts. For example, they could describe the role of light and materials in blocking light to make shadows (even though they could not readily explain those roles). Also, their capacity to describe and explain key concepts was moderated by the provision of key vocabulary. The first probes offered almost no scaffolding; students had to generate responses from scratch; however, after we recorded those initial responses, we gave students a list of key vocabulary and invited them to respond to the same probes. Their responses were much richer when key vocabulary could be consulted.

Engagement. Our initial forays into the qualitative data tell us that this approach promotes high degrees of engagement for both students and teachers. Data from teacher feedback through the daily logs, our own observations, punctuated by quotes from our interviews tells us that, without a doubt, whatever else the intervention did, it engaged participants. For students, this meant that students wanted to read on and work on, showing very little of the boredom that often sets in when students are working within an organized curriculum. A comment by one of the participating teachers in filling out her daily log epitomizes teacher reaction to the intervention:

> I loved the way they used the vocabulary outside of the science time. They were always seeing light sources and objects that were blocking the light. I felt they got the vocabulary and used it with ease. The students had an excellent understanding of light and light sources at the end of the unit. (Teacher 1, personal communication, June 2, 2015)

Insight 4: Fundamental Principles[2]

Our fourth insight is that a *set of fundamental principles* undergirds our approach and accounts for whatever impacts it have. Some of these principles are new to us—meaning we probably would not have discovered them without moving our work to include novice readers. The other principles reinforce insights we gathered in our work with students in grades 2–5 (see Billman & Pearson, 2013; Cervetti et al., 2006). Whether old or new, it is this set of principles that we propose that teachers, schools, and curriculum designers need to embrace as they set out to guide students in becoming careful readers, writers, talkers, and doers—through both first-hand and second-hand investigations to acquire new knowledge.

Principle #1. Children, including novices who are English learners, bring a lot to the classroom. Finding ways to bridge from the resources they bring to the science and literacy goals we want them to achieve is our number 1 challenge and responsibility.
When we began our journey into integrated curriculum for beginning learners, we were uncertain whether strategies used to bridge prior knowledge and learning goals

with older students would work with students still learning to read. We learned that it is not necessary to wait until middle school to begin students' journey into deep knowledge of science. If we design units that strategically build on the important resources children bring to school, namely oral language, knowledge of the natural world, rudimentary reasoning skills, and emerging print-based practices, we can—and on the basis of our evidence, we should—begin this journey from the very first day of formal schooling. What are those resources? There are several.

Children enter school already using language as a tool to learn and talk about the world around them. Acquiring language and the rudiments of domain related language begin at the earliest of ages and, in fact, establish the groundwork for future networks of specialized language (Clay, 1998; Dewey, 1990/1957; Lindfors, 1999). In most cases, each child's experiences with adults and knowledgeable others contribute to learning language and building knowledge frameworks that help her categorize experiences and ways of interacting and communicating with others. These language structures and vocabulary serve as the foundation for engaging in the work of learning the specialized vocabulary and ways of talking in school-based learning opportunities.

Children enter school with substantial knowledge about the natural world.
We know that a classroom of students will include a range of individual and group differences in background and experiences. However, all children come to school with foundational knowledge of the natural world (Duschl et al., 2007) and with curiosity and a general enthusiasm for learning about the world. This rudimentary knowledge along with experiences and dispositions for learning, serve as a springboard for developing the literacy and language tools to satisfy this need to know. Ideally, the tools of written language (reading and writing) are then put to work in helping students expand their knowledge. It's just the synergy needed to create a virtuous cycle between literacy and science learning.

Children are capable of complex thinking and reasoning about what they know.
Thinking skill development is linked to knowledge development; not surprisingly children think more facilely (synthesize ideas, draw inferences, connect evidence to claims) about topics they know a lot about—and less facilely about those they don't (Duschl et al., 2007; Glaser, 1984; Metz, 1995). It stands to reason that, engaging students in discipline related learning opportunities helps them develop more sophisticated reasoning skills and then consciously use these skills to build knowledge of the practices and conceptual knowledge of the domain.

In short, what we have learned in our current work with first graders aligns with what the field has known should be conventional everyday practices in classrooms. Granted, in spite of this knowledge, educators have thoroughly failed to implement what we know or are far too cautious in challenging students to learn what is within their grasp. The findings from this project add to the already strong case from the field for making a commitment and curricular move to provide challenging learning opportunities for children. Importantly, we now have the policy backing we need from both the *Next Generation Science Standards* (2013) and the *Common Core State Standards* (2010) to take the risk that must be taken for all of our students, including language minority students.

Principle #2. Participation in a disciplinary community is key to acquiring disciplinary expertise and literacy.
If experience and text are equal partners in acquiring and codifying knowledge for a discipline, then community is a discipline's medium of survival and vitality. Fundamentally, knowing science (or any subject) also includes knowing how to participate in the community with other scientists. That means not only having a conceptual understanding of the discipline but also knowing how to ask the questions, test the hypotheses, collect the data, engage in the discourse in ways that foster deeper understanding of the discipline.

As we indicated in Principle 1, children are already engaged in participating in their worlds and already use language to communicate with those around them to build understanding. Learning the language of science—even in grade 1—requires opportunities to engage in activities that require authentic use of the language to read, write and talk about science. This situated practice builds new knowledge, fosters connections, and adds to children's repertoires of participation (Wells, 2002). So when our young English learners are learning to read and learning how to do science, they are also learning how to participate in a learning community, both providing to and receiving from the community resources in the form of ideas, observations and language.

Principle #3. Throughout the K-12 curriculum, students should acquire literacy (reading, writing, and oral language) expertise while in the pursuit of disciplinary knowledge and inquiry skill.
We first discovered this principle for ourselves (Billman & Pearson, 2013; Cervetti et al., 2006) in our work with the 2–5 *Seeds and Roots* curriculum, and we had hoped that our findings for first grade would demonstrate that it applied equally as well to the earliest stages of schooling. Even in first grade reading, we asked, would reading, writing and language be better viewed as tools rather than goals.

The conventional wisdom in teaching reading, as we suggested at the outset of this essay, is that first students learn to read, and later they read to learn—a position that we thoroughly oppose. First, we know that an undue focus on developing the skills of reading that is not coupled with understanding/learning may lead children to misunderstand the nature and purposes of reading along with how reading and writing can support personal interest development (Levy, 2009; Nolen, 2001). Second, we also know that early education can improve children's reading and writing skill levels, but it is increasingly clear that a focus on skills needs to be coupled with a quest for knowledge. In that respect, we think and have seen in our work with older elementary school children that having opportunities to use language to talk, read, and write about science positively impacts literacy learning (Cervetti et al., 2012; Wang & Herman, 2005). Neuman and Roskos (2005) has captured the essence of this seemingly counter-intuitive idea that a quest for knowledge situates, rationalizes, and enhances skill learning. She claims that we have underestimated the role of knowledge in understanding of early literacy and early literacy development:

> Children want to learn about their worlds. As they acquire knowledge, they become fascinated with the tools of communication—reading and writing—and what they can do with them. In play, for example, young children use both real and pretend writing and reading to enhance the drama and realism of the pretend situation. They want to master the tools of literacy, as when writing down the football facts that interest them, and they'll often seek help from more proficient writers and readers who serve as spontaneous apprentices to help them learn about written language and how to use it for various purposes. After all, literacy development is not just a matter of learning a set of technical skills. It is a purposeful activity involving children in ways of making, interpreting, and communicating meaning with written language. (p. 5)

Principle #4. Knowledge and comprehension form a virtuous cycle of learning for all students.

We have known for a long time (see Anderson & Pearson, 1984), since the early days of the cognitive revolution in fact (Rumelhart & Ortony, 1978), that knowledge drives comprehension. More recently, we have become aware of the other side of that equation—that comprehension drives knowledge development (see Cervetti & Hierbert, 2015; Neuman & Roskos, 2005); we actually learn new ideas when we read. Both claims are accurate because the two constructs—knowledge and comprehension—form a symbiotic relationship, a virtuous cycle, if you will, with learning as the link between the two. It works like this:

> The more you know (K), the better you comprehend.
> The better you comprehend (RC), the more you learn.
> The more you learn (L), the more you know.
> The more you know (K), …

Knowledge is cumulative. Some science educators insist that first-hand investigations should precede second-hand (reading about science) investigations; just as surely there are reading educators who insist that if reading comes first, the hands on science will be all the better. Importantly, we have discovered through our design-based approach to research and development, that there is no magic formula for situating or ordering literacy versus science activities. Rather, whatever experience comes second benefits from the experience that came first. This is a desirable outcome because it validates the *knowledge is cumulative* principle. We believe that some day we may possess more nuanced research-based knowledge about how to optimally sequence first-hand versus second-hand investigations. It might be, for example, that for very obscure topics, second-hand should come first. As of now, however, we have very little research-based guidance to help us make such decisions. Even so, we can take some solace in the consistent observation that has surfaced in our work with grade 1 and well as grades 2–5: Whatever comes second benefits from whatever comes first.

Texts can be a part of the virtuous cycle. Texts can be designed and sequenced to support the K-RC-L-K cycle. Our approach to reading is built on a synergistic design of texts coupled with instruction that supports students through a Gradual Release of Responsibility toward reading all of the unit texts more independently. The five books are read using a combination of the Read-Aloud, Shared Reading, and Partner Reading approaches (Burkis & Yaris, 2014). Teachers lead or guide students' interactions with books that come early in the unit; then, students read books that come later in the unit with a partner. The readability parameters and design of each book align with the instructional mode of reading—e.g., read-alouds are longer while partner reads are shorter; read-aloud texts introduce key vocabulary that will be used in subsequent texts. By structuring texts and reading instruction in this way, students have multiple experiences with the content and vocabulary of the unit before they are responsible for reading unit texts more independently. The approach worked because texts were designed and sequenced to propel students through the K-RC-L-K cycle; as such, the sequencing of information and the reading mode supported a gradual release of responsibility to the students, to the extent that they were able to handle the partner read with a minimum of teacher support. In the words of one of the teachers in our CIT,

> My low readers felt so successful. One student was so proud of himself for reading the book that he wanted to go to the front of the room and read it to the class. Again I love the books and so do the students. They were engaged but finished very quickly. I had each group answer the questions for me to check for understanding. (Teacher 2, personal communication, June 11, 2015)

Knowledge always represents a glass half full. One of the factors often singled out by school personnel to explain the lack of reading comprehension among vulnerable readers is their lack of knowledge about the content of the books and articles they are asked to read: "If only my students knew more about X or Y, then I'd be able to teach them to read about Z more effectively" … This perspective assumes that the knowledge glass kids bring is half empty. While we are sympathetic with the concern that many students don't benefit from a rich set of experiences before they come to schools for a more formal education, we believe it is counterproductive to even entertain this line of reasoning. When all is said and done, what Pearson said

20 years ago (a reflection on what he said in 1978) about the relationship between knowledge and comprehension still rings true:

> This, by the way, this is exactly the point that I tried to make 16 years ago in *Teaching Reading Comprehension* (Pearson & Johnston, 1978)—that a teacher's job is always to bridge from the known to the new. There really is no other choice. Children are who they are. They know what they know. They bring what they bring. Our job is not to wish that students knew more or knew differently. Our job is to turn students' knowledge and the diversity of knowledge we encounter in a classroom into a curricular strength rather than an instructional inconvenience. We can do that only if we hold high expectations for all students, convey great respect for the knowledge and culture they bring to the classroom, and offer lots of support in helping them achieve those expectations. (Pearson, 1996, p. 272)

Always assume that the glass is *at least* half full.

Rather than offering a conclusion or a coda, we will end with this last principle. And it is fitting way to end a chapter in a book addressing ways of nurturing human rights for students in developing countries. So we, as researchers looking for ways of promoting understanding and knowledge through the lenses and tools of reading, writing, and language, end our essay with a simple declaration: The river that leads to new knowledge for our students runs through—and is nurtured by—the reservoirs of knowledge that those students bring, both individually and collectively, to our schools. To fail to privilege the precious resources stocked full in that reservoir will only prolong the inequities and widen the gap between the educational haves and have-nots. The solution to eradicating the inequities and closing the gap requires that we place great trust in children's experience, curiosity, and quest for knowledge about the world in which we all live.

ACKNOWLEDGMENTS

The research reported here was supported by the Institute of Education Sciences, U.S. Department of Education, through Grant R305A130610 to the Regents of the University of California. The opinions expressed are those of the authors and do not represent views of the Institute or the U.S. Department of Education.

NOTES

1 By design and intention, we use the term first hand investigations to refer to what most science educators would call "hands on" activities, where students are involved in learning concepts by DOING. We use the term second hand investigations to designate READING about those ideas, maybe even about someone else DOING something. Think of second hand as either eyes on or minds on activities. The goal is to use both first and second hand to achieve greater synergy.

2 Many of the ideas in this section first appeared in Billman and Pearson (2013).

REFERENCES

Anderson, R. C., & Pearson, P. D. (1984). A schema-theoretic view of basic processes in reading comprehension. In P. D. Pearson, R. Barr, M. L. Kamil, & P. Mosenthal (Eds.), *Handbook of reading research* (pp. 255–291). New York, NY: Longman.

Billman, A. K., & Pearson, P. D. (2013). Literacy in the disciplines. *Literacy Learning: The Middle Years, 21*(1), 25–33.

Burkis, J., & Yaris, K. (2014). Break through the frustration balance vs. all-or-nothing thinking. *Reading Today, 32*(2), 26–27.

Cervetti, G., Pearson, P. D., Bravo, M. A., & Barber, J. (2006). Reading and writing in the service of inquiry-based science. In R. Douglas, M. Klentschy, & K. Worth (Eds.), *Linking science and literacy in the K-8 classroom*. Arlington, VA: NSTA.

Cervetti, G., Barber, J., Dorph, R., Pearson, P. D., & Goldschmidt, P. (2012). The impact of an integrated approach to science and literacy in elementary school classrooms. *Journal of Research in Science Teaching, 49*(5), 631–658.

Cervetti, G. N., & Hiebert, E. (2015). The sixth pillar of reading instruction: Knowledge development. *The Reading Teacher, 68*, 548–551.

Cervetti, G. N., Tilson, J., Goss, M., Castek, J., & Jaynes, C. (2008, October). *Teaching and learning vocabulary in the context of science.* Presentation at the annual meeting of the California Reading Association, Sacramento, CA.

Chall, J. S. (1983). *Stages of reading development.* New York, NY: McGraw-Hill Book Company.

Chall, J. S. (1996). *Learning to read: The great debate* (3rd ed.). Fort Worth, TX: Harcourt Brace College Publishers.

Clay, M. M. (1998). *By different paths to common outcomes.* York, ME: Stenhouse.

Dewey, J. (1990/1957). *The school and society; The child and the curriculum.* Chicago, IL: University of Chicago.

Droop, M., & Verhoeven, L. (2003). Language proficiency and reading ability in first-and second-language learners. *Reading Research Quarterly, 38*(1), 78–103.

Duke, N. K. (1998). 3.6 minutes per day: The scarcity of informational texts in first grade. *Reading Research Quarterly, 35*(2), 202–224.

Duke, N. K. (2000). For the rich it's richer: Print experiences and environments offered to children in very low-and very high-socioeconomic status first-grade classrooms. *American Educational Research Journal, 37*(2), 441–478.

Duschl, R. A., Schweingruber, H. A., & Shouse, A. W. (2007). *Taking science to school: Learning and teaching science in grades K-8.* Washington, DC: National Academies Press.

Echols, J. C., & Hosoume, K. (1999). *Ant homes under the ground.* Berkeley, CA: Great Explorations in Math and Science (GEMS), Regents of the University of California.

Elleman, A. M., Lindo, E., Morphy, P., & Compton, D. L. (2009). The impact of vocabulary instruction on passage-level comprehension of school-age children: A meta-analysis. *Journal of Research on Educational Effectiveness, 2*, 1–44.

Garcia, G. E. (1998). Bilingual children's reading. In M. Kamil, P. Mosenthal, P. D. Pearson, & R. Barr (Eds.), *Handbook of reading research* (pp. 813–834, Vol. 3). Mahwah, NJ: Lawrence Erlbaum and Associates.

Glaser, R. (1984). Education and thinking: The role of knowledge. *American Psychologist, 39*(2), 93–104.

Kintsch, W. (1988). The role of knowledge in discourse comprehension: A construction-integration model. *Psychology Review, 95*(2), 163–182.

Lesaux, N. K., & Geva, E. (2006). Synthesis: Development of literacy in language minority learners. In D. L. August & T. Shanahan (Eds.), *Developing literacy in a second language: Report of the National Literacy Panel* (pp. 53–74). Mahwah, NJ: Lawrence Erlbaum Associates.

Levy, R. (2009). Children's perceptions of reading and the use of reading scheme texts. *Cambridge Journal of Education, 39*(3), 361–377.

Lindfors, J. W. (1999). *Children's inquiry: Using language to make sense of the world.* NewYork, NY: Teachers College Press.

Macmillan/McGraw-Hill. (2011). *Treasures.* New York, NY: Author.
Metz, K. E. (1995). Reassessment of developmental constraints on children's science instruction. *Review of Educational Research, 65*(2), 93–127.
National Governors Association Center for Best Practices & Council of Chief State School Officers. (2010). *Common core state standards for English language arts and literacy in history/social studies, science, and technical subjects with appendices A–C.* Washington, DC: Authors.
Neuman, S. B. (2001). The role of knowledge in early literacy. *Reading Research Quarterly, 36*(4), 468–475.
Neuman, S. B., & Roskos, K. (2005). Whatever happened to developmentally appropriate practice in early literacy. *Young Children, 60,* 1–6.
NGSS Lead States. (2013). *Next generation science standards: For states, by states.* Washington, DC: The National Academies Press.
No Child Left Behind Act of 2001, P.L. 107-110. 20 U.S.C.
Nolen, S. B. (2001). Constructing literacy in kindergarten: Task structure, collaboration, and motivation. *Cognition and Instruction, 19*(1), 94–142.
Pearson Education, Inc. (2011). *Scott Foresman reading street.* New York, NY: Author.
Pearson, P. D. (1996). Reclaiming the center. In M. Graves, P. van den Broek, & B. M. Taylor (Eds.), *The first R: Every child's right to read* (pp. 259–274). New York, NY: Teachers College Press.
Pearson, P. D., & Johnston, D. D. (1978). *Teaching reading comprehension.* New York, NY; Holt, Rinehart and Winston.
Rumelhart, D. E., & Ortony, A. (1978). The representation of knowledge in memory. In R. C. Anderson, R. J. Spiro, & W. E. Montague (Eds.), *Schooling and the acquisition of knowledge.* Hillsdale, NJ; Lawrence Eralbaum Associates.
Stahl, S. A., & Fairbanks, M. M. (1986). The effects of vocabulary instruction: A model-based meta-analysis. *Review of Educational Research, 56*(1), 72–110. doi:10.3102/00346543056001072
Storch, S. A., & Whitehurst, G. J. (2002). Oral language and code-related precursors to reading: Evidence from a longitudinal structural model. *Developmental Psychology, 38*(6), 934–947.
Van Daalen-Kapteijns, M. M., Elshout-Mohr, M., & De Glopper, K. (2001). Deriving the meaning of unknown words from multiple contexts. *Language Learning, 51*(1), 145–181.
Wang, J., & Herman, J. (2005). *Evaluation of seeds of science/roots of reading project: Shoreline science and terrarium investigations.* Los Angeles, CA: CRESST.
Wells, G. (2002). The role of dialogue in activity theory. *Mind, Culture, and Activity, 9*(1), 43–66.

P. David Pearson
Graduate School of Education
University of California, Berkeley
USA

Alison K. Billman
The Lawrence Hall of Science
University of California, Berkeley
USA

MEGAN M. GOSS, HELEN MIN AND CARRIE STROHL

3. DESIGNING STANDARDS-DRIVEN CURRICULUM TO SUPPORT

Access to Science for All Students

INTRODUCTION

The national debate regarding the adoption of more uniform standards across the United States has reignited our discussion of the relationship between educational expectations, student learning and teacher pedagogy. Since their inception, standards have served as guideposts for establishing learning outcomes for students, and, more recently, for educators working to meet expectations that are aligned with accountability testing. For curriculum developers, standards are a critical component because all curricula that aim to be adopted must list expected standards for every lesson. Therefore, the curriculum design process necessarily includes highlighting as many standards as possible. Although this has routinely been a minimum benchmark for all adoptable curricula, it often turns out to be an ineffective or even counterproductive effort at providing quality and equitable 'standards-driven' curriculum (Hamilton et al., 2012; Harris, 2012). The question then becomes, "What does it actually mean for curricula to be truly 'standards-driven'?"

To us, designing standards-driven curriculum means attending to deep content knowledge (i.e., depth over breadth) and to the practices of the discipline that are part of both the Next Generation State Standards (NGSS) (NGSS Lead States, 2013) and the Common Core State Standards for English Language Arts and Literacy (CCSS ELA) (National Governors Association Center for Best Practices & Council of Chief State School Officers, 2010). Ultimately, our task is to generate a rich and rigorous curriculum that not only meets the applicable standards, but also ensures all students' access to learning, engagement and success in science. In so doing, we aim to support classroom teachers who serve a student population with diverse language and learning needs (Aud et al., 2013; Zeichner, 2012), while also infusing an authentic portrayal of the nature of science (Schwartz, Lederman, & Crawford, 2004). We present our work in the context of the US; however, the issue of language access is universal, as our international colleagues have demonstrated (Babaci-Wilhite, 2013; Skutnabb-Kangas, 2008). Although we remain humbled every day by the demands of creating 'standards-driven' curriculum, we believe that we have found a few important factors that inform our work and make reaching this high bar possible: disciplinary literacy as an essential spine of unit planning and execution.

Z. Babaci-Wilhite (Ed.), Human Rights in Language and STEM Education, 35–52.

This foundation guides us in making science curriculum that approaches the demands for excellence established by the newest standards and by those that we ourselves set to achieve through our work.

This chapter focuses on designing effective standards-driven curriculum for middle school science teachers and, perhaps more importantly, for middle school science students. Adolescent learners pose a set of unique design challenges for curriculum developers. For instance, the reality that many students encounter a slump in academic achievement in middle school (Guthrie & Davis, 2003; Stevens, 2003; Fields, 2014) can be explained, in part, by a lack of opportunity to develop the disciplinary literacy skills required to contend with increasingly complex content that students encounter in and beyond secondary school. In response, many researchers (e.g., CAAL, 2010; Draper & Siebert, 2009; Heller & Greenleaf, 2007; Lee & Spratley, 2010; Moje, 2008; Shanahan & Shanahan, 2008, 2012) call for pedagogical approaches that more intentionally frame content-area instruction in discipline-specific ways. Heeding this call, the NGSS and CCSS recommend a shift towards science curricula infused with the skills, strategies and experiences that resemble those used in the discipline of science.

Although a subset of education researchers (e.g., Kesidou & Roseman, 2002; Pearson et al., 2010) recognize that improving classroom science instruction relies on attending to the disciplinary literacy demands of science, until now, this shift did not permeate the field of science education widely. For instance, although science educators have long since urged changes in how students engage with both content and processes in the science classroom, the reality of what students do and experience each day in classrooms hasn't changed significantly (NRC, 2007, 2012). In general, some science has become more active than in the past, likely a result of the push for "inquiry" that came from an earlier wave of curricular reforms. However, on the whole, one very important—in fact, crucial—aspect has not changed significantly: the kinds of discourse opportunities that are provided for students in science remain stagnated.

Science, in particular, is a discipline defined by the persistent push against what is known; in practice, scientific thinking is based on constantly asking questions and demanding more evidence. As Duschl and Osborne (2002) explained, it is essential for a curriculum to strive to provide students with good activities and ideas about how scientists do their work; however, they also contend,

> Equally important…is the need to educate our pupils and citizens about how we know and why we believe; to expose science as a way of knowing…such a shift requires a focus on 1. how evidence is used in science for the construction of explanations, and 2. on the criteria used in science to evaluate the selection of evidence and the construction of explanations. (p. 40)

We wholeheartedly agree with the descriptions of what a rich and impactful curriculum could and should look like. In the following sections, we describe how these descriptions inform our curricular design process.

DISCIPLINARY LITERACY AS A GUIDING ORIENTATION FOR CREATING RESPONSIVE AND ENGAGING MIDDLE SCHOOL SCIENCE CURRICULUM

A disciplinary literacy emphasis provides a guiding orientation for creating curriculum at both the level of a cohesive unit as well as at the level of an individual lesson to achieve the goal of pushing science teaching beyond the status quo. In this kind of curriculum, students learn science by experiencing the ways of thinking that characterize the actual work of science. To do this, the center of gravity moves away from the teacher and discovery is placed into the minds of the students (Jiménez-Aleixandre & Erduran, 2007; Osborne, 2010). As Shanahan and Shanahan (2014) pointed out, disciplinary literacy holds the promise of inviting students to join the discipline as a member of the scientific community. We accomplish this by reframing student talk, embedding authentic reading and writing experiences, and employing the central discourses of all science—constructing explanations from evidence and engaging in argumentation (NRC, 2012).

Disciplinary literacy relies on disciplinary knowledge as well as an awareness of disciplinary traditions used to construct that knowledge (Shanahan & Shanahan, 2008). Practicing scientists engage in literacy for a range of purposes, such as explaining their findings, conducting research, connecting to the work of others, or sharing ideas with the public. The increasing diversity of available texts and the accumulation of scientific evidence requires a scientifically literate person to be able to critically interpret an author's stance and purpose, distinguish relevant from irrelevant evidence, and use complex information to make decisions. Some consider this repertoire of skills a fundamental civil right (e.g., Greenleaf & Hinchman, 2009; Lee, 2004), while others highlight the social justice aspects of developing disciplinary literacy in young adults (see Moje, 2007 for a review). Students across the globe learn in contexts wherein the language of instruction differs from the language of the home (UNESCO, 2011; Babaci-Wilhite, 2013; Skutnabb-Kangas, 2008). The specialized language of science becomes a 'common language' that all students can learn in order to participate in discipline-specific learning.

The start of our work on this project can be traced back to the *Seeds of Science/ Roots of Reading*®, a joint research and development project between the Lawrence Hall of Science and the Graduate School of Education at the University of California, Berkeley. This initial work sought to find the natural 'sweet spots' or synergistic commonalities between the practices of science and literacy (Cervetti et al., 2006, 2007). We founded all of our current curricular work on principles that capitalized on those synergistic sweet spots and have, so far, been met with very positive results (Cervetti et al., 2012). The first premise we followed is that students learn about the natural world through experiencing it in a "firsthand" way (e.g., direct observation) as well as in a "secondhand" way (e.g., reading about it) (Palincsar & Magnussen, 2001). We also explicitly taught students how to make predictions and inferences, to ask questions and access prior knowledge, and to summarize findings from text and experience, all of which are powerful cognitive strategies for learning and doing science and developing advanced literacy. From this strong foundation, we began to think about how we might expand these principles so that they were more suited to the specific needs of middle school students.

Our pivot towards an updated approach came at approximately the same time as the many researchers in both science and literacy education were calling for a shift towards more disciplinary-specific practices within content area classrooms. This shift also corresponded with how the CCSS initiative was implemented to address the challenge of "college and career ready" literacy in technical subjects like science and math, as well as the inclusion of overlapping core practices in the NGSS (for an illustration, see Stage et al., 2013). We recognized that the middle school environment (where students are more ready to pull many strategies together, and where they could begin to seriously take on the kinds of sophisticated, intellectual talk, reading and writing that are associated with scientific argumentation) would be the perfect place to pull all of the literacy-science threads together under one umbrella. The natural umbrella we found ourselves under—the one that tied all of the formerly disparate pieces together—was the disciplinary literacy practice of argumentation.

Argumentation as a Hallmark of Our Approach

In our curriculum, we aim to create a classroom culture in which argumentation is used to collaboratively construct science knowledge. In other words, we want the authority over knowledge to shift from residing solely in the teacher to being shared with students through a process of structured interaction in which complex science ideas are contemplated and shaped. Although middle school age students are fully capable of engaging in everyday argumentation, they do so differently than adults (Felton & Kuhn, 2001) and from scientists. Infusing carefully orchestrated argumentation sequences in our units provides a holistic and tangible ideal representation of science. Therefore, when we first conceive of a unit, we use an iterative process in which we break down the disciplinary core ideas to determine the conceptual build that will best convey what students will learn. We then consider the authentic contexts and topics that would provide argumentation opportunities that the particular discipline would require. Finally, we create an overarching science question, along with claims and evidence with which students can grapple over the course of a unit. In so doing, students can probe and answer a rich and meaningful question or address a difficult and relevant problem in a way that reflects what scientists in the sub-disciplines (geology, chemistry, astronomy, etc.) actually do.

An essential characteristic of this overarching structure is to pose questions as genuinely open-ended (i.e., without a single "correct" answer), so that the focus of interaction is on the co-construction of knowledge through investigation, analyzing data, and culling evidence from multiple sources to support a claim. For example, in one physical science unit in which students learn about phase changes, we propose the question: *Why is this lake on [the moon] Titan continually changing?* Students learn that the lake seems to disappear and reappear every seven years, and are asked to consider the claims that, *It is changing because it is evaporating?* or *It is changing because it is condensing and freezing?* Although the consensus within the scientific community is that the lake is evaporating, scientists have debated this in the recent past. The question and claims are legitimate and compelling enough to launch students into an authentic need to learn more about phase change, both at the larger physical scale, and at the molecular level. By the end of the unit, it doesn't matter if some students believe that the lake keeps 'disappearing' and 'reappearing' because it becomes a solid, because there still remains some evidence for this possibility. What is most important is for students to gather and evaluate evidence about the question of Titan's lake. Eventually, they take a final stand and make an argument about which claim is strongest and why, based on the evidence they have collected and the conceptual understanding they have gained throughout the unit. This structure is woven into the fabric of each unit to instantiate the disciplinary literacy that is unique to and essential for participation in science.

After establishing the overarching argumentation structure for a unit, we attend to the more fine-grained opportunities that students need in order to participate with confidence in the challenging cognitive work of scientific argumentation. At the lesson or activity level, our pedagogical approach is to embed the purposeful use of argumentation, while also supporting students' developing understanding of the practices involved. In this way, students learn, over time, both *how* an argument takes shape as well as *what* specific parts make up an argument (e.g., what constitutes a claim or supporting evidence). This dual understanding—teaching students about both the structure and function of an argument—gets to the heart of the disciplinary literacy focus that guides our work. To illustrate this approach, we now unpack several examples from our curriculum.

BUILDING STUDENT CAPACITY TO ENGAGE IN ARGUMENTATION

Argumentation is a practice that takes a considerable amount of dedicated classroom time for students to internalize the associated norms and exhibit the strategies, skills and habits that allow them to read, write, speak and evaluate scientific ideas with confidence and authority. However, research shows that students come to class prepared to make cases and consider opposing views, as they have been doing this kind of thinking in authentic ways from a very young age (Bricker & Bell, 2008). The goal of our curriculum is to offer students opportunities to learn the strategies and skills needed to improve their capacity to think about evidence while simultaneously providing deep content learning. This is surprisingly difficult to accomplish without first addressing the traditional classroom culture in which the focus of science is on conveying content information (i.e., facts) to students. In many science classes, there is an urge to pass on the canon of scientific knowledge and an emphasis on students getting this information 'right.' Therefore, students rarely encounter moments when their interpretations of ideas are given respectful attention. Even more rare is the opportunity to receive and evaluate 'messy' evidence. To foster authentic argumentation opportunities, our curriculum asks students to think critically and evaluate actual evidence. Because argumentation happens in the scientific community for a range of purposes: to provide others with evidence about new discoveries, to explain their thinking about the natural world, and to hold conversations about these important ideas, it is important for students to also participate in varying kinds of argumentation.

Practicing Oral Discourse Skills with an Evidence Sort

This first example of how we weave intentional argumentation activities into our curriculum is one that invites students to practice oral discourse skills, while also introducing a way to distinguish between relevant and irrelevant evidence. The following activity is taken from an Earth science unit about geologic formations. The topic authentically sets a purpose for argumentation through the introduction of a mysterious 'disappearing' city. The teacher unveils the following description:

> This is a mystery about a city that disappeared. In 1749, workers were digging in the countryside of Italy and accidentally uncovered a large object, which turned out to be the walls of a building. Ancient cities are often buried over time, so this wasn't so unusual. What was strange about this city was that much of it was buried under solid rock, not layers of dirt. Also strange was the fact that people were found inside their houses! Over the next 150 years, archaeologists continued to dig in this site, uncovering an entire city.

After seeing actual images of remains and ruins, students are asked to consider the same question that scientists asked after these remains and ruins were discovered, *What destroyed this city?* After students share their initial ideas, they are then presented with one possible claim *The city was destroyed by a sudden volcanic eruption.* They receive a set of cards, each printed with one of the following statements:

- There are active volcanoes found near where the city was found.
- Many bodies were found in the middle of doing everyday actions. For instance, one man was found pulling bread out of the oven.
- A new island was formed in 2006 by a volcanic eruption in the Pacific Ocean over 3,000 miles away.
- The city was buried under layers of ash and rock.
- There were over 16,000 people found buried in the city.
- Many people still live near Mt. Vesuvius today because the land is very good for growing crops.

Students read each statement and consider whether the statement would make a strong piece of evidence to support the claim, *The city was destroyed by a sudden volcanic eruption*. They rely on Argumentation Sentence Starters for Oral Discourse as a support for their partner discussion, such as:

I think this piece of evidence supports this claim because…
I don't think this piece of evidence supports this claim because…
I agree/disagree because…

We have found that sentence starters can be a friendly way to invite contributions to a large-group discussion such as an argument while also providing support for using the academic discourse associated with argumentation. Middle school students especially require a supportive classroom environment that encourages them to appropriate the language of science (Brown & Spang, 2008), while also providing a series of fading scaffolds (McNeill et al., 2006).

After students discuss the evidence in pairs, the teacher leads a whole class discussion about the possible evidence, prompting students to explain their thinking about why and how each piece of evidence does or does not support the claim. Students are also encouraged to think about the relationship between pieces of evidence, with questions such as, "If you had this evidence and nothing else, would you be convinced about this claim? Why or why not?" and "Is there another piece of evidence that you could combine with this evidence that would make you feel more strongly that the claim is being supported?" The teacher introduces the terms 'relevant' and 'irrelevant' and discusses the role of relevant evidence in an argument. The lesson concludes by connecting this practice back to what scientists actually do: weighing evidence against other evidence and a claim in order to test out the strength of both the evidence and claim, putting evidence together in ways that make sense, and listening to and agreeing and disagreeing respectfully with their peers.

Practicing oral discourse skills with an evidence sort builds students' awareness of the *structure* of an argument (e.g., a claim, with supporting evidence), but it also gives them a firsthand experience with the *function* of an argument in science. The product or artifact is not the completed evidence sort, but the talk itself. Given the low-stakes forum (peer-to-peer interaction), embedded supports (sentence starters and practice of the language of argumentation), streamlined content (preselected evidence) and real-world example (the mystery city was Pompeii), students can take small steps toward building a repertoire of discourse "moves." They might simply agree or disagree with or build on a statement made by someone else, or they may make a full-blown reasoned argument. Furthermore, the activity also serves another purpose—it provides students with background knowledge that they can draw on when investigating the relationship between rock layers and formations on Earth.

Through activities like the evidence sort, students can learn to see the oral argumentation as a medium through which scientists share ideas, disagree with each other, and actively engage in a discussion. Thus, students understand the role and importance of oral argumentation in a scientific community. The nature of this activity also allows students to practice reasoning and explaining the logic behind their ideas without focusing on a correct answer, tapping into the natural ability that even young students have to reason (McNeill, 2011). When students contemplate whether and how evidence relates to the claim itself, they will also realize how each piece of evidence relates to each other, and how this relationship builds support for the claim. In this way, they may begin to understand the complex nature of reasoning and the relative importance of evidence; one singular piece of evidence does not offer much support for a claim—a scientist must consider the relationships between individual pieces of evidence and the claim and must be able to explain how these relationships potentially build an argument (Jiménez-Aleixandre & Erduran, 2007; Driver et al., 2000; Duschel & Osborne, 2002). The next example addresses in more detail how we embed practice with reasoning into the practice of argumentation.

Developing a Strong Written Argument Using the Reasoning Tool

In contrast to the above example, this next example focuses more explicitly on building students' capacity to write cohesive arguments by attending to the essential component of reasoning. Reasoning is an integral part of creating sound, logical, and convincing arguments in science (Able, 2008; Bricker & Bell, 2008; Duschl & Osborne, 2002; McNeil, 2011; Osborne, 2010; Simon et al., 2002). In a written argument, reasoning is the "glue" that connects a claim with evidence; this connection is made more apparent to the reader through the use of cohesive linguistic tools. It is a particularly difficult component of an argument for students to develop because the writer must balance the demand of synthesizing the content with the act of composing the written structures that communicate the argument effectively. In this activity, students are introduced to a Reasoning Tool, a graphic organizer that serves as a scaffold for providing students with explicit practice with focusing on developing reasoning in written arguments.

Consider the everyday claim, *Vegetables are good for you.* Common evidence to support this claim might be, "Vegetables contain fiber and vitamins, such as Vitamin C." Without reasoning, this argument is incomplete; the reader is left wondering, "How are fiber and vitamins good for me?" Reasoning is the "so what" about a scientific argument. This type of everyday example provides a perfect opportunity to use the Reasoning Tool to explain why the evidence matters. For instance, in the context of a life science unit, students may have already learned some things about the role of fiber and vitamins in the human body. Through a guided discussion, the teacher models how to add reasoning statements to the argument using the Reasoning Tool.

Once students have the grasp of the Reasoning Tool, it can be applied to arguments about more complex scientific phenomena, such as the unique camouflage of a Luna Moth (*Actius luna*). Observe the photograph below and consider the following argument:

> Luna moths have camouflage to help them avoid predators. The Luna moth has green, yellow, and brown coloring, brown spots, and wings are curved and

Evidence	This evidence matters because...	Therefore, ...
Vegetables contain fiber.	Eating vegetables that contain fiber can reduce the risk of heart disease, obesity, and type 2 diabetes.	Vegetables are good for you. (Claim)
Vegetables contain Vitamin C.	Eating vegetables and fruits that contain a lot of Vitamin C can help heal wounds and protect against colds and allergies.	

Figure 3.1. Reasoning tool for everyday argument about vegetables

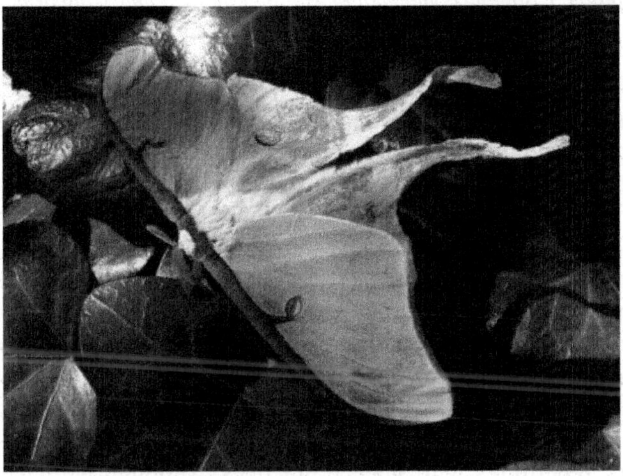

Figure 3.2. Luna Moth image and argument

shaped like a long brown stripe on top. The wings are curved and shaped like leaves. The wings look like they are piled on top of each other.

One possible claim to answer the question, *Why are Luna Moths camouflaged?* is *Luna moths have camouflage to help them avoid predators.* Several pieces of observational evidence are offered, but the argument doesn't provide the *reasoning* behind the evidence. Using the Reasoning Tool, a teacher guides students to make the connection between the claim and the evidence more explicit, thus improving the argument. Although the purpose of the Reasoning Tool is to practice reasoning through an argument, it also reinforces the structure of argument as consisting of a claim that answers a question about the natural world, evidence (information about the natural world that supports the claim), and reasoning, the tie that binds the argument together into a cohesive whole. Using the Reasoning Tool, students can write a more complete (and convincing!) argument.

From a disciplinary literacy orientation, the above example highlights how core science ideas (e.g., organisms interact in a system, species have characteristics that enable survival), discrete science knowledge (e.g., Luna moths are green and yellow

Evidence	*This evidence matters because...*	*Therefore, ...*
It has green and yellow coloring, brown spots, and a long brown stripe on top.	The coloring makes the moth look the same as the colors found on the tree on which the moth is sitting. The tree is brown, green, and yellow and has brown spots. The brown stripe on the moth looks like a stick on the tree. This helps the moth blend in.	Luna moths are camouflaged to look like leaves *so when they sit on trees* , they can avoid predators.
The wings are curved and shaped like leaves. The wings are piled on top of each other.	The leaves on the tree that the moth is on are layered on top of each other. The shape of the leaves is similar to the shape of the moth's wings.	

Figure 3.3. Sample completed reasoning tool for Luna Moth argument

with brown spots), the nature of scientific inquiry (e.g., scientists make observations that can be used as evidence) and authentic literacy practices (e.g., reading, writing, talking to support argumentation) come together seamlessly to create a collection of experiences that simultaneously supports the acquisition of discipline specific skills. This is truly a standards-driven approach. Essential to all of the above example activities is a proclivity on the teacher's part to build a community of science learners. Our approach to embedding this tenet is to explicitly build a culture of argumentation. The autonomy and ability to thoroughly participate in a series of activities where students discuss and make arguments about difficult conceptual material is made possible because they have been preparing for this kind of sophisticated work all along, throughout the unit (and those units that come before), which is why we embed many opportunities, such as evidence sorts and reasoning tools to gradually build students' repertoires with science practices. Although we have attempted to typify our approach to curriculum design using disciplinary literacy as a guiding orientation, the whole is definitely greater than the sum of the parts. Therefore, we offer one last example of a key culminating experience that serves as a cornerstone of each and every unit we design.

A Culminating Argumentation Experience: The Science Seminar

In the following example, students prepare for and engage in a whole-group argumentation discourse routine called a Science Seminar. We offer the science seminar series as an exemplar for this chapter because it contains virtually all components of argumentation that are essential in science, while at the same time holding students to rigorous standards for comprehending and explicating content. The Science Seminar provides a unique, egalitarian argumentation opportunity; all students face a new and unique scientific problem to which they must apply the content of a unit. Ideally, the topic of the Science Seminar has several plausible explanations so that students are able to feel the freedom to explore the problem without feeling constrained by reaching one 'right' conclusion. The teacher's role is to guide students to independently interpret the evidence they encounter and take a stance on an engaging science topic with their peers. Repeatedly, we have seen students of varying language abilities and degrees of participation thrive in this student-centered space. For instance, when we piloted the use of science seminars in multiple classrooms, we often observed students with language barriers, feelings of social anxiety or other potential inhibition, eager to engage in these discussions. Evidence from field notes and interviews with teachers who piloted our curriculum also noticed similar patterns. As one teacher stated:

> … every child could contribute and you talk about engagement – every child, even my lowest, lowest kids…I have this one boy who is so low, I mean he is language learning disabled…but he could spit out how we take in molecules from digested food…and I'm standing there going 'are you kidding me?' It just blows away everything you ever thought about those few who have such severe learning disabilities. (Teacher Interview, Fall, 2013)

In a way, the Science Seminar provided a level playing field for students of varying backgrounds and abilities, which is why we aim to seamlessly weave argumentation into the fabric of our units in ways that scientists (and science teachers) can get behind.

The Seminar itself typically occurs at the end of a unit, but is predicated by one or two days of preparation wherein students gather and weigh evidence that will be used to construct an argument. After the Seminar, students produce a written argument that reflects how their thinking has evolved and solidified over the course of the sequence. The ability to engage in a series of activities based on difficult conceptual material is possible because the students have been preparing for this kind of sophisticated work all along, throughout the unit (and those units that come before). During the unit, they learned content, but they also learned and practiced the structural and functional pieces of argumentation. Students we interviewed generally expressed that the format allowed them to learn and participate. One student explained:

> In the science seminar…it was really nice to be able to have everybody just listen to you, and you could speak freely because I know I'm like really shy sometimes, and sometimes I don't want to talk. But it was easier when it was something like that that we did… when we have discussions, I feel like it's—if I have like thoughts or questions, I feel like it's easier to ask like my peers first and like they can tell me what they know, and then if I don't get that answered, I can ask like the teacher. (Student Interview, 2014)

It is worth noting that this is a long-term investment, and it often takes years of practice for students to gain proficiency with all of the rich and complex pieces of argumentation, including a classroom culture of collaboration.

This Science Seminar example below is drawn from an Earth-science unit about wind and water currents. In this sequence, students evaluate evidence, discuss their thinking and finally prepare to write an argument that answers the question, *Why does the Atacama Desert get so little precipitation?* In the Atacama Desert Seminar, the science underlying the argument pertains to how average precipitation levels in a climate region depend on the amount of water vapor carried in the wind. The amount of water vapor in the air is affected by several factors, including air temperature, surface ocean temperature, and regional topography. In the Atacama Desert of South America, a specific pattern occurs called the rain shadow effect. When prevailing winds that are rich in water vapor hit a mountain range they are forced up to a higher, colder elevation. The air's capacity to hold water vapor quickly decreases with the temperature, and almost all the water vapor precipitates as rain on the windward slope. Once the wind reaches the other side of the mountain range, it has very little water vapor left. Hence, the leeward side of a mountain tends to be dry and have a desert-like climate. This is a complex natural phenomenon that students begin to grasp more deeply as they use argumentation to make sense of multiple forms of evidence.

Evidence provided to students before the seminar can come in many forms that are authentic to real science practices–graphs, tables, visual data, models, factual background reading and written testimonials. For instance, to prepare for the Atacama Desert Seminar, students first view and interpret (with guided instruction and discussion) a series of visual representations that offer them an initial understanding of the context and provide data showing that the Atacama Desert has received less than 8 inches (20 cm) in the past 33 years, making it one of the driest places on Earth. In addition, they are presented with the following claims, which help to guide and frame further analysis of evidence that they are provided:

> Claim 1: Prevailing winds on the Pacific coast cause extremely low precipitation in the Atacama Desert.
>
> Claim 2: The location of mountain ranges causes extremely low precipitation in the Atacama Desert.
>
> Claim 3: Surface temperatures of the ocean cause extremely low precipitation in the Atacama Desert.

Students spend the next day trying to answer the question about why the Atacama Desert is so dry by interpreting more evidence, evaluating the three competing claims, and reasoning about the possible explanations. To begin this difficult work, students work in pairs and read a map showing the average ocean temperatures and the following brief explanation of how ocean temperature affects precipitation:

> The amount of ocean water that evaporates varies with surface temperature. Cold ocean water evaporates more slowly than warm ocean water, forming less water vapor. Water vapor can condense into precipitation. Therefore, less water vapor leads to less precipitation. Ocean water that evaporates is carried in the direction of prevailing winds.

Students also receive a diagram showing how the rain shadow effect takes place, as well as data about winds, the average surface temperature of the nearby ocean, and other evidence that plays a role in determining the best answer to the question about what makes the Atacama desert so dry. All students are expected to analyze all of the available evidence, so that during the Seminar, all participants will be familiar with and can follow the thinking of their peers. Students we interviewed perceived how collaboration led to greater content understanding. For instance, one student noted:

> When we're working with a partner, you can kind of explain what you're thinking to them and also, if they don't get something, you can help them and also kind of clarify your own thoughts about that when you're explaining it to them. (Student Interview, 2014)

As students sort through multiple forms of evidence in preparation for the seminar, they consider which pieces of evidence they feel are more valuable for making a strong and convincing argument about the topic at hand. It is of the utmost importance for students to have the autonomy to build their own arguments by choosing the evidence that they consider to be the most valuable and convincing.

Argumentation is a very personal experience; while students should be guided and supported in learning the best ways to present and think about evidence and create cohesive arguments, it is also essential that they are provided with time to gain skills and perspectives that reflect their own understandings and beliefs. In the Science Seminar sequence, all evidence and thinking is valued, and there is no explicit effort to preference one set of evidence or one line of thinking over another. Because of this, students are free to focus on getting to know and making a strong case for one piece of evidence, or to tie various pieces together to form a logical argument. This careful orchestration of knowledge and practice ultimately results in a discussion that is open, inclusive and welcoming to students of all language backgrounds and levels. Many students noticed that the structure of the seminar increased participation. For example, a student stated:

> I think I like the conversation… if you're just in class, and you, and you're not having a conversation, and you raise your hand, the teacher might miss you, but it's easier in a conversation because… you have class time devoted to doing the conversation. So, you have more time, there's more people who will get a chance to talk. (Student Interview, 2014)

So, how does the actual Science Seminar work? To set up the actual Seminar, students are divided into two groups and seated in two concentric semicircles. Students seated in the inner circle are the first to address the competing claims by presenting relevant evidence, while the students seated on the outer semicircle actively listen and take notes. A volunteer begins by reviewing evidence connected to one of the possible claims. Thereafter, students engage in an authentic opportunity to agree or disagree productively and build knowledge collaboratively. After about ten minutes, students switch places and the students who were listening, then get to add their thinking to the discussion. At times, students may look to the teacher for authoritative input during a class discussion, but we encourage the teacher to guide the conversation back to the students themselves. Not only does this tactic result in students seeing their peers as intellectual resources, but it also cultivates a robust, egalitarian learning community. However, it is often useful to provide students with some general prompts that support the language of argumentation, such as:

"Do you agree or disagree with that idea? Why?"
"What other evidence could support the claim?"
"Is there another way we could explain this claim?"
"What other claims could you make? Based on what evidence?"

Although it takes time to build the culture of argumentation we mentioned previously, we have found through our firsthand experience of piloting units in actual classrooms that students respond well to these prompts. Both language frames and the language of argumentation increases students' adeptness with oral argumentation, which has been shown to supports students' abilities to write more structured arguments (Reznitskaya et al., 2001); this is what students do after participating in the Science Seminar. Because writing an argument about content is as much a function of good, clear writing as it is of knowing the content itself, students conclude each unit with a final piece of argument writing. The final written argument provides the teacher with a valuable tool for assessing students' learning through examining their uses of scientific language.

What we have described above details one example of the core disciplinary practice of argumentation in science and how it looks in the culminating experience in one middle school curricular unit. The Science Seminar sequence is often the high point of a unit, because it is structured to support student autonomy and taps into middle school students' seemingly innate tendencies towards wanting to discuss and productively argue many sides of an issue. It is critical that students are provided with the tools to acquire and practice argumentation skills, as these are embedded in the field of science itself. Students should be able to negotiate the meaning of complex text, identify evidence, and use these skills in the service of oral and written scientific argumentation. Colombi and Schleppegrell (2002) describe these types of skills as "advanced literacy." They state, "students need to move beyond basics to do science, history and other subjects; to construct arguments and critique theories; and to integrate print, visual, interactional, and electronic means of developing and sharing knowledge" (p. 2). The curricular approach we have outlined here is one avenue by which students may begin to acquire these skills. What we advocate for is that all students should have access to such instruction in order to have the necessary supports over time to develop these language skills. This access should be considered a fundamental purpose of education and right for those students whose needs we aim to serve (Lee, 2004; Babaci-Wihlite, 2013).

CONCLUSION

The NGSS has taken a strong stance in advocating for students to experience more literacy-related activities such as reading and writing scientific arguments in the science classroom. In addition, they describe a need for students to experience the nature and practices of science while learning science. It is our belief that if you plan for, attend to, and instill disciplinary literacy into every aspect of students' experiences in the classroom, you can accomplish the goals established by the NGSS (and the CCSS as well). On one level, the planning needed to do this is rather fine-grained and precise: you must expose students to the kinds of texts scientists in the discipline actually read—graphs, tables, computer readouts, more traditional texts, etc.—and then provide them with guidance and practice so that they eventually have confidence that they understand the best ways to interpret and comprehend these texts. You must also support students as they build an understanding of the components and structures of scientific arguments. This kind of planning focuses mainly on helping students to build an understanding of the structures involved in participating in science, and even in each specific field of science.

On another level, the planning is more overarching and woven into the fabric of the content itself. This is where we design to provide a culture of argumentation for all students through careful planning of the curricular direction. It is through these experiences that students begin to learn and then become adept at participating in the functional aspects of argumentation. Providing students with authentic and interesting scientific questions to consider gives a context where both deep content learning and deep immersion in argumentation can happen, as students are driven to learn more about the content and to collect evidence to better understand and then support or refute relevant claims. As students progress through a single unit, then a series of units that are built upon a foundation of disciplinary literacy and argumentation, they learn more about both the structure and function of arguments and argumentation in science. When it is embedded successfully and seamlessly, disciplinary literacy becomes the way for students to both learn science and to articulate this learning. When we collect evidence that our approach works with all students, it becomes clear why science curriculum design is a complex (and rewarding) endeavor.

REFERENCES

Able, V. (2008). When scientific knowledge, daily life experience, epistemological and social considerations intersect: Students' argumentation in group discussions on a socio-scientific issue. *Research in Science Education*, *38*, 67–90.

Aud, S., Wilkinson-Flicker, S., Kristapovich, P., Rathbun, A., Wang, X., & Zhang, J. (2013). *The condition of education* (NCES 2013-037). Washington, DC: National Center for Education Statistics.

Babaci-Wilhite, Z. (2013). A study of escalating debates on the use of a global or local language in education. In D. B. Napier & S. Majhanovich (Eds.), *Education, dominance and identity* (pp. 121–132). Rotterdam, The Netherlands: Sense Publishers.

Bricker, L. A., & Bell, P. (2008). Conceptualizations of argumentation from science studies and the learning sciences and their implications for the practices of science education. *Science Education, 92*(3), 473–498.

Brown, B. A., & Spang, E. (2008). Double talk: Synthesizing everyday and science language in the classroom. *Science Education, 92*(4), 708–732.

Carnegie Council on Advancing Adolescent Literacy. (2010). *Time to act: An agenda for advancing adolescent literacy for college and career success.* New York, NY: Carnegie Corporation of New York.

Cervetti, G., Pearson, P. D., Bravo, M. A., & Barber, J. (2006). Reading and writing in the service of inquiry-based science. In R. Douglas, M. Klentschy, & K. Worth (Eds.), *Linking science and literacy in the K-8 classroom.* Arlington, VA: NSTA.

Cervetti, G. N., Pearson, P. D., Barber, J., Hiebert, E. H., & Bravo, M. A. (2007). Integrating science and literacy: The research we have, the research we need. In M. Pressley, A. K. Billman, K. H. Perry, K. E. Reffitt, & J. M. Reynolds (Eds.), *Shaping literacy achievement.* New York, NY: Guildford.

Cervetti, G., Barber, J., Dorph, R., Pearson, P. D., & Goldschmidt, P. (2012). The impact of an integrated approach to science and literacy in elementary school classrooms. *Journal of Research in Science Teaching, 49*(5), 631–658.

Colombi, M. C., & Schleppegrell, M. J. (2002). Theory and practice in the development of advanced literacy. In M. Schleppegrell & M. C. Colombi (Eds.), *Developing advanced literacy in first and second languages: Meaning with power* (pp. 1–19). Mahwah, NJ: Lawerence Erlbaum.

Davis, E. A., Palincsar, A. S., Arias, A. M., Bismack, A. S., Marulis, L. M., & Iwashyna, S. K. (2014). Designing educative curriculum materials: A theoretically and empirically driven process. *Harvard Educational Review, 84*(1), 24–52.

Draper, R. J., & Siebert, D. (2009). Content area literacy in mathematics and science classrooms. In S. R. Parris, D. Fisher, & K. Headley (Eds.), *Adolescent literacy, field-tested: Effective solutions for every classroom* (pp. 105–116). Newark, DE: International Reading Association.

Driver, R., Newton, P., & Osborne, J. (2000). Establishing the norms of scientific argumentation in classrooms. *Science Education, 84*(3), 287–312.

Duschl, R. A., & Osborne, J. (2002). Supporting and promoting argumentation discourse in science education. *Studies in Science Education, 38*, 39–72.

Felton, M., & Kuhn, D. (2001). The development of argumentive discourse skill. *Discourse processes, 32*(2–3), 135–153.

Fields, R. (2014). *Towards The National Assessment of Educational Progress (NAEP) as an indicator of academic preparedness for college and job training.* Washington, DC: National Assessment Government Board.

Greenleaf, C. L., & Hinchman, K. (2009). Reimagining our inexperienced adolescent readers: From struggling, striving, marginalized, and reluctant to thriving. *Journal of Adolescent & Adult Literacy, 53*(1), 4–13.

Guthrie, J. T., & Davis, M. H. (2003). Motivating struggling readers in middle school through an engagement model of classroom practice. *Reading & Writing Quarterly, 19*(1), 59–85.

Hamilton, L. S., Stecher, B. M., & Yuan, K. (2012). Standards-based accountability in the United States. *Education, 3*(2), 149–170.

Harris, D. M. (2012). Varying teacher expectations and standards: Curriculum differentiation in the age of standards-based reform. *Education and Urban Society, 44*(2), 128–150.

Heller, R., & Greenleaf, C. (2007). *Literacy instruction in the content areas: Getting to the core of middle and high school improvement.* Washington, DC: Alliance for Excellent Education.

Jiménez-Aleixandre, M. P., & Erduran, S. (2007). Argumentation in science education: An overview. In M. P. Jiménez-Aleixandre, S. Erduran, & M. Pilar (Eds.), *Argumentation in science education: Perspectives from classroom-based research* (pp. 3–27). Rotterdam, The Netherlands: Springer.

Kesidou, S., & Roseman, J. E. (2002). How well do middle school science programs measure up? Findings from Project 2061's curriculum review. *Journal of Research in Science Teaching, 39*(6), 522–549.

Kuhn, D. (2010). Teaching and learning science as argument. *Science Education, 94*(5), 810–824.

Lee, C. (2004, Spring). Literacy in the academic disciplines and the needs of adolescent struggling rwzzzeaders. *VUE, 3*, 14–25.

Lee, C. D., & Spratley, A. (2010). *Reading in the disciplines: The challenges of adolescent literacy.* New York, NY: Carnegie Corporation.

McNeill, K. L. (2011). Elementary students' views of explanation, argumentation, and evidence, and their abilities to construct arguments over the school year. *Journal of Research in Science Teaching, 48*(7), 793–823.

McNeill, K. L., Lizotte, D. J., Krajcik, J., & Marx, R. W. (2006). Supporting students' construction of scientific explanations by fading scaffolds in instructional materials. *The Journal of the Learning Sciences, 15*(2), 153–191.

Moje, E. B. (2007). Developing socially just subject-matter instruction: A review of the literature on disciplinary literacy teaching. *Review of research in education, 31*(1), 1–44.

Moje, E. B. (2008). Foregrounding the disciplines in secondary literacy teaching and learning: A call for change. *Journal of Adolescent & Adult Literacy, 52*(2), 96–107.

National Governors Association Center for Best Practices & Council of Chief State School Officers. (2010). *Common core state standards for english language arts and literacy in history/social studies, science, and technical subjects.* Washington, DC: Authors.

National Research Council. (2007). *Taking science to school.* Washington, DC: The National Academies Press.

National Research Council. (2012). *A framework for K-12 science education: Practices, crosscutting concepts, and core ideas.* Washington, DC: The National Academies Press.

NGSS Lead States. (2013). *Next generation science standards: For states, by states.* Washington, DC: The National Academies Press.

Oliveira, A. W., Akerson, V. L., & Oldfield, M. (2012). Environmental argumentation as sociocultural activity. *Journal of Research in Science Teaching, 49*(7), 869–897.

Osborne, J. (2010). Arguing to learn in science: The role of collaborative, critical discourse. *Science, 328*(5977), 463–466.

Palincsar, A. S., & Magnussen, S. J. (2001). The interplay of firsthand and text-based investigations to model and support the development of scientific knowledge and reasoning. In S. Carver & D. Klahr (Eds.), *Cognition and instruction: Twenty-five years of progress* (pp. 151–194). Mahwah, NJ: Lawrence Earlbaum.

Pearson P. D., Moje E. B., & Greenleaf, C. (2010). Literacy and science: Each in the service of the other. *Science, 328*(5977), 459–463.

Reiss, M. J., Millar, R., & Osborne, J. (1999). Beyond 2000: Science/Biology education for the future. *Journal of biological education, 33*(2), 68–70.

Reznitskaya, A., Anderson, R. C., McNurlen, B., Nguyen-Jahiel, K., Archodidou, A., & Kim, S. (2001). Influence of oral discussion on written argument. *Discourse Processes, 32*(23), 155–175.

Schwartz, R. S., Lederman, N. G., & Crawford, B. A. (2004). Developing views of the nature of science in an authentic context: An explicit approach to bridging the gap between nature of science and scientific inquiry. *Science Teacher Education, 88*, 610–645.

Shanahan, C., & Shanahan, T. (2014). The implications of disciplinary literacy. *Journal of Adolescent & Adult Literacy, 57*(8), 628–631.

Shanahan, T., & Shanahan, C. (2008). Teaching disciplinary literacy to adolescents: Rethinking content-area literacy. *Harvard Educational Review, 79*(1), 40–59.

Shanahan, T., & Shanahan, C. (2012). What is disciplinary literacy and Why does it matter? *Topics in Language Disorders, 32*(1), 7–18.

Simon, S., Erduran, S., & Osborne, J. (2002). *Enhancing the quality of argumentation in school science.* Paper presented at the Annual Meeting of the National Association for Research in Science Teaching, April 7–10, 2002, New Orleans, LO.

Skutnabb-Kangas, T. (2008). Human rights and language policy in education. In N. Horberger (Eds.), *Encyclopedia of language and education* (Vol. 1, pp. 107–120). New York, NY: Springer.

Stage, E., Asturias, H., Cheuk, T., Daro, P. A., & Hampton, S. B. (2013). Opportunities and challenges in next generation standards. *Science, 340*, 276–277.

Stevens, R. J. (2003). Student team reading and writing: A cooperative learning approach to middle school literacy instruction. *Educational Research and Evaluation, 9*(2), 137–160.

UNESCO. (2011). *Competency framework for teachers*. Paris: United Nations Educational, Scientific and Cultural Organization.

Zeichner, K. (2012). Educating teachers for cultural diversity in the United States. *Teacher Education in Plural Societies* (RLE Edu N): *An International Review*, 141.

CURRICULAR RESOURCES

The Learning Design Group. (2014). *Engaging in argumentation with a science seminar: Regional climate in the Atacama Desert. Strategy guide to developing literacy practices in science*. Berkeley, CA: Lawrence Hall of Science, University of California at Berkeley.

The Learning Design Group. (2014). *Reteaching loop: Practicing oral discourse skills. Strategy guide to developing argumentation practices in science*. Berkeley, CA: Lawrence Hall of Science, University of California at Berkeley.

The Learning Design Group. (2014). *Reteaching loop: Using the reasoning tool to develop a strong written argument. Strategy guide to developing argumentation practices in science*. Berkeley, CA: Lawrence Hall of Science, University of California at Berkeley.

Megan M. Goss
The Lawrence Hall of Science
University of California, Berkeley
USA

Helen Min
The Lawrence Hall of Science
University of California, Berkeley
USA

Carrie Strohl
University of California, Davis
USA

PART II

EQUITY AND CRITICAL PEDAGOGY IN TECHNOLOGY AND HUMAN RIGHTS EDUCATION

JABARI MAHIRI AND JEREMIAH SIMS

4. ENGINEERING EQUITY

A Critical Pedagogical Approach to Language and Curriculum Change for African American Males in STEM

CARNARIES IN THE MINESHAFT

Skinner (2008) argued that a community's health can be gauged by how well black men and boys are doing within it. Similarly, Noguera (2008) suggested that black males can be viewed as canaries in a mineshaft in that they reflect and react to the highest levels of toxicity in U.S. society. "With respect to health, education, employment, income, and overall well-being," Noguera noted, "all of the most reliable data consistently indicate that black males constitute a segment of the population that is distinguished by hardships, disadvantages, and vulnerability (2006, p. 11). These disadvantages and vulnerabilities are even more pronounced in STEM education and careers.

According to reports from the Level Playing Field Institute (Scott, 2010; Scott & Martin, 2012), the STEM opportunity gap for African American students is profound. For example, just 43% of African American students reach proficiency in 5th grade science compared to 80% for Asians and whites. By 6th grade with respect to math proficiency, forty-six percentage points separate African American (35%) from Asian students (81%). Throughout middle and high school, proficiency rates in math and science continue to decline ensuring that African American students are less likely to access and be successful in rigorous college preparatory coursework in these subjects. Predictably, very few graduate from college with degrees and career opportunities in STEM, the fields with seven of the top 10 fastest-growing occupations between 2010 and 2020 (Bureau of Labor Statistics, 2009).

These statistics are troubling for African American students generally, and though not disaggregated by gender, they are even more problematic for African American males who have the highest school dropout rates of any demographic category. For example, data on black males and public education in all 50 states in the U.S. indicate that they remain at the bottom of high school graduation rates in all but 13 states, and in those 13 states, Latino males are on the bottom (Schott Foundation, 2015). Clearly, there is a need to dramatically change the language, curriculum, and pedagogy utilized for learning STEM as well as other academic subjects as a human right in education for African Americans and, ultimately, for all students.

Z. Babaci-Wilhite (Ed.), Human Rights in Language and STEM Education, 55–70.

In 2011 Sims, referred to as Mr. J in this chapter, took charge of developing a unique, out-of-school education program aimed at transforming the language and pedagogy of curricular approaches to learning STEM in order to counter some of the problematic constraints on African American male academic achievement in these subject areas and related careers. Mahiri was a key contributor to the design, implementation, and research of this initiative since its inception. The program has three core classes – math, computer science and mobile apps (CS), and the rhetorical analysis of manhood (RAM) – that are taught on two Saturdays each month throughout the school year to African American males in each of the three grades of middle school.

Hosted and funded by a neighborhood-based non-profit organization located in the San Francisco Bay Area, the participants are mainly drawn from under-resourced, under-performing public schools in several Northern California, urban municipalities. The informal acronym of the program is MAN UP (Male Aptitudes Nurtured for Unlimited Potential). With origins in African American vernacular, "MAN UP" has several meanings. One is to rise to the occasion to complete a task. Another is to take care of one's responsibilities. A third is to prepare oneself for a fight. The team of instructors led by Mr. J (RAM), Mr. K (math), and Mr. S (CS), along with Mahiri developed and implemented the program to leverage all three meanings, in part, by seamlessly incorporating African American language, culture, and daily life experiences into STEM learning. Within this framework we designed learning experiences that uniquely nurtured the participants' aptitudes – their innate or acquired capacities and talents – for STEM.

We consciously and collaboratively instituted a critical pedagogical approach as a deliberate way to develop the students' identities with and competencies in STEM as well as their capacities to make socially just applications of STEM. In line with this approach, we saw the linguistic, theoretical, and discursive components of rhetorical analysis as a tool kit that was central to the students' development because it afforded them with critical perspectives to not only absorb STEM, but to cognitively wrestle with the applications and ethics of STEM knowledge and skills. Although RAM classes were key to this approach, critical pedagogical practices were woven throughout the teaching and learning of all three courses.

The authors of this chapter decided to formally study the impact of this critical pedagogical approach for learning and making identity connections to STEM for the young African American male participants during the 2013–14 academic year. Although MAN UP served a total of 43 students in 6th, 7th, and 8th grades, we focused on the 17 students in the first year of MAN UP (fourteen 6th graders and three 7th graders) to capture how they responded to and developed through the program during their first year of participation. Data collected on the first year MAN UP cohort included pre and post surveys, individual and focus group interviews with the students; interviews with the instructors of the science/technology and math classes; participant observation, video recordings, and field notes on class activities and field trips; and the students' writing in online reflective journals.

By the end of the 2014 academic year, we found that the first year cohort had significant increases in key STEM competencies as well as in their personal identification with STEM content and careers. Additionally, we found substantial increases in their understanding and implementation of socially just applications of STEM. Perhaps the overall impact of MAN UP was summed up by one student who commented, "There aren't a lot of African American males in STEM, but opportunities can open up because of what we are doing here. We are raising expectations."

MINING STEM POTENTIAL

Much has been written about the problems of a digital divide between African Americans and others, and similar considerations also attend to STEM disparities. In this section, however, we review scholarship that offers perspectives and prospects for ameliorating these issues through more comprehensive understandings of effective ways to build on the cultural and intellectual assets of African American youth. These understandings of STEM come from research and scholarship from the social sciences, from multicultural education, and from critical pedagogy. In combination, they offer counter considerations to the pathological framing of African American males in education, in society, and in the media.

Academic performances of African Americans and other marginalized groups can be characterized in terms of opportunity gaps rather than achievement gaps. Since opportunities in a society are selectively constructed and socially structured, they can be re-constructed through a more culturally connected language of instruction and designs of curriculum that make educational achievement a right for all students. In line with this, scholarship reviewed in this section reveals ways that African American students develop identities with and competencies in STEM subjects when exposed to pedagogies that are linguistically and culturally relevant and that connect STEM learning to socially beneficial applications.

Connecting Cultural-Linguistic Identities and STEM Competencies

Steele's (2010) work on individual and group stereotype threats provides an expansive lens for viewing how identity contingencies and other socially constructed obstacles work to mitigate the academic achievement of African Americans. Stereotype threats are felt in specific situations in which stereotypes associated with one's individual or group identity are prevalent. Mere awareness of the stereotype can be distracting enough to negatively affect a person's performance in the domain connected to it.

Negative stereotypes of African Americans' academic achievement in STEM subjects can be debilitating, but their affects can also be circumvented by a variety of interventions. One mitigating approach that has been identified is providing critical feedback along with opportunities for analysis of the larger societal context and structures that motivate stereotypes (Steele, 2010). In this regard, research is also beginning to show how some African American students are achieving in STEM by developing abilities to better manage or "flip" the negative impacts of stereotype threat (McGee & Martin, 2011). Another approach extends from how teachers can work to foster feelings of identity safety in how classroom discourse and culture are constructed to access rigorous curriculum content (Mahiri, 2011). Additionally, students' authentic lived experiences must be valued and incorporated into the overall learning of STEM (Emdin, 2010).

Many scholars have shown the importance of connecting positive academic identities to the cultural-linguistic identities and lived experiences of students themselves. For example, Nasir (2011) and Tate (1995) demonstrated ways that positive academic identity development is at the root of academic success in math for African American students. Nasir's (2011) research found a positive reciprocal correlation between positive academic and cultural identity development and increased math performance for African American students. Lee (2005) illustrated the importance of attending to specific English dialects of urban students to directly build on their cultural-linguistic competence for learning curriculum content. Lee (2008) further showed the significance of accessing cultural identity to effectively teach African American students by synthesizing findings from numerous research studies supporting critical connections between language, identity, and learning.

Essentially, linking learning to students' cultural-linguistic diversity and competences is the cornerstone of approaches to multicultural education (Banks, 2007) and culturally relevant pedagogy (Ladson-Billings, 1995). Banks noted, "The more we know about a student's level of identification with a particular group and the extent to which socialization has taken place within that group, the more accurately we can predict, explain, and understand the student's behavior in the classroom" (2007, p. 27). Similarly, the central idea of culturally relevant pedagogy is that if learning structures and stimuli are grounded in a cultural-linguistic context that is familiar to students, there will be greater potential for cognitive expansion and knowledge growth (Allen & Boykin, 1992; Ladson-Billings, 1995). Gay (2000, 2010) expanded upon the framework of culturally relevant pedagogy by delineating methods of culturally responsive teaching that specifically used cultural characteristics, experiences, and perspectives of ethnically diverse students to increase their learning. Five essential components of this approach are developing a knowledge base about cultural diversity, including diverse curriculum content, demonstrating caring and building learning communities, cross-cultural communication, and cultural congruity in instruction. In working to build the African American middle schoolers' identification with STEM, the MAN UP instructors employed each of these components throughout the program.

Connecting STEM Learning and Social Justice

Social justice as an educational framework is integral to critical pedagogical approaches to teaching and learning (Gorski, 2013; Giroux, 2011; Duncan et al.,

2008; Smyth, 2011). These approaches attempt to equip marginalized students with the theoretical lenses necessary to disrupt the inequitable conditions they face and work to navigate. Extending from the work of Freire (1971) and further informed by scholars like Giroux (2001) and McLaren (1994), critical pedagogy seeks to facilitate this transformation by empowering students to locate and develop their own critical, agentive voices and productive capabilities. A central goal of critical pedagogy is to disrupt the "banking model" of education – the positioning of learners as empty vessels to be filled with ideas and information that continues their oppression.

Limitations of the banking model outlined by Freire have been addressed in STEM subjects as the "absorption" model. Linn and Eylon (2011) noted hallmarks of the absorption approach in science instruction as requiring students to listen to lectures, read textbooks and complete exercises, and conduct experiments or investigations following step-by-step procedures. Fundamental aspects of the absorption approach do not build on individual interests and intuitions or cultural-linguistic backgrounds and competencies of learners. Linn and Eylon (ibid) argued that when these constraints are circumvented by a more active and agentive knowledge integration approach, then "everybody can learn science" (ix).

Proponents of critical pedagogy argue, however, that opening access to STEM subjects and careers to everybody, by itself, is not enough. They feel that though access to high quality STEM instruction is a civil and human right, a larger critique of the positioning and roles of STEM in the very processes of oppression and marginalization must accompany increasing access to STEM content. They argue against attempts to limit these kinds of critiques to the social sciences and humanities and advocate that these perspectives are just as imperative to the learning of STEM subjects. Blikstein (2008), for example, suggested that knowledge and use of digital technology, and particularly the design of new digital devices, derived from math and science are necessary tools for fulfilling Freire's vision of humanization and societal transformation.

Studies conducted in after school contexts (Vakil, 2014) and within school contexts (Norris, 2014) found significant ways that critical pedagogical approaches enhanced African American and Hispanic middle school students' social justice perspectives in conjunction with their learning STEM subjects. Moses and Cobb's (2001) work with the Algebra Project earlier connected these kinds of efforts to the socio-political implications of how math (and algebra specifically) is both a gateway to and gatekeeper of STEM content and careers.

African American males more than any other group have been marginalized the most by these sySTEMic forces in education generally and in the learning of math specifically (Darder, 1997). To counter this, Gutstein's (2005) argued for a version of math education that affords hyper-marginalized students a nutritive educational space where they can not only learn math, but also where they can learn to apply math to issues that inform and affect their socio-political and socio-economic realities in the larger societal context. In this regard, Gutstein noted, "mathematical literacy, as a form of functional literacy distinct from critical literacy, serves the needs of capital accumulation in the United States" and further that "A reconceptualization of the purpose of mathematics education is needed – one that includes envisioning mathematical literacy as critical literacy for the purpose of transforming society, in its entirety, from the bottom up toward equity and justice, for all students whether from dominant or oppressed groups" (2005, p. 28).

METHODS

The central question for this study was: What is the impact on African American middle school males of a critical pedagogical approach for developing identification with, competencies in and socially just applications of STEM? We used qualitative methods for collecting and analyzing data guided by the work of Denzin and Lincoln (2003). We also utilized the work of Spradley (1979) with respect to conducting interviews with the students and their instructors. In addition to interviews, we made audio/video recordings, and took field notes on the study participations' various learning activities during their Saturday classes and fieldtrips. We also provided opportunities for the first year cohort to express considerations about their learning and social experiences as African American males in STEM.

The program took place every other Saturday from September 2013 to June 2014 for a total of 22 sessions. Once for each of the two semesters, there was an additional Saturday session devoted to a field trip to places like the Exploratorium in San Francisco and the Chabot Science and Space Center in Oakland. Each Saturday session lasted five-hours from 9:30am to 2:30pm. In addition to various whole group activities, the three 75-minute classes on math, computer science, and the rhetorical analysis of manhood took place simultaneously with each of the three grade levels rotating to the instructor for each course with content pitched to the appropriate grade.

Site

The host organization negotiated space for the program to take place at an urban charter middle school that was centrally located among the various municipalities in which the majority of participants lived. This site had classrooms equipped with mobile desks on wheels, smart boards, and there were several large open spaces suitable for group activities. Additionally, there was a set of 20 chrome books available for use by the instructors. A few participants in MAN UP also attended the charter school that was the site for this Saturday program, but the program had no formal relationship with the school.

Participants

Although there were 43 participants in the program across the three grade levels, the focal participants for this study were the 17 students in their first year of the MAN UP program in sixth grade along with their instructors. The first year cohort was selected to explore the impact of the program during their first year of participation. A key demographic characteristic of these 17 first year students was that 80% qualified for free or reduced lunch and thus considered to be from low-income families. Eighty percent also attended Title 1 public schools. Sixty percent were from single parent homes, and 60% had parents or guardians that did not have a college degree. The average GPA of participants was 2.99 because the program's selection process focused on students who were already achieving some measure of success in school.

The instructors for the three Saturday classes were men. Mr. J, an African American who directed the program and taught the rhetorical analysis of manhood (RAM) classes, was also a doctoral student in education. The instructor for the computer science and mobile apps (CS) classes, who the students called Mr. S, was Iranian and also a doctoral student in education. Additionally, he had a master's degree in engineering. The instructor for the math classes, who the students called Mr. K, was an African American who was also a public school math teacher.

Data Collection

Data collected on the focal cohort included pre and post surveys on STEM identification and program satisfaction as well as academic content assessments through pre and post concept inventories. Also, individual and focus group interviews were videotaped with the students and transcribed to gain insights into how they were experiencing the program across the school year with particular emphasis on their work on the central projects in each class. Interviews with the instructors of the math and computer science classes were videotaped and transcribed to get their perspectives on student learning linked to the curriculum. Participant observation, video recordings, and field notes on class activities and field trips were also utilized for the duration of the study.

Data Analysis

Across all data sources we looked for evidence of the focal participants' developing identification with, competencies in, and critical application of key aspects of STEM that were taught in their three courses and the other activities of the MAN UP program. Consequently, our fundamental approach was to analyze the entire range of data sources through the overlay of these three categories of identification, competencies, and critical applications. Since we wanted to see the students' initial and on-going *development* in these three categories with respect to STEM, we analyzed for each category through the various data sources collected during the first and second semesters. Therefore, we were able to identify, code, and analyze evidence of development of the focal participants in all three classes across the entire academic year of the study.

FINDINGS

Our findings on the focal students in the MAN UP program are reported in the three central categories that were delineated in the research question for this study. Over the course of the 2013–14 academic year, we found that the focal cohort had developed significant increases in their personal identification with STEM subjects and careers as well as increases in key STEM competencies. Additionally, we found substantial increases in their understanding and implementation of socially just applications of STEM. Evidence for these findings came from all of our data sources, and five of the main sources were the pre and post identification and program satisfaction surveys, the pre and post concept inventories, focal group interviews, interviews with the math and CS instructors, and the participants' work on final projects in each of their three classes that was documented with videos.

Developing Identification with STEM

We felt that creating a space for the students to develop a positive STEM identity was not only exigent, but also a prerequisite for increases in STEM competencies. And while opportunities for positive STEM development were intentionally created in all three courses, the rhetorical analysis of manhood (RAM) course was a primary space for conversations around African-American male identities vis-à-vis STEM education. So, along with pre and post STEM identification surveys, year-end focus group interviews, and interviews with the instructors of the other two courses, the manhood course attempted, to develop and assess shifts in the focal students' identification with STEM as evidenced by their confidence in applying STEM to important issues and in their expressed desires to major in STEM, and also to, ultimately, pursue STEM careers.

Pre and Post STEM Identification Surveys

A pre-survey was administered during the program's first session in September of 2013, and a post-survey with the same questions plus additional ones on participants' satisfaction with the program was administered during the final academic session in May of 2014. The first prompt on these surveys asked students to "Please draw a scientist, technologist, engineer and/or a mathematician. (Please label your drawing)." On the pre-survey 14 of the 17 focal students drew a balding, bearded, middle-aged European American male in a white lab coat. None of the respondents drew a person of color. One student drew a European American woman, and the remaining two drew extraterrestrial aliens. On the post surveys, 16 of the 17 focal students drew STEM professionals that were clearly African-American males. Only one drew a picture of a balding, bearded, middle-aged European American male in a white lab coat. In fact, eight of the students drew pictures of themselves.

The pre-survey also asked students whether or not they saw themselves as future STEM professionals, and only three of the 17 students from the focal class agreed. On the post-survey, 16 or 17 agreed that they saw themselves potentially as STEM professionals. For example, they indicated considerable increases in their aspirations for pursuing STEM in high school, college, and careers in the post-survey in contrast to the pre-survey. Clearly, the focal students initially held rigid conceptions of what STEM practitioners looked like, but over the course of the program, their perceptions dramatically changed to seeing people who looked like them, including actually picturing themselves as STEM practitioners.

Additionally, in response to post-survey questions about overall satisfaction with the program, 15 of the 17 participants indicated that their abilities to identify with STEM subjects and careers was highly impacted by having instructors who were themselves role models of STEM practitioners. For example, one student responded, "I benefited mentally seeing role models of my color skin." Another noted how important it was "getting help from teachers who inspired me and were my role models." A third respondent wrote on his post-survey that "they helped me see myself as someone who could do well in math and science." The participants also indicated satisfaction with their own perceptions of having increased their identification with STEM through completion of the program and "getting together with kids my age and race to do this work."

The instructors further reinforced the participants' identification with STEM by intentionally calling them "applied STEM practitioners" throughout the program. Interviews with Mr. S, the CS instructor, and Mr. K, the math instructor were conducted by Mr. J at the close of the 2013–14 academic-year. Mr. K felt that positioning and referring to the participants as applied STEM practitioners was important to their identity shifts. "I feel like this approach worked," he said in his interview. "They responded to it and tried to own up to the responsibility of their titles of applied mathematicians… They were owning their math identity."

Connecting STEM Identities in the Courses

In all three MAN UP courses emphasis was placed on positioning students as "applied STEM practitioners." This was seen, in part, in the work on final projects that the students completed for each class. For math class the focal students created a symbolic mathematical representation in the form of a crest or insignia using linear equations to represent them as young scholars to the outside world. The symbol they created was a mallet breaking a large wall. At the year-end celebration, the class described the mathematical processes they mastered to create their insignia along with its symbolic implication. Speaking for the group, one student explained, "We created a mallet because in this program we are creating symbols using math… If that doesn't smash stereotypes about us, I don't know what will… We like the mallet because rubber is malleable… We are, too."

In computer science the focal class contributed to the creation of a mobile application to raise environmental awareness that was led by the 8th graders. The focal cohort also contributed to its creation. Though the focal class did not lead this project, their understanding of their identities as STEM producers rather than consumers was reinforced through it. In focus group interviews, for example, one focal student reported that he now "feels connected to the producer role" because in developing this app he was "doing the producing instead of just talking about it." Another focal student commented that this process had "convinced" him that can and he will be a video game engineer that "creates [produces] good video games with positive African American male role models, unlike GTA [Grand Theft Auto]."

Similarly, the final project in the rhetorical analysis of manhood class to curate a production of diverse digital narratives modeled the museum installation called "Question Bridge Black Male" also reinforced STEM identities in the focal students in a variety of ways. Mahiri, who served as one of the advisors for the development of Question Bridge Black Males <http://questionbridge.com/> introduced the MAN UP instructional team to this exhibit, and the program eventually took all of the students on a field trip to see it at a local museum.

In RAM the goal of addressing issues of identity was explicit, and the students brought what they were learning about rhetorical analysis to the "re-production" of a Question Bridge as a final class project. The focal students videotaped themselves and students in the other two classes addressing questions of black male identity and life experiences as modeled in the museum exhibit. When asked why they chose this project, students talked about how they wanted to smash stereotypes around African American males in STEM and how they felt that using technology to counter stereotypes of African American males was analogous to "performing their argument."

Developing Competencies in STEM

In addition to developing the participants' STEM identities, the MAN UP program also focused on developing specific STEM competencies required for college and careers. To assess shifts in math learning, we used a core "concept inventory" based on California State Standards and the Common Core State Standards. Different concept inventories were developed for CS and RAM that reflected specific content addressed in those classes. Like the pre and post surveys, the concept inventories (except for the RAM class) were administered at the beginning and end of the 2013–14 academic-year.

Pre-post Concept Inventories

The math course aimed to prepare students to continue to be successful in grade-level math and also provide foundational skills and dispositions necessary to be successful as they advanced to higher levels of math. The math concept inventory captured the focal participants' pre and post knowledge in three content areas: ratios and proportions, geometry, and statistics. The focal students demonstrated growth across all three content areas with a 14% increase in the number of correct answers on the concept inventory. Interestingly, those students whose family income qualified them for free or reduced lunch made the highest gains of the group.

Their responses in focus group interviews and on the program satisfaction surveys indicated that they also saw increases in their math competencies. For example, all but one reported that they had learned a lot and that the program had improved their math skills. They spoke highly about ways that Mr. K's math course affected them. Many noted that they were ahead of the math they were learning in school as a result of participation in MAN UP. As one student said, "When I went to my math class, everyone was struggling except me because I had already learned it here," Another shared, "I always really liked math, but Mr. K helped me explore math and what it is and how to use it instead of just learning it."

Similar developments in STEM competencies were achieved in the computer science/mobile apps class. Mr. S taught students how to design and build mobile apps using App Inventor programming language. However, beyond learning to program, the course focused on big ideas of computing including abstraction, design, recursion, simulations, and the limits of computation. The mobile apps concept inventory consisted of 8 items and asked students to evaluate their skill level and ability to complete skills tasks, based on a 3-point Likert scale ranging from "not at all true" to "very much true." The students showed significant gains in their knowledge and skills with increases on every item including understanding how to create a storyboard and implementation plans for apps, knowing what a mock-up is, understanding how to create a design rationale, and understanding how to use app programming software.

The participants' interview and satisfaction survey responses confirmed that they felt they were developing important STEM competencies. For example, 16 of the 17 students agreed that the program taught them a lot about computer science and that they now felt that computer science was interesting and fun. As one student noted, "I've never had a computer science class before. This was all new to me. Mr. S was so patient, and he really knows his stuff. I'm beginning to think that I really can be a video game designer."

Finally, the rhetorical analysis of manhood class offered a unique space for seeing how the first year cohort was developing competencies in STEM as well as critical language skills. This course explored conceptions of manhood generally, and conceptions of African American manhood specifically in terms of definitions; individual, group, and societal perceptions; and the power relationships and media representations that influence all of these considerations. The students learned to use principles of rhetorical analysis to understand and critique the various considerations of manhood surrounding African American males and honed their presentation and technical skills through the Question Bridge like video project.

The initial concept inventory for RAM was administered at the beginning of the second semester as opposed to the start of the first semester for the other two classes, and the post concept inventory was administered at the end of the program. The 19 questions on this inventory sought to ascertain students' conceptions of manhood. Data from these questions indicated that many students initially understood manhood as a static positionality that was predicated exclusively on the ability to provide both financial and physical security for women and children. This static conception of manhood shifted over the course of the second semester to a more malleable understanding of manhood as connected to being a caretaker, but also having responsibilities for improving the larger community. This shift was captured in an interview response of one of the 6th graders who noted, "Manhood is understanding that you have a responsibility to your family and your community, and your planet."

Across all three classes and across the various sources of data, there were clear indications that the MAN UP program had significant impacts on developing competencies as well as identities in connection with STEM subjects and careers. As one participant concluded in a focus group interview, "It's a great way to spend my Saturdays and stay off the streets because I bettered myself and my STEM skills."

Developing Socially Just Applications of STEM

Mr. J, Mr. K, Mr. S, and Mahiri worked collaboratively to design the courses and activities such that the students would not only be learning STEM, but also doing and applying STEM to equity and social development issues that the participants identified as important. In math this was realized through Mr. K connecting learning to larger societal inequities on one hand while centering it in the lived experiences of the students on the other. A key aspect of his pedagogical approach was to engage the students in critical dialogues about how math is crucial to engagements in personal and cultural group contexts beyond classwork in schools. Starting with simple examples like how frequent attempts have been made by unscrupulous merchants to short-change some African Americans based on assumptions that they might not be able to accurately count their change to the roles of math calculations for understanding the complexities of wealth accumulation as it is affected by things like credit scores, credit card debt, differential interest rates, taxes, stocks and bonds, and other kinds of math driven investment instruments.

One of the many examples of Mr. K's focus on socially just applications of math was his unit on the role of fast food in health disparities suffered by African Americans. The guiding question for the unit was, "Is there a connection between poor health and poverty?" Students could choose any topic for which they felt they could use math to facilitate understanding and positive change. They decided to use statistical analysis to arrive at a list of healthy food alternatives for low-income people. They researched the amount of calories, fats, saturated fats, proteins, sugar, and sodium in the foods that they normally ate because according to one student "we all eat, pretty much, the same junk." These projects culminated in PowerPoint presentations to the class, but they also hoped to share their findings at their respective schools as well as at community events so that they could in the words of one focal student, "educate their brothers and sisters how important diet is." Ultimately, the focal students saw their experiences learning math in the MAN UP program as "better and more challenging" because it was "math in real-life situations" that was "more hands on and more interaction than regular class and more active."

In the computer science class, a critical pedagogy approach to engaging the participants in mobile app development was taken throughout the school year. CS integrated technology and computer programming with political and societal issues to expand students' thinking on how technology can be leveraged to address the big problems of our day, and how contributions to community justice can be made through the design and creation of technology. Over the course of the school year, Mr. S noted that he saw a definite shift in students' perspectives about thee roles and values of technology in contributing to social justice. Initially, his students felt that the extent of help they could provide to their communities was to devise ways for people to find resources or organizations that were able to help people in need. Over time, however, the focal students began to see themselves as the help. For Mr. S, this shift in the students to become socially conscious, applied STEM practitioners was one clear way that the critical pedagogy approach was working.

One of a number of examples of this was in the focal class' extensive participation in the 8th grade class' design and construction of a mobile app to raise environmental awareness for middle and high school students. The app was designed and played as a football game wherein the running back gained yardage with each right answer selected regarding sustainability, global warming, and environmental awareness.

As noted earlier, Mr. J's implementation of critical pedagogy in the RAM class was key to how the students were facilitated to cognitively wrestle with the applications and ethics of STEM knowledge and skills in conjunction with their emerging STEM identities. By continuously tackling different notions of manhood that are deeply entrenched in our national ethos, the students were positioned to also critically question other axiomatic conceptions. They did this by working through the rhetorical triangle in order to identify logical, emotional, and ethical arguments that were being put forth. Mr. J taught the students how everything can be considered as an argument, even the spatial positioning of their class rooms at their home schools.

This approach began with the very first assignment in the RAM course in which the students searched Google Images for the word manhood and black manhood. They, then, were told to select one of the first 15 images that popped up for each search and to deconstruct the implicit argument underlying the image using the rhetorical triangle as an analysis tool and paying close attention to the notions of European male vis-à-vis African American male that were being promulgated. Eventually, they began applying this critical framing to their STEM courses. So, where STEM had initially been neutral and axiomatic, the students began to ask more critical questions by applying the tools of rhetorical analysis to their STEM learning. They wanted to know: STEM for what, and, STEM for whom?

The impact of his approach on student learning is partially captured in the final class project to create a Question Bridge like video documentary (noted earlier in this chapter) that would reveal the diverse views on young African American men on the positive possibilities and intricate dimensions of manhood. Mr. J guided the focal cohort on storyboarding the project by using design principles learned in the mobile applications course. Before filming their classmates, the students developed a critical focus for the project utilizing elements of the rhetorical triangle and the circle of critical praxis that were central to the RAM curriculum. Once they concretized the arguments that would undergird the documentary, they then developed questions to ask the students who would be featured in the video.

This project exemplified several core components of the MAN UP program simultaneously. It reflected the utilization of digital technology in the production of video documentary in conjunction with developing the students' understanding of how to use STEM to achieve social justice goals. Importantly, there was substantial cross-fertilization of development of the focal students across all three courses such that they were reinforced through work in math, computer science, and the rhetorical analysis of manhood to increasingly see themselves as capable producers of positive social change. Essentially, they learned to see themselves as young men who could use STEM to make the world a better place. Or in MAN UP vernacular, they became applied STEM practitioners.

DISCUSSION

Deciding to take a critical pedagogical approach to learning, identity, and skill development of African American middle school students was considered to be radical by the host organization of the MAN UP project. However, the three instructors and Mahiri were convinced that connecting STEM learning to socially just applications would be crucial to the effectiveness and success of the program, and we believe this is reflected in the findings we reported. We were also clear on the possibilities of our approach to STEM learning from the work of scholars like Moses with the Algebra Project (1992), Gutstein's work on mathematical literacy as a critical literacy for increasing equity in society (2005), and Blikstein's work on the use of science and digital technology as necessary tools for fulfilling Freire's vision of humanization and societal transformation (2008). This work argues, and we think the MAN UP program demonstrates, that mere access to STEM subjects and careers is not enough. Instead, critiques of the roles of STEM in the processes of oppression must accompany considerations of the potential for STEM to be used to mitigate oppression as part of the process for opening wider access to STEM as a civil and human right.

Throughout the program we were careful not to push students into thinking that they needed to be future STEM professionals in order to be successful. Instead, we encouraged them to ask questions as to why STEM education is important. What is it about STEM education that has prompted all of the attention about racial and demographic differences in outcomes? Rather than encouraging them to simply identify with STEM majors and careers, we invited them to ask critical questions about the roles of STEM in larger societal processes and problems. Instead of passively receiving STEM knowledge, they were encouraged and equipped to relate whatever knowledge they received to their daily lives and the lives of other marginalized people. In so doing, they would be able to flip the script and co-construct along with their instructors new possibilities for STEM and for themselves.

CONCLUSION

The findings from this study provided evidence that there was indeed a shift in identity taking place for the focal class. This is important because we know from the work of Nasir (2011) and others that there is a positive correlation between the development of an academic identity and increased levels of math competency, for example. Our findings illustrate connections between identity development and increased competency in computer science and the creation of mobile apps also. Additionally, our findings suggest that our critical pedagogical approach was important in helping the students understand ways to apply STEM for social justice purposes.

REFERENCES

Allen, B. A., & Boykin, A. W. (1992). African-American children and the educational process: Alleviating cultural discontinuity through prescriptive pedagogy. *School Psychology Review, 21*, 586–596.

Banks, J. A., & McGee Banks, C. A. (Eds.). (2004). *Handbook of research on multicultural education* (7th ed.). San Francisco, CA: Jossey-Bass.

Blikstein, P. (2008). Travels in Troy with Freire: Technology as an agent of emancipation. In P. Noguera & C. A. Torres (Eds.), *Social justice education for teachers: Paulo Freire and the possible dream* (pp. 205–244). Rotterdam, The Netherlands: Sense Publishers.

Darder, A. (1998). *Schooling as a contested terrain.* Claremont, CA: Institute for Cultural Studies in Education.

Denzin, N. K., & Lincoln, Y. S. (Eds.). (2003). *Collecting and interpreting qualitative materials* (2nd ed.). Thousand Oaks, CA: Sage.

Duncan-Andrade, J., & Morrell, E. (2008). *The art of critical pedagogy: Possibilities for moving from theory to practice in urban school.* New York, NY: Peter Lang.

Emdin, C. (2010). *Urban science education for the hip-hop generation: Essential tools for the urban science educator and researcher.* Rotterdam, The Netherlands: Sense Publishing.

Gay, G. (2000). *Culturally responsive teaching: Theory, research, and practice.* New York, NY: Teachers College Press.

Gay, G. (2010). *Culturally responsive teaching: Theory, research, and practice* (2nd ed.). New York, NY: Teachers College Press.

Giroux, H. (2001). *Theory and resistance in education: Toward a pedagogy for the opposition* (rev. ed.). Westport, CT: Bergin & Garvey. (Original work published 1983)

Gorski. P. C., Zenkov, K., Osei-Kofi, N., Sapp, J., & Stovall, D. (2013). *Cultivating social justice teachers: How teacher educators have helped students overcome cognitive bottlenecks and learn social justice concepts.* Sterling, VA: Stylus Publishing.

Gutstein, E. (2005). *Reading and writing the world with mathematics: Toward a pedagogy for social justice* (Critical Social Thought). New York, NY: Routledge.

Kapor Klein, F., & Scott, A. L. (2011). *The tilted playing field: Hidden bias in Information Technology workplaces*. Retrieved from http://www.lpfi.org/sites/default/files/tilted_playing_field_lpfi_9_29_11.pdf
Ladson-Billings, G. (1995). Toward a theory of culturally relevant pedagogy. *American Educational Research Journal, 32*, 465–491.
Lee, C. D. (2005). Culture and language: Bi-dialectical issues in literacy. In P. Anders & J. Flood (Eds.), *The literacy development of students in urban schools*. Newark, DE: International Reading Association.
Mahiri, J. (2011). *Digital tools in urban schools: Mediating a remix of learning*. Ann Arbor, MI: University of Michigan Press.
McGee, E. O., & Martin, D. B. (2011). "You would not believe what I have to go through to prove my intellectual value!": Stereotype management among academically successful Black mathematics and engineering students. *American Education Research Journal, 48*(6), 1347–1389.
McLaren, P. (1994). *Life in schools: An introduction to critical pedagogy and the foundations of education*. New York, NY: Longman.
Moses, R. P., & Cobb, C. E. (2001). *Radical equations: Civil rights from Mississippi to the Algebra Project*. Boston, MA: Beacon Press.
Nasir, N. S. (2011). *Racialized identities: Race and achievement among African American youth*. Stanford, CA: Stanford University Press.
Noguera, P. (2008). *The trouble with lack boys: And other reflections on race, equity, and the future of public education*. San Francisco, CA: Jossey-Bass.
Norris, A. (2014). Make-her-spaces as hybrid places: Designing and resisting self-construction in urban classrooms. *Journal of Equity & Excellence in Education, 47*(1), 63–77.
Scott, A. L. (2010). *Dissecting the data: The STEM education opportunity gap in California*. Retrieved from http://www.lpfi.org/sites/default/files/dissecting_the_data_-_STEM_ed_opportunity_gap_lpfi_report.pdf
Scott, A. L., & Martin, A. (2012). *Dissecting the data 2012: Examining STEM opportunities and outcomes for underrepresented students in California*. Retrieved from http://www.lpfi.org/sites/default/files/dissecting_the_data_2012_final.pdf
Skinner, R. (2008). *Black Caucus highlights disparages and suggests philanthropic solutions*. Retrieved from http://www.atlanticphilanthropies.org/news/black-caucus-highlights-disparages-and-suggests-philanthropic-solutions
Smyth, J. (2011). *Critical pedagogy for social justice*. New York, NY: Continuum.
Spradley, J. P. (1979). *The ethnographic interview*. Belmont, CA: Wadsworth.
Steele, C. (2010). *Whistling Vivaldi: And other clues to how stereotypes affect us*. New York, NY: W. W. Norton.
Tate, W. F. (1995). Returning to the root: A culturally relevant approach to mathematics pedagogy. *Theory into Practice, 34*(3), 166–173.
Vakil, S. (2014). A critical pedagogy approach of engaging urban youth in mobile app development in an after-school program. *Equity & Excellence in Education, 47*(1), 31–45.

Jabari Mahiri
Graduate School of Education
University of California, Berkeley
USA

Jeremiah Sims
Graduate School of Education
University of California, Berkeley
USA

INGA BOSTAD AND ALED DILWYN FISHER

5. CURRICULUM AND SOCIAL CHANGE IN EDUCATION FOR A SUSTAINABLE FUTURE?

Ecophilosophy, Critical Inquiry and Moral Dilemmas

INTRODUCTION

How can we create a transformation in education through curriculum or social change that retains and inspires the value of being a part of nature, and learning to take care of nature and each other?[1] The view that education is key in achieving a sustainable society has become mainstream. Education for sustainable development (ESD) has been reiterated as vital in numerous documents and declarations. This builds on nearly half a century of developments in environmental education (EE).

The position now occupied by ESD has not gone unquestioned. Fundamentally, is today's conception of sustainable development depoliticised and detached from questions of value? Does it equate narrowly-defined, discrete 'environmental,' 'economic' and 'social' interests in an uncritical fashion? Have socioeconomic perspectives become equated with economic growth, and, if so, what kind of knowledge is it crucial to look for in education for sustainability?

The depoliticisation of sustainability means various unexamined normative premises that promote market interests and neoliberalism become taken for granted and neutralised in liberal consensus thinking. It is therefore crucial to scrutinise these normative premises. This includes questions about the conditions labour market demands, which privilege instrumentalist decision-making skills and positivistic certainty, place on contemporary environmental debates. This is a question of the possibilities of broader reflection on ecological issues in education.

Nonetheless, an uncritical return to the earlier premises of EE is no more appealing than the unaltered continuation of ESD. Guiding assumptions that education *in, for* and *about* 'nature' would automatically lead to environmental consciousness and action require critical scrutiny. What *kind* of education *in* nature is relevant? What knowledge *about* nature is it necessary to impart? What role do teachers have in this? More fundamentally, what ideas about 'nature' does EE assume? Do these promote a romantic, dualistic ideal of a separation between human society and pristine wilderness? Such an ideal clashes with a socio-ecological approach that recognises the interconnectedness of humans, and the physical systems in which we are situated, as well as the ways in which knowledge, skills, and values are also ecologically embedded. We attempt to avoid a romanticised conception of a

Z. Babaci-Wilhite (Ed.), Human Rights in Language and STEM Education, 71–90.

revival of the knowledge people once held about 'nature,' which can overlook basic dilemmas we face today, and shy away from an understanding of systemic causes of complex socio-ecological crises.

In this chapter we examine and argue for an education that gives children and, indeed, all students relevant and vital knowledge and information about complex socio-ecological challenges, while, at the same time, providing them with space for reflection over values and ethics. Furthermore, we will explore how an open and pluralistic education can quickly become relativistic, or prioritise normatively-loaded conceptions of 'democracy' and 'pluralism' over ecology and ecological arguments. How can we give students the knowledge basis they obviously require without placing too many constraints on debates about socio-ecological problems? What kind of 'pluralism' should we promote?

We will also illustrate how ESD frequently exhibits an instrumental perspective. This critique will be used to examine weaknesses in contemporary education and promote alternative models of ecological pedagogy based on the following premises: firstly, space for real critical reflection, where normative problems are seen as part of the dissemination of basic science and knowledge; secondly, a reconnection with the physical world through understanding and insight in ecology, biology, and the natural sciences generally; and, thirdly, that ecological issues are seen as distinct, but simultaneously inextricably linked with, socioeconomic relations and politics.

EDUCATION IN AN ECOLOGICAL PERSPECTIVE

A Holistic View of Education

We share the premise that education is the key to social change in general, and a sustainable future in particular. Nonetheless, we base this on a holistic conception of education. Education is a living, progressive process, broader and more comprehensive than formal characteristics of schooling, exams, and educational institutions. Education continues throughout life, both through actions and thought, imagination and understanding, theory and practice. This means education is not necessarily easily controlled – the idea of a predefined learning outcome is problematic. An increase in the number of classes in mathematics or the natural sciences, does not automatically lead to an increase in knowledge. This view, which divides the world into discrete subjects where isolated knowledge is to be accumulated, underlies an atomised, instrumentalist education system that works against broader ecological considerations, or reflection across disciplines.

Using a holistic approach, education becomes a larger, less clearly-bounded *process* that covers everything from formal education, to informal information work, campaigning and political debate, art and culture. Nevertheless, we recognise the continuing centrality of formal education in this process. 'Formal' state-based education offers enormous opportunities for influence, but also major challenges. The aforementioned instrumentalist approach affects the entire pedagogic process, and to what extent people have the opportunity to be, or actually are, open to alternative influences. Furthermore, the education system tends to become more specialised and outcome-orientated as one travels through it. The challenges associated with EE at university level are therefore different than at preschool. This does not mean people become 'lost' to ecological awareness at a certain point as they mature, but recognises that earlier stages can significantly shape later outcomes. Indeed, research on early childhood suggests that many characteristics become ingrained before children enroll in 'formal' institutions. This shows the complex relationships between different stages of the education system and the educative environments we find ourselves in, reinforcing the need to consider these holistically.

History of Environmental Education

The early objectives of EE, linked to the 1960s' explosion in environmental consciousness, were "education for nature," promoting limits to population and consumption; and "education in nature," including outdoor and experiential education (Kopnina, 2014a, p. 124), based on the view that "environmental attitudes are shaped by distinct experiences… acquired in early childhood" (Kopnina, 2014b, pp. 73, 75–76). This was partly a departure from "science-oriented" nature education dominating early post-war approaches, which assumed scientific knowledge of environmental challenges, and their technological solutions, would instil environmental consciousness (Kopnina, 2014b, p. 75). However, EE's early institutionalisation and subsequent internationalisation through the UNESCO-UNEP International Environmental Education Programme (1975), still retained guiding assumptions "that if people *knew* about their relationship to the environment, then taking action… would be logical. This reflected the era's optimism in public advocacy and technology, that environmental problems were governed by isolable and controllable linear relationships (Cockerill, p. 635).

The 1980s witnessed the growth of broader approaches building on the more experiential style, including "earth education" and action research partnering pupils and communities in local problem-solving. These addressed both environmental and socio-economic questions (Kopnina, 2014b, p. 76). ESD represents the mainstream manifestation of these trends. However, more critical strands emerged, including "environmental citizenship," representing a "political model" of EE linking socio-environmental injustice to power and social change (Kopnina, 2014b, p. 76).

As Kopnina (2014b, p. 76) summarises, "focus has shifted to integrating environmental concerns… with the socio-economic factors that shape the concept of development." Similarly, Kyburz-Graber et al. (2006) chart how traditional approaches have given way to more transdisciplinary perspectives that examine how an "environmental issues" are defined within socio-ecological settings as "only partly questions of scientific evidence" but also of "interests, needs, values, interpretations, conditions and social contexts." This requires learning environments "in which learners have the opportunity to explore, analyse and interpret human actions in real-life situations," promoting "local and socially relevant knowledge," and reflection on value systems. Such a socio-ecological approach "includes the experiences and perspectives of the people involved" (Kyburz-Graber et al., 2006, pp. 101–104). This critical form of the integrative trend in EE can inform a critique of mainstream integrative approaches in ESD.

Critique of Education for Sustainable Development

ESD is often presented as an advancement of EE, with some EE organisations changing their aims, and even their names, to reflect the new focus on sustainable development (Cockerill, p. 636). From 2005 onwards, this integrative approach is seen in the declaration of the UN's Decade for Education for Sustainable Development. However, many observers reject conflation of EE and ESD. Critics suggest that while the former focussed primarily on intrinsic environmental protection, the latter takes an instrumental stance in which human needs are preeminent, seeking 'balance' between the 'environment,' 'society' and 'economy,' and assuming these can be easily reconciled (often through technological fixes) without fundamentally altering existing practices. Indeed, this purported anthropocentricism is not only an *inter*species bias of human over non-human, but obscures contemporary global, neoliberal cultural and socioeconomic *intra*species power relations, prioritising "the (mostly Western) idea of economic growth as the greatest good" (Kopnina, 2014b, p. 77) alongside other Northern-centered values and assumptions about appropriate social institutions. ESD therefore lacks critical reflection over the type of development actors it promotes– namely, 'rational,' autonomous market actors – privileging short-term, utility-maximising and individualistic behaviour to the exclusion of other aspects of development and long-term effects on biodiversity and future generations. Thus, while sustainable development simultaneously promotes liberal democratic values of pluralism and public deliberation in its human development-inspired understanding of the means and ends of development, it retains deep-rooted normative parameters for acceptable forms and objects of discussion and participation.

There is therefore a methodological paradox apparent in ESD's professed pluralism – a "fuzziness" that appears to open for the possibility of hearing alternative, more critical perspectives, but also potentially "masks the implicit anthropocentric agenda of sustainable development discourse" (Kopnina, 2014b, p. 79). Furthermore, there is an unresolved tension between a more instrumentalist emphasis on "education for environment or for sustainability" and "open, democratic or pluralistic approaches cautioning about educational indoctrination." While much pedagogy exists on a continuum between these poles, the former assigns value to "the environment itself, with its protection as the educational aim," while the latter gives priority to "social values, such as democratic representation," rather than substantive environmental outcomes (Kopnina, 2013, p. 125). This echoes a tension in ecology between the relative certainty of some forms of environmental knowledge, particularly at more general or abstract levels, and the uncertainty of socio-ecological responses, particularly given the importance of local, contextualised knowledge in complex socio-ecological systems. This is seen in attempts to integrate more general, quantitative accounts of climate change with experiential and qualitative understandings of change in order to aid adaptation.

Fundamentally, as Bonnett (1999, p. 315) addresses, "if there are… ecological imperatives deriving from the laws of nature, then clearly certain policies are proscribed." This could risk reducing 'democracy' to simply a means to entrenching these imperatives (Ibid). However, such imperatives themselves may be uncertain, as much ecological knowledge is, especially as complex interrelationships are at play over long timeframes. Indeed, knowledge is never pure and uncontested, but intersected by power relations, disciplinary divisions and cultural framing (ibid, p. 316). For example, the term 'climate change' is an example of the way language can be called on to frame knowledge so as to serve particular interests. Indeed, ethnographic research undertaken in Norwegian communities found high levels of knowledge about the physical dynamics of climate change were distorted by forms of "socially-organized denial" designed to fit, according to Norgaard (2006, p. 347), both Norwegian's nature-loving self-image and petroleum economic interests (seen in interpreting Norwegian oil and gas as part of the solution to climate change).

This debate directly confronts assumptions about the 'neutrality' of rationality itself, and whether purportedly rational 'democratic' processes will prevail, leading to sustainable outcomes; as Bonnett (1999, p. 321) asks, "can education afford to be procedurally neutral when so many other powerful influences in society certainly are not?" Bonnett (1999, p. 323) endorses "sustainability as an attitude of mind" that "seeks awareness of as many facets and significances of nature of possible," one in which democratic values are unpackaged to reveal their *liberal* democratic biases that combine certain kinds of "public articulation, explanation and evaluation" with a "calculative element" inherent in promotion of the market. This allows for wider ideas of democracy that encompass other forms of value, including those of the larger biotic community, and alternative forms of expressing democratic participation, including broader *economic* democracy, and forms of ownership and resource use beyond market and property-based exploitation.

As Kopnina (2014c, pp. 37, 41) details, this critical approach to democracy is necessary for understanding how power relations affect nominally neutral processes. Agreeing that we cannot guarantee ecological awareness through value-neutral deliberative democratic pedagogy, she asks how proponents of a plural approach would react if the human-centred values of these processes, such as gender and racial equality, were also up for discussion, stressing a pluralism that goes beyond *human* pluralism to encompass *non-human* views and values (Kopnina, 2014c, p. 42). Kopnina (2014b, p. 81) advocates a return to the focus "on the protection of the natural environment as the essential basis of long-term human welfare" in foundational EE approaches, combined with "strategic education emphasizing active environmental citizenry" and a focus on "the anthropogenic causes of severe environmental problems." This more systemic orientation is supported by Huckle (2010, pp. 135, 140), who calls for an ESD that confronts "structures and processes shaping the global political economy," calling for global citizenship education that permits an exploration of models of global democracy that "challenge current economic thinking by promoting economic democracy alongside political and cultural democracy." Indeed, recognition is needed of how a form of 'democracy' that allows deliberation over unsustainable ideas is ultimately undemocratic in undermining the fundamental physical prerequisites of life and, thusly, participation in democracy, especially as ecological crises affect groups differentially and, consequently, their capacity to participate. The question is how to promote in practice critical reflection over what is 'democratic' and 'undemocratic' inherent in global citizenship education.

Problems of Lococentrisism

How does an appreciation of the need for a more critical ecological pedagogy relate to pedagogical theory? One suggested approach is through place-based learning, building on earlier 'education in nature' approaches. However, the assumption that place-based learning improves ecological awareness can become fetishised as "lococentricism." As Garrard (2010, pp. 233, 236) points out regarding place-based literary ecocriticism, proponents often offer anecdotal descriptions of educational settings and student experiences, rarely examining whether students attitudes have changed. Naturally, measuring attitudinal dynamics involves a number of normative and methodological difficulties – not least avoiding the instrumentalist approach to measurement of educational attainment prevalent today. – but Garrard is surely right in doubting the extent to which forms of lococentric EE engender behavioural change. Furthermore, Garrard (2010, p. 239) suggests lococentric ecocriticism is "an obsession locally peculiar to American environmentalism," and how "the local" itself can be problematic normatively as well as epistemologically, not least in an ever more globalised world characterised by transformed notions of place.

Similar conclusions apply to pedagogy more broadly, and might be discernible in Scandinavian attitudes to *friluftsliv*: a folk concept, translated literally as "open air life," stressing experiences in and enjoyment of nature as crucial to Scandinavian culture and way of life. Examining outdoor education in Sweden, Sandell and Öhman (2010, p. 121) highlight a divergence between instrumental and transformative approaches to *friluftsliv*:

> On the one hand, outdoor recreation and education helps to strengthen prevailing mainstream societal development by improving popular health and the knowledge acquired at school, as well as promoting environmentally friendly behaviour... On the other hand, an undercurrent of questioning, counterculture and social critique with regard to the development strategy

> of modern urban industrial society is linked to the theme of outdoor life for intrinsic value purposes.

Their research into outdoor activity centres, "all weather schools," and outdoor learning within the radical environmental movement suggests "outdoor activities can generate situations in which personal moral relations to nature are created," for example where encounters with animals inspire "a moral reaction – a spontaneous feeling of care" (ibid). Sandell and Öhman (2010, pp. 124–127) identify six education potentials for encounters with nature:

- "An experience-based meaning of nature" encourages meaning-making "in spite of educational contexts rooted in other traditions;"
- "A relational ethical perspective" that complements value-based environmental ethics;
- "A fourth perspective" to sustainable development engendering "a comprehensive existential perspective that originates from aesthetic and emotional relations with nature;"
- An opportunity for urban dwellers to experience "human ecology in practice;"
- A contribution to "sensing the quality of a simple life," which has "an important ideological role" in developing critiques; and Stress on "democracy, identity and dwelling," encouraging reflection over "linkages between sustainable development, mobility and a sense of place."

The authors admit, nevertheless, that "we know very little about the continuity of this learning" in other contexts, "whether a strong attachment to the local environment also creates a strong environmental commitment in general," or "whether a concern for nature also leads to a concern for social and economic sustainable issues" (Sandell & Öhman, 2010, p. 127) In other words, it is difficult to chart effects on pupils' subsequent behaviour. Indeed, despite widespread public support for *friluftsliv*, Scandinavian countries cannot feasibly claim to be world leaders in sustainability; Norway is a major oil and gas exporter, while it is unclear whether there are causal links between certain policies (like Denmark's renewable energy investments) and *friluftsliv*. Furthermore, the examples used in this Swedish study are not part of the core curriculum in public schools. Although the authors recognise their aforementioned six conclusions about education in nature as potentials, and might argue all schools should adopt the "all-weather" approach, their findings may have diminished relevance in the immediate situation within the education sector.

The romanticisation of place and locality can therefore become a mirror-image of the failings of the displacement of the modernist approach it seeks to critique at the other extreme. Insisting education *must* start by "approaching the world through affectivity and not cognition" (Ritchie, 2013, pp. 34, 37) thus might restrict ways students can develop ecological awareness through a critique and reapplication of reasoning; indeed, it ignores one of the origins of modern environmentalism in scientific observation and modelling of human effects on ecological systems, which often had limited experiential content. Simply replacing one one-size-fits-all pedagogic approach with another overlooks research that uncovers a more complex, multi-level picture of how we conceive of 'the environment' (often thinking "more environmentally" at the global level) with multiple, conflicting subjectivities (Garrard, 2010, p. 241). As Garrard (2010, p. 241) concludes, there "remains a widespread, but largely untested and untheorised assumption that education about the environment… delivered through the environment… will automatically be education for the environment."

The nuanced phenomenological approach endorsed by Howard (2008, pp. 302, 304) recognises "we must move beyond science while being inclusive of scientific knowledge" through a "cross-curricular conversation" that recognises ecological problems' social production. Ecological interrelations are embedded in cognitive frames *and* biological materiality, requiring us to learn to "interpret our bodily felt experience of the claims of the biotic community" (Howard, 2008, p. 305). This might be taken further to embrace the kind of multi-level, intersectional and inter-subjective considerations outlined above. Ultimately, rather than discussing pluralism only in terms of the *values* taught, we might also discuss a pluralist approach to *techniques* used that combine aspects of place-based study, critical reflection, scientific inquiry, culture, and more. Rather than EE being one niche of curricula, ecological concerns must be integrated at all levels in numerous different ways that embrace many different entry points and forms of learning.

Again, this points to the need to radically reform the education system as part of the radical reforms needed for societies in general. Practically, as Howard (2008, p. 308) notes, transdisciplinary approaches require "time for quiet reflection, and paying close attention," conditions that are rare in modern learning environments of "fragmented schedules divided by bells, and standardized testing." As Kyburz-Graber et al. (2006, pp. 111–112) put it, while critics may suggest critical environmental education leaves teachers and pupils "overstretched," global ecological crises demand "educational settings in which students and teachers learn to interact with complex issues;" as socio-ecological approaches "do not do us the favour of being disciplinary, well-structured and easily accessible," learning must encourage "thoughtfulness and reflection."

A QUESTION OF CRITIQUING KNOWLEDGE?

Teaching Critical Ecological Thinking

Kopnina (2013, p. 125) points out that while instrumentalist checklists exist to measure "sustainability competencies," there are "no significant studies of moral reasoning about sustainable development and environment." She quotes general research on moral reasoning that suggests "the presence of information about the impact of ecological damage on the environment, especially a more "wild" environment, elicited more ecocentric reasoning, while the presence of a social commitment elicited more non-environmental moral reasoning," with 'anthropocentrically'- and 'ecocentrically'-inclined individuals unlikely to find "common solutions" through deliberation given their "different moral voices." Kopnina (2013, p. 126) speculates as to whether the pluralist approach may again represent "pluralism of one species" and reduce environmental concerns to "'one of many' sustainability facets." By examining the use of vignettes in classroom discussions to explore moral reasoning, Kopnina finds that ESD's pluralism is positive when encouraging pupils to "tackle their moral reasoning and… distinguish between conflicting interests within the dominant discourse," but provides no guarantee that instrumentalist 'anthropocentric' perspectives do not end up dominating. This "'plurality' refers to humans only" to the disadvantage of perspectives emphasising ecosystemic integrity. Her conclusion is to push the critical side of these discussions further into a deconstruction of notions of "development" (Kopnina, 2013, p. 130). This requires "pedagogical guidance" if children or students are to "be moved beyond acknowledging and accepting contradictions to actually making ethically informed decisions" (Kopnina, 2013, p. 131). Indeed, one of Kopnina's vignettes, regarding how a poor farmer should manage an unnamed species of trees on his land, is presented in such a way that we receive far more information about the farmer and his social situation and nearly nothing about the importance of the trees in question to the broader ecosystem, or indeed any social or cultural role they play (Kopnina, 2013, p. 128). This speaks to the previously-addressed need to have an understanding of pluralist classroom discussion that provides as much information about socio-ecological contexts as possible in order to ensure a truly democratic debate where socio-ecologically informed moral reasoning is possible.

However, there is an issue regarding how one defines views as either (or more) 'ecocentric' or 'anthropocentric.' Moral reasoning may be more complex than this (often false) dichotomy implies. For example, Kopnina (2013, p. 127) quotes research that suggests "only biospheric altruism leads to sacrifice rather than quality-of-life solutions to environmental problems," asking:

> Does lumping together ethical issues as distinct (and sometimes mutually opposing) as poverty eradication and environmental protection bring the torrent of 'unsustainability' under the protective umbrella of generalized altruism, or does it cleverly mask the strategy of powerful elites to peruse economic development at the cost of both social and ecological catastrophe?

This passage seems to equate 'ecocentric' ('biospheric altruistic') stances with 'sacrifice,' constructing sharp distinctions between 'sacrifice' and 'quality-of-life' that perhaps few thinkers who identify with ecocentricism would accept. Implying poverty eradication and environmental protection are distinct and sometimes mutually opposing is also not a straightforward empirical statement, but a normative assertion. The two may well be distinct in existing economic paradigms, but this is because of the way material wealth is systematically prioritised over ecological interests in current systems in precisely the way ecocentric accounts criticise. Thus, if we define 'ecocentricism' in terms of 'sacrifice' and 'anthropocentricism' in terms of 'quality of life' and 'poverty eradication,' we can potentially ignore not only nuances in moral reasoning, but the critical forms of moral reflection that are vital to a broader socio-ecological account of sustainability, including an examination of what is meant by 'quality of life,' 'poverty,' 'sustainability' and the other normatively loaded phrases Kopnina employs. The labelling of support for ecological protection to secure human survival and flourishing, for example due to family or community ties, as 'anthropocentric' can confuse *intra*species belonging with *inter*species, human-first chauvinism, and might obscure *intra*species (human) inequalities under *inter*species differences. Holmes Rolston III's (1998, pp. 140–141) non-dichotomous approach recognises "the possibility of valuation" is a result of evolution, and value is thus both "anthropogenic" and "biogenic," existing along a continuum where "value increases in the emergent climax but is continuously present in the composing precedents." Value "fans out from the individual to… the matrix," 'intrinsic' and 'instrumental' values become "local details of value embedded in global structures," and subjects must situate themselves "within the limits of decentralized community," where "*systemic* value" describes duties arising "in an encounter with the system that projects and protects these member components in biotic community." This contingent value approach, encouraging longer-term valuation as part of an interconnected whole, can provide a basis for exploring ecophilosophy and education.

The importance of teaching critical thinking is therefore clear. Kyburz-Graber et al's (2006, p. 107) work on socio-ecological school projects found reflection on values, and the interaction between 'nature' and 'society,' requires "much more professional attention than was expected," particularly as teachers themselves clung to established practices even as their students displayed more critical tendencies. In the case of reflexive classroom discussion of biotechnology, students privileged natural scientific knowledge, expecting what they imagined to be invariably male, neutral-scientific researchers to offer technical solutions to societal issues. Even where preliminary stages included discussions about principles, these were often discarded later as pupils looked for a single, 'correct' scientific answer. Partly due to teachers' priorities to create debate for its own sake, the students' discussions thus often mirrored techno-optimistic public discourses around biotechnology (Kyburz-Graber et al., 2006, pp. 109–110). Again, the need for teaching and encouraging critical thinking through providing as many different perspectives as possible, and opportunities for exploring these, comes through.

Values, Science and Ecology

One key aspect of doing this is examining the links between values, science, and ecology. As we have seen, several challenges linked to creating a critical, investigative classroom are found in the need for a single 'correct scientific answer.'

In addition, research policy is marked by an unclear, superficial multidisciplinarity, where the humanities and social sciences are not fully appreciated.[2] It is important to see ESD in a larger context where direct and indirect control and monitoring structures characterise education, and make it difficult to find time and resources for more reflective pedagogy. International rankings, prioritising fixed outcomes and grades, could easily overshadow the development of critical skills, attitudes and time-consuming learning to understand moral, societal and environmental dilemmas. While the demand for improved knowledge in the natural sciences is increasing, the need for knowledge critique is greater than ever. Most of the facts and information considered relevant today will be less relevant in the decades to come, making basic skills more important. Questions of how to generate hypotheses and test them, differences between justifications and claims, the certainty of scientific evidence, and how to argue coherently and sufficiently for ethical norms and values are relevant here.

The positivist ideal could be seen as distorting both natural and social science given how questions about ethics, morals, norms and values are seen to transgress requirements for repeatability, objectivity and methodological stringency. Positivism's much-debated relationship with ethics has its roots in Weber's value-free ideal:

> In the three "capitols" of logical positivism (Vienna, Cambridge and Uppsala), there was agreement that a scientific ethics is a logic impossibility. Of course, there were also several different views on ethics in the Vienna circle. However, there was generally agreement that moral norms and value statements could not be concluded from factual statements, and that factual statements are the only ones that can be true or false. *This* is the only thing scientists can say anything about. (Tranøy, 2002, p. 57)

The ideal of value-free science has itself been criticised for being normative, just as the ideal of sustainable development is normatively framed. In accordance with logical positivism, most scientists are trained to think of values as "subjective, biased, emotional, even irrational," and of science as a value-free activity. For Hargrove (2000, pp. 114, 116), this "anti-value indoctrination" has become "the single most serious inhibition to the application of environmental ethics in public policy."

Hargrove shows how utilitarianism and logical positivism have also influenced the economic system, excluding forms of value that cannot be converted into numerical quantities. This means values are not just deprioritised in education, but that education is based on positivistic accounts that actively encourage students to *discount* values, biases and feelings:

> Like utilitarianism before it, positivism reduces values to the arbitrary, subjective, irrational feelings of individuals independent of their social context...By defining values as feelings, value studies can become factual studies about how people feel. Counting the numbers of people who feel one

> way or the other permits quantification, which is then presented as objective value information. (Hargrove, 2000, pp. 122–123)

Even more important is Hargrove's (2000, p. 123) conclusion that the apparent basis of modern social science as "the study of the rational choices of individuals made self-interestedly or selfishly… in the context of moderate scarcity," does not need to be justified, precisely because it is "arbitrary, subjective, and emotional." The reassertion of values through critical reflection on socio-ecological issues, including the examination of this value basis for modern science, is, consequently, vital.

ECOPHILOSOPHY AND EDUCATION

Dewey

The core of education, according to Dewey and others, is preparation for social participation. However, this preparation assumes a non-dualistic concept of knowledge. Education must not only bridge the gap between the practical and theoretical, between individual life experiences and group rules, but encourage the freedom to take responsibility for one's surroundings. Social control, or social goals for school, are connected by Dewey (2007, p. 244) to increased freedom:

> But the essence of the demand of freedom is the need of conditions which will enable an individual to make his own special contribution to a group interest, and to partake of its activities in such ways that social guidance shall be a matter of his own mental attitude, and not a mere authoritative dictation of his acts.

The pedagogic lesson we can draw from this is, according to Dewey:

> Not… that teachers would find their own work less of a grind and strain, if school conditions favoured in the sense of discovery and not in that of storing away what others pour into them; nor that it would be possible to give even children and youth the delights of personal intellectual productiveness – true and important as are these things. It is that no thought, no idea, can possibly be conveyed as an idea from one person to another. (Dewey, 2007, p. 132)

Freedom becomes, for Dewey, the role personal thought and reflection play in the learning process, the ability to anticipate different consequences and respond to them. A democratic society is one that opens for greater variation of mutual and shared interests, in contrast to those that see socialisation as preservation of shared customs (Dewey, 2007, p. 376).

The school as a genuine, practising democracy is emphasised both descriptively and normatively by Dewey. All social environments have a formative effect; the everyday environment of school forms the mental and emotional dispositions for individuals' actions through engaging them in activities that awaken and reinforce certain impulses (Dewey, 1997, p. 51). We never raise pupils directly, says Dewey, but indirectly through the totality of relationships to do with the activities that characterise a living and active human being. A theory of knowledge corresponding to the social environment must therefore be developed – one that is driven by a method for sharing experience, where knowledge is always in close contact with a spectrum of human life in community.

According to Dewey, education and schooling should aim to create a personal interest in taking care of democracy, meaning a community where there is a balance between what one gives and receives, "a spirit of companionship and shared activity" (2007, p. 288), and where the connection between the learning that happens inside and outside school or formal training is clear and thoughtful. This interest is created through two different dimensions that relate to the philosophical and political implications of Dewey's theories on democracy and education (Sjøden, 1997, p. 28). *Firstly*, the long-term ideal of a participatory democracy where everyone is able to develop their capacity to participate in society, where private interest is evaluated against community interests based on the consequences for the society as a whole. *Secondly*, a dimension that relates to the importance of developing an interest and awareness for others, and their needs and interests, meaning that school, through communication and relationships, is a richer place than the private sphere. At the same time as we consider these, according to Dewey, a narrow, moralistic ethics is responsible for "failure to recognize that all aims and values which are desirable in education are themselves moral" (2007, p. 288). There is also a conflict between natural sciences and the humanities, which have experienced an unfortunate mechanical and instrumental split in school curricula. This conflict is the result of a dualist philosophy for Dewey (2007, p. 225), which encourages an alternative philosophy regarding the connection between nature and humanity: "Mind and the world are regarded as two independent realms of existence having certain kinds of contact with each other". The mechanical philosophy rests additionally on the illusion that technology represents nature as it is, that concepts reflect reality, and method, the subject matter (Dewey, 2007, p. 230). "The consequences are that such a mechanical understanding of nature has… resulted in a belief that we can predict and control the happening of events, ignoring the qualities of event" (Dewey, 2007, p. 230). Experience, meanwhile, according to Dewey (2007, p. 231) "knows no division between human concerns and a purely mechanical physical world."

It is not necessary to waste time with contradictory definitions of knowledge; according to Dewey (2007, p. 286), it is sufficient to say that there is "knowledge gained at first hand through the exigencies of experience which affects conduct in significant ways." In general education, we need to distinguish between understanding, critical reflection, assessment and exams, and not try to base our assessments and exams on the premise that these constitute the sum of learning outcomes. Simultaneously, the moral dilemmas described in textbooks and curricula are not necessarily transferable to the social context in which schools find themselves, as well as not necessarily leading to social change.

An ethical or moral theory that is satisfied with being well argued for on paper, at a linguistic, philosophical level, is, for Dewey, an incomplete ethics. A theory is also incomplete if, first and foremost, it seeks to intervene in 'the other's' world. Moral categorical imperatives are radically different from so-called technological imperatives. The latter neither demand, forbid or permit. They *enable*, for better or worse. In summary; developing environmentally-attune education requires encouraging the freedom to take responsibility for one's surroundings. This freedom is a substantive freedom and not a formal or empty one; it is developed in a value-based and morally open-minded social context. It is the attitude of pedagogy as a discovery of knowledge, skills and values that is the heart of Dewey's insight, and not the storing of what others pour into us. Education must be open for greater variation of mutual and shared interests, where private interests are evaluated against community interests based on consequences for society as a whole. Such a view tessellates with Holmes Rolston III's levels of value (while broadening community to the *biotic* community), which can be used for developing Dewey's insights in a socio-ecological perspective.

Natural Science, Ecophilosophy and Environmental Education – Næss and Freire

As we have discussed earlier, labour market concerns demand an instrumental approach to knowledge about ecology, at the same time as this presents problems for paradoxes, value conflicts and norm hierarchies. To also avoid the failures of earlier EE traditions that stressed fact-based, 'hard' scientific approaches, a different approach to the interplay between 'natural science' disciplines and values must be developed, and how this relates to the broader issues affecting both education systems and the wider educative environment in a holistic approach.

Sjøberg has shown in several works how natural sciences have been deprioritised in both OECD countries generally and Norwegian primary schools particularly because teacher training has too little emphasis on natural sciences, textbooks are written by authors without a strong academic background, and the number of lessons has been reduced and characterised by a descriptive approach (1998, p. 83). This is somewhat paradoxical given the aforementioned proliferation of empiricist, positivist approaches, inspired by the natural sciences, in the social sciences; and the previously-described reliance on technical-scientific knowledge in ethical discussions around environmental issues where values and critical thinking have been downplayed. However, it speaks somewhat to the consequences of the high-level of specialisation, disciplinary separation and assessment-focused instrumentalism encouraged by the educational system. Sjøberg underlines the need to see environmental issues as a part of both natural and social studies. Natural sciences should first and foremost be taught as a "problem-filled, human activity" (Sjøberg, 1998, p. 18). This relates to the Freirean concept of using "generative themes in a problematising way," taking relevant issues "captured from people's experience and presented… in a form that encourages [participants] to see it critically." The key here becomes the choice of "codification," or "the medium… used to capture the essence of everyday issues" presented (Ledwith, 2011, p. 101).

Similarly, for Næss and Jickling (2000, p. 48), there is an ongoing process of preparing students to examine their premises and subsequent consequences so as to clarify their thinking, articulate how they feel, and not the least to identify real differences and dilemmas. The most basic tool in this approach is to use actual situations and dilemmas in politics, the media, history or the community: as Næss and Jickling (2000, p. 49) put it, "with deep prejudices you must use some example of how you would behave in a particular situation." When doing this, one often finds there is an underestimation of the cognitive value of feelings, as "what people say in favour of economic growth is sometimes highly emotional" (Næss & Jickling, 2000, p. 53). Transferred to the teacher, her role is to make the children see previously unappreciated connections between actions and values and train them in this, including to "use their body language and encourage personal relations, so as to keep the children's imagination and motivation intact" (Næss & Jickling, p. 54). For both Næss and Freire, dialogue is therefore key. Indeed, in Freirean thinking, without dialogue, there can be no communication, which is the basis for "true education" and critical inquiry through listening to others' narratives, recognising difference and making complex issues "simple but not simplistic" (Ledwith, 2011, pp. 105–107). This stress on dialogue is the root of seeing education as a holistic *praxis*, where there is "a dynamic between action and reflection" in which "theory informs action and action generates theory" (Ledwith, 2011, p. 106).

According to Næss, scientific knowledge is always a value question;[3] we have long had enough technical ecological knowledge and it is a blind alley to ask for more; in Næss and Jickling (2000, p. 55) word's, "wisdom is what we need." Indeed, Næss and Jickling (2000, p. 55) distinguished between "scientific knowledge," "research," and "value priority," suggesting we must "undermine the prestige of" the former in favour of the latter two. He suggested seeing teachers as *researchers*. For Næss, researchers "may never pretend to have contributed to scientific knowledge;" teachers should "talk as much about what we don't know as about what we do know." Early education should consequently "have very few finished products," preferring "many tools and very many natural things which the children can use," including "patches of free nature" that are allowed to grow without preplanned intervention (Næss & Jickling, 2000, p. 57). Similarly, the very young should be dissuaded from "trying to be perfect;" instead, they should be allowed to "try to classify bad habits and say: 'There is a bad habit here which really has some influence, maybe I could change that. And then I keep some of the others'" (Næss & Jickling, 2000, p. 59). On this more mutual relationship between teacher and pupil, just as Næss emphasised the possibility of teachers entering into dialogue with even the youngest pupils around issues of basic values and worldviews, Freire understood teachers and pupils as "co-learners" that seek to create "critical, inquiring and responsible citizens" capable of radical action and reflection. As such, "this is the interface of praxis at which the knowledge and theory of the educator come together with the everyday experience of the people" (Ledwith, 2011, p. 102).

Næss also addressed questions of pedagogy regarding how stages of formal education are embedded in broader community structures. He supported slowly introducing "relations of humans to nature with the historical background—not from the history of humans, but history in general," rather than going straight to terms like ecology, stressing questions of "'what does this feel like?' Instead of: 'What is this?'" (Næss & Jickling, 2000, p. 56) He noted that children often have a "rudimentary philosophy of life" before they enter school that is often lost in the first years of school as they are "coerced to listen" (Næss & Jickling, 2000, p. 56). In addition, Næss insisted on a close relationship between education and action. He argues that teachers should not shy away from activism, but be careful not to lead students:

> You may say: "If we have a particular deep kind of priority and value, what then follows from this?" And if you always use these qualifying "ifs" then you are on the right side… It is completely right for you to do everything you can to have the community adopt your policy. But, as a teacher, this should always be done with qualifying "ifs." "If you have the following value priority, then…" (Næss & Jickling, 2000, p. 61)

This is, again, similar to the Freirean concept of education as either "liberating" or "domesticating." A pedagogy of liberation seeks a process of "conscientisation… whereby people become aware of the political, socioeconomic and cultural contradictions that interact in a hegemonic way to diminish their lives," breeding an awareness and critical insight that engenders collective action (Ledwith, 2011, p. 100).

CONCLUDING REMARKS

How can we test, critically and holistically, an education where the measure of success is not measurement? We have to not only ensure children can go out and experience nature and use their bodies, but also realise that the education system is disconnected from nature and ecological challenges; this is an enormous challenge, and it is daunting to consider how we can explore this complicity with creating confusion.

We must avoid making education for a sustainable future yet another educational technology, where children are required to learn by heart and repeat predefined factual knowledge. We have seen how well-justified facts about ecology and scarcity of resources are crucial in education and training, but also how a single correct answer can hinder an understanding of what we cannot know with certainty. There are sizeable challenges that are beyond the purview of this article, but this article reveals a clear need for tackling them.

Together, the perspectives of Dewey, Næss and Freire and others give further clarity to the holistic view of education, and build on the socio-ecological approach that seeks pedagogic practices in participatory, transdisciplinary settings where multiple forms of knowledge are examined alongside values and shared understandings. We have built this holistic, socio-ecological view by examining the failings of value-laden ideas of ESD and the shortcomings of its assertion of a particularised form of pluralism that excludes more critical socio-ecological perspectives; the need to avoid a mirror-image of this through privileging lococentric or any other single approach, encouraging a pluralism of both values and techniques; the way this requires a reform of the educational system and a new approach that promotes critical thinking, the reassertion of values into pedagogy and the avoidance of oppositional value categories, instead favouring a more contingent and embedded value theory; how these insights can be developed in educational settings through Dewey's view of education in a broader, holistic approach; and Næss and Freire's insights into the exploration of ecological issues through generative themes grounded in experience, the importance of dialogue, and the erosion of strict educator-educated binaries so as to encourage a critical educational praxis that is action-orientated.

We have put an emphasis on showing the need for a new education for a sustainable future, both through revealing the deep-lying scientific and depoliticised basis of both the discourse and teaching, but also through showing how one can promote ecological awareness and capacity for action through systematic and holistic teaching and learning environments grounded in co-learners' experiences. Just as Gadamer portrays aesthetic judgements as a form of alienation, where the experience is filtered or is something other than what it is immediately understood as, so can ecological awareness and actions also be inspired of peoples' experiences of being a part of nature. To experience alienation and belonging, fear for natural disasters, and unpredictable natural experiences, can also promote an unease or disturbance that awakens deeper questions of where we come from, and why we wish to protect our surroundings, and what we wish to protect, for an uncertain future.

Ultimately, a socio-ecological approach and ecophilosophical perspectives challenge education in terms of its ontology, epistemology, and methodology. These challenges which debate not just what is taught, but how and where we teach, are interlinked. The ontological basis of the entire education system must change in order to meet this challenge by recognising the ecological embeddedness of human society, and the socio-ecological context of all pedagogy. This requires confronting the following epistemological dilemmas we have outlined:

- *Certainty dilemma* – the environmental movement emerged partly as a response to growing scientific certainty regarding environmental degradation. But ecology recognises that even aspects of natural science are dynamic and uncertain. This is particularly the case when taking into account inter-subjective experience as part of a socio-ecological approach. This requires a renewed emphasis on critical thinking and values.

- *Place dilemma* – EE stresses place-based, experiential learning, and local knowledge as key to understanding socio-ecological approaches. At the same time, the interconnectedness of global ecological systems means some aspects of the ecological crisis manifest themselves elsewhere, and more general, abstract forms of knowledge are sometimes vital for understanding environmental phenomena, particularly when projecting into the future (as computer modelling of climate scenarios shows). In other words, some aspects of the ecological crisis are difficult to experience directly. This speaks to the need for a plurality of not just values and information discussed, but methods used to explore these values and the nature of the information we tend to receive in formal educational settings.

These, in turn, manifest themselves in the following methodological pedagogical dilemmas:

- *Supplement-replacement dilemma* – how can we contribute to ecological awareness through formal education when it is not possible to completely control classroom outcomes, let alone broader educational ones? What should the priorities be? Accepting a holistic view of education and the more honest, humble role of the teacher assumed by Næss and Freire helps to explore this tension.
- *Instrumental-pluralist dilemma* – how can an open, pluralist style that encourages debate and reflection be squared with the need to impart certain basic ecological information? How do we avoid relativism? How do we establish a value basis for classroom discussion with creating too many prescriptions for children's opinions? The idea of sustainability as a state of mind is helpful here in seeking to expose co-learners to numerous different perspectives, as well as using the insights into different levels of value and critical thinking to explore these.
- *Short versus long-term dilemma* – how can we accommodate holistic, reflexive approaches to education within existing curricula based on separate, discrete subject areas and the educational system? What role do teachers, as individuals and a professional group, have in this? What educational reforms are needed? What practical programmes for achieving these changes are available? Understanding education as a socio-ecologically embedded and contextualised praxis is vital here.

Nonetheless, addressing this requires significant changes in pedagogical practice and the educational system. It would be naïve to assume that these changes can be brought about in a linear, straight-forward reform process; instead, it will be up to various educational actors, the co-learners themselves, to develop a praxis of education for a sustainable future through a variety of complimentary approaches.

NOTES

1 As formulated in the Norwegian Kindergarten Act 2005-06-17-64, Section 1: "Children shall be able to develop their creative zest, sense of wonder and need to investigate. They shall learn to take care of themselves, each other and nature. The children shall develop basic knowledge and skills."

2 See for instance Martha C. Nussbaum (2010): *Not for profit. Why Democracies Need the Humanies* Princeton University press.

3 The expression "all living beings are a part of the same whole" is neither a norm nor a description, it is an intuitive reaction to what it is to be human. According to Næss's ecosophy or deep ecology, this experience can be socially repressed or underdeveloped, and when it is argued for, one must bring forward a norm or reason for this experience. The deep ecological understanding has been criticised by among other Peder Anker for diluting the differences between natural things – rivers are given equal status to dogs, who are themselves placed on the same level as humans. Part of this critique was countered by Næss through his ranking of norms, in addition to the division between identity (belonging) and norms (good, evil).

REFERENCES

Bonnett, M. (1999). Education for sustainable development: A coherent philosophy for environmental education? *Cambridge Journal of Education, 29*(3), 313–324.

Cockerill, K. (2013). A failure reveals success: A comparative analysis of environmental education, education for sustainable development, and industrial ecology education. *Journal of Industrial Ecology, 17*(5), 633–641.

Dewey, J. (1997). *Democracy and education: An introduction to the philosophy of education.* New York, NY: Free Press / Simon & Schuster.

Dewey, J. (2007). *Democracy and education: An introduction to the philosophy of education.* Sioux Falls, SD: NuVision.

Garrard, G. (2010). Problems and prospects in ecocritical pedagogy. *Environmental Education Research, 16*(2), 235–245.

Hargrove, E. (2000). Toward teaching environmental ethics: Exploring problems in the language of evolving social values. *Canadian Journal of Environmental Educaton, 5*(1), 114–133.

Howard, P. (2008). Ecology, phenomenology, and culture: Developing a language for sustainability. *Diaspora, Indigenous, and Minority Education, 2*(4), 302–308.

Huckle, J. (2010). ESD and the current crisis of capitalism: Teaching beyond green new deals. *Journal of Education for Sustainable Development, 4*(1), 135–142.

Kopnina, H. (2013). An exploratory case study of Dutch children's attitudes towards consumption: Implications for environmental education. *Journal of Environmental Education, 44*(2), 128–144.

Kopnina, H. (2014a). Education for sustainable development (ESD): Exploring anthropocentric–ecocentric values in children through vignettes. *Studies in Educational Evaluation, 41*, 124–132.

Kopnina, H. (2014b). Revisiting Education for Sustainable Development (ESD): Examining anthropocentric bias through the transition of environmental education to ESD. *Sustainable Development, 22*(2), 73–83.

Kopnina, H. (2014c). Education for Sustainable Development (ESD) as if environment really mattered. *Environmental Development, 12*, 37–46.

Kyburz-Graber, R., Hofer, K., & Wolfensberger, B. (2006). Studies on a socio-ecological approach to environmental education: A contribution to a critical position in the education for sustainable development discourse. *Environmental Education Research, 12*(1), 101–114.

Ledwith, M. (2011). *Community development: A critical approach* (2nd ed.). Bristol, CT: Policy Press.

Næss, A., & Jickling, B. (2000). Deep ecology and education: A conversation with Arne Næss. *Canadian Journal of Environmental Education, 5*, 48–62.

Norgaard, K. M. (2006). "We Don't Really Want to Know": Environmental justice and socially organized denial of global warming in Norway. *Organization and Environment, 19*(3), 347–370.

Ritchie, J. (2013). A pedagogy of biocentric relationality. *New Zealand Journal of Educational Studies, 48*(1), 34–49.

Rolston III, H. (1998). Challenges in environmental ethics. In M. E. Zimmerman, J. B. Callicott, G. Sessions, K. J. Warren, & J. Clark (Eds.), *Environmental philopsophy: From animal rights to radical ecology* (2nd ed., pp. 124–144). Upper Saddle River, NJ: Prentice Hall.

Sandell, K., & Öhman, J. (2010). Educational potentials of encounters with nature: Reflections from a Swedish outdoor perspective. *Environmental Education Research, 16*(1), 113–132.
Sjøberg, S. (1998). *Naturfag som allmenndannelse : en kritisk fagdidaktikk.* Oslo, Norway: Ad notam Gyldendal.
Sjøden, N. (1997). Forord. In J. Dewey (Eds.), *Demokrati og utdanning.*
Tranøy, K. (2002). Arne Næss i 90. Filosofi, positivismestrid, etikk. *Norsk Filosofisk Tidsskrift, 17*, 56–63.

Inga Bostad
Faculty of Law
Norwegian Center for Human Rights
University of Oslo
Norway

Aled Dilwyn Fisher
Faculty of Law
Department of Public and International Law
University of Oslo
Norway

LANETTE JIMERSON AND PAGE HERSEY

6. LOCALIZING HUMAN RIGHTS EDUCATION THROUGH TECHNOLOGY

Two Literacy Based Examples

INTRODUCTION

From the civil rights movement to #BlackLivesMatter California has been a geographic space of activism and struggle for social justice. These movements make visible how human rights activism is always local and global. Issues such as police brutality, worker's rights, and immigration impact students' lives and as such human rights education is fundamental to any education that seeks to develop a democratic citizenry. Furthermore, human rights education curricula that connect students to local, regional, and global struggles for justice through the use of technological tools can deepen students' ability to critically read the word and the world.

As literacy educators within K-12 schools, we believe that a key component of literacy development is the understanding of literacy, including digital literacy, as a right of education. Literacy is broadly defined to include both traditional school literacies and the literacies youth develop outside of formal schooling. In this vein we position literacy, and thus language, as a right in education (Babaci-Wilhite, 2015) and a vehicle for addressing education as a human right. Babaci-Wilhite frames language as a right in education as the intentional use of students' first language, out-of-school literacies, and cultural knowledge as a foundation for language learning within traditional schooling contexts. Such a foundation is critical to students' ability to take full advantage of their human rights, both their rights to education and their local and global rights to economic and social stability.

In this chapter we present two literacy-based examples of how educators might leverage technological tools to localize human rights education. In the first classroom, ninth grade students investigate the promise and limitations of the pursuit of happiness in relation to the Universal Declaration of Human Rights (UDHR) Articles 23 and 25, the writing of philosopher John Locke and several canonical texts. Students engage their linguistic and cultural knowledge to leverage technological tools in order to gather data, collect interviews, communicate their experience, and the experience of friends and family members' quest for human rights. In the second classroom, middle school special education students examine police brutality through blogging. Much like the research of Mahiri (2011), blogging

Z. Babaci-Wilhite (Ed.), Human Rights in Language and STEM Education, 91–108.

provides the ability of youth to bridge their out-of-school linguistic knowledge to the genre of blogging in order to speak against experiences of societal violence both locally and globally.

Implications of these two studies highlight the effectiveness of transformative classroom spaces wherein students leverage technological tools to make personal and local connections to human rights.

BACKGROUD OF HUMAN RIGHTS EDUCATION

Flowers (2000) defined Human Rights Education (HRE) as simply, "all learning that develops the knowledge, skills, and values of human rights" (p. 6). She described the distinction between learning *about* human rights and learning *for* human rights, concluding that a successful HRE program needs both components. "Learning *about* human rights is largely cognitive, including human rights history, documents, and implementation mechanisms" (p. 10). Learning *about* human rights creates a common vocabulary and provides a context for an understanding of the rights all humans share. Flowers defined the other side of a successful HRE program as learning *for* human rights, which means understanding and embracing the principles of human equality and dignity and the commitment to respect and protect the rights of all people. It has little to do with what we know; the "test" for this kind of learning is how we act (Flowers, 2000, p. 13).

Lohrenscheit (2002) added detail to Flowers's (2000) differentiation between learning *about* and *for* human rights. The author suggested that learning *about* human rights emphasizes "knowledge, understanding and values" while learning *for* human rights places greater emphasis on "respect, responsibility and solidarity" (p. 177). The learning context is crucial to the implementation of HRE – language, culture, and needs of students must be recognized. Later scholarship has also pointed to an element of learning *through* human rights, which "includes learning and teaching in a way that respects the rights of both educators and learners" (United Nations, 2011). By participating in open dialogue and inquiry, it is possible for teachers and students to critically engage with meaningful issues to learn *about, for*, and *through* human rights (see Table 6.1).

RATIONALES FOR HUMAN RIGHTS EDUCATION

What does HRE have to offer our students, schools, and communities? One study found that HRE programs resulted in reduced incidences of bullying and truancy and created a calmer learning environment (Vasagar, 2010). Covell, Howe, and Polegato (2011) concluded that a fully implemented HRE program "buffers the effects of social disadvantage by promoting school engagement" (p. 203). Human rights education is "based on the premise that an educated citizenry is the greatest guarantee and ultimate sanction of human rights" (Tarrow, 1990, p. 12). Moves

Table 6.1

Approach	*Students learn*
Learning *about* human rights	• History and origins of human rights movement • Specific UN instruments and frameworks • Major players in human rights movement • Common human rights issues and controversies (Lohrenscheit, 2002).
Learning *for* human rights	• How to participate in human rights movements at local level and beyond • Solidarity with connected human rights movements • Collective action • Tensions between personal interest and larger good within the human rights movement (Lohrenscheit, 2002).
Learning *through* human rights	• In an environment of mutual respect • In a democratic learning space where all voices are encouraged and heard • From teachers and each other • By exercising human rights principals in the learning environment

toward curricular change and the potential benefits of HRE programs have prompted scholars to call for the support of HRE at the national level (Cardenas, 2005; Jennings, 2006; Hornberg, 2002; Stone, 2002; Suarez, 2007). Reardon (1995) proposed that HRE is particularly suited to address the challenging global issues our students face.

The comprehensive nature of HRE means it can be adapted to fit into any subject area at any grade level. "The mutability of HRE is its strength" (Bajaj, 2011, p. 507). Possibilities for overlap exist between HRE and peace education, education for social justice, citizenship education and other disciplines. "Human rights education can support such efforts by providing explicit attention to the larger international human rights movement as well as providing content that defines social justice specifically from a human rights perspective" (Jennings, 2006, p. 289). Though HRE is "defined by the universal framework of international (and sometimes regional) standards, the specific topics and their applications depend upon local and national contexts" (Tibbitts, 2008, p. 2). Local problems rarely exist in a vacuum; HRE is well suited to address the transnational nature of issues facing local communities. HRE also provides a vehicle for the development of students as global citizens who can work to address injustice locally and transnationally.

TECHNOLOGY, CRITICAL LITERACY, AND TEACHING THROUGH HUMAN RIGHTS EDUCATION

A democratic education should provide opportunities for students to critically question the construction of knowledge and examine the different lived experiences of various groups of people. Literacy instruction in classrooms is well situated to support students in developing the ability to both deconstruct and construct "critical texts that can be used in the struggle for social justice" (Morrell, 2005, p. 315). The requirement that students are both able to understand and construct texts can be the vehicle (HRE) for teaching *through* human rights education. As students "see language and literacy learning as political acts [and] realize literacy as tied to power relations in society" (Morrell, 2005, p. 315) they can then engage as citizens in the struggle to create change not only for communities but for marginalized communities across the globe. This view of language and literacy provides students with opportunities to "take advantage of the immense learning opportunities available at home [and] in communities" (Babaci-Wilhite, 2015, p. 15) to create change. Thus teaching *through* HRE is most successful when initiated through students' lived experience as, "a lived experience that becomes integrated into a student's understanding of the world is more likely to produce lasting behavior modifications" (Norlander, 2014, p. 71).

Norlander, (2014) in her research on the intersection of HRE and technology, argues that one aspect of grounding a student's lived experience is providing "awareness of the relevant tools for subsequent informal learning to occur" (p. 71). The use of technological tools to support students' initial and subsequent learning can "enhance the fundamental project of education" and "change the way people relate to one another" (p. 72) including how students understand the implications of human rights for not only their community but for national and global communities as well. "The use of digital tools is envisioned not as a discrete skill or a particular type of literacy but rather as a way students can become conscious creators and producers of cultural spaces in processes of literate production" (Schwartz et al., 2014, p. 184). Technological tools such as cell phones and communicative collaboration platforms like Twitter, Youtube, and Wordpress allow for the rapid transmission of ideas, the collective construction of knowledge, and a speaking against human rights violations—such as the recent movement #BlackLivesMatter—at low cost. Technological tools allow educators to go "beyond the limits imposed by a traditional classroom" (p. 73) and speak to the democratic purposes of education. The use of tools, such as text messaging and online posting, builds on students' out-of-school literacy practices and highlights the value of "students' everyday experience for analytical and personally responsive academic inquiry" (Schwartz et al., 2014, p. 192). In fact without educators who teach *through* human rights and integrate the use of technological tools into their instruction, the objectives of a democratic education are impossible (Mahiri, 2011). As Norlander (2014) notes, "Education is an indispensable component of activism… The integration of information, training and action enables engaged transformative learning" (p. 76).

THE PURSUIT OF HAPPINESS AND LOCALIZING HUMAN RIGHTS EDUCATION

Pacific East College Academy (PECA) is located in Aloe, a suburban community, and serves approximately 300 students in grades 9–12 that are bussed in from the neighboring, low-income communities that surround Aloe. PECA student demographics mirrored that of the 21st century urban school. Seventy-seven percent of the students are Latino, 17 percent African American and 8 percent Pacific Islander. The student demographics are in stark contrast to the population of Aloe, which is 81 percent Caucasian. In addition to the racial disparity the median family income of Aloe, $82,000, was double that of the median family income of Spirit residents where more than 80% of the students lived.

PECA began as a small charter school connected to a local university however the imagined collaboration between the university and the school slowly disintegrated over the last five years. One of the co-authors secured a teaching position within Aloe as a result of a request to the university attached to the charter. The goal was to to teach at the school in order to better understand the implications of the new Smarter Balanced Assessment Consortium (SBAC) Writing Performance Tasks based on the Common Core State Standards, which the university had a key role in developing. The new online assessments shift away from traditional literature as the focus and expand to include a range of topics from environmental issues to art, thus students ability to engage through technology with issues such as human rights and global decision-making is a necessary skill in 21st century English classrooms.

In a freshman humanities course, there were 26 students, with the majority of students identifying as Latino, and an equal percentage of the remaining students identifying as Tongan, African American, or East Indian. All of the students resided in the communities surrounding Aloe, with the exception of one student who lived nearly 53 miles away, whose parents worked in an adjacent community. The demographics of the class were significant as it is this demographic of students—educationally and socially disadvantaged—that are most impacted by high-stakes assessments. The SBAC assessment not only expected students to write on a variety of topics from science to popular culture, the use of technology to administer the assessment meant that students needed to gain technical proficiency. Students would need to efficiently cut and paste text, highlight text, and take digital notes through a commenting system for use during the construction of their essay. The expectation that students would draw from the abundance of information accessed through technology to develop and support claims, arguments, and communicate information marked a significant shift.

Given these two foci—the content of the assessment and the technical skills of using technology—there were two primary teaching goals: develop students' technical abilities and investigate the strategies, concepts, and dispositions necessary for students to demonstration competency. To reach these two goals curriculum was designed that would provide authentic opportunities for learning and application and sought donors to purchase digital tools for the students.

THE PURSUIT OF HAPPINESS UNIT

Developed over the 2014 winter break the unit focused on students' understanding and experience of the pursuit of happiness. This focus came from a deepened understanding that much of what students were asked to do within the classroom—construct questions of various cognitive levels, engage in text-based discussion, draft extensive written responses—was misaligned to the learning from lived experience in the community. The community in which students lived created economic hardships. These hardships impeded students and their families' pursuit of happiness, an aspect highly lauded as the goal of education. To address this disconnect the Pursuit of Happiness unit was designed as an opportunity to investigate the disconnect as well as use technological tools to expand students' engagement within and outside of their community so they could communicate their expectations of education and financial opportunities.

The grounding text for the unit was John Locke's (1905) *An Essay Concerning Human Understanding* and the UDHR. Students studied several paragraphs within book three of Locke's *An Essay Concerning Human Understanding* and compared the language to the UDHR and Declaration of Independence. John Locke's philosophy on the pursuit of happiness and the rights and responsibilities of citizens in their pursuit of happiness aligns with teaching *about* and *for* human rights (Lohrenscheit, 2002). Central to Locke's argument is the need for all men to take into account the lived experience of others and to make decisions that honor everyone's right to experience both social and financial stability, or thinking *for* the right of every human to have particular living conditions. Locke uses food and financial wealth as key examples of how mankind should temper their own decision-making; a key focus of UDHR is the way in which decision-makers at all levels provide or constrain human rights. Interrogating the UDHR and the Declaration of Independence provided a global and national frame of human rights—particularly in terms of access to food and child labor.

Along with these grounding texts students also read *The Alchemist* by Paulo Coelho (1993), *Of Mice and Men* by John Steinbeck (1937), *Romeo and Juliet* by Shakespeare, and research studies on the psychological and physiological effects of experiencing happiness. The inclusion of research articles was necessary to ground some of the UDHR points that in an initial reading may seem trivial such as Article 24: the right to rest and leisure and Article 27: the right to freely participate in the cultural life of the community.

In addition to these key texts all of the students had their own personal digital devices, which were funded by an entrepreneur.[1] Students were given a $350 limit and were able to select the device of their preference as long as they submitted a rationale and a budget that included insurance. Many of the students selected

Chromebooks—which had the most capabilities and structure as a standard laptop; a few students selected Kindles based on a belief that Kindles were better devices for playing video games; and one student paid the difference in order to select an iPad based on its popularity as a high-value and benefit item within her family and community.

The Pursuit of Happiness and Teaching through Human Rights Education

The Pursuit of Happiness unit drew on several routines established in the class and the opportunity to teach *through* human rights was drawn from a weekly routine called out-of-school literacy. Every Monday students were required to share in pairs what they had learned for the first time or had learned more over the weekend. Students could talk about sleeping, playing video games, getting angry, watching a movie or sports. The routine also required students to connect their learning to a discipline that could be studied in college. This latter part of the routine—connecting daily lived experience to a discipline—was designed in order to highlight the construction of knowledge undergirding all that we do as humans and provide dialogue on the expanded range of topics the new assessment asked student to engage. Once students shared in pairs, students were selected to share with the entire class. Additionally, at the end of each semester students were required to develop an informational presentation on one of the literacies they had shared with a partner using both technological tools and disciplinary language. In these Monday share-outs, a number of times one male student detailed picking crops with his father. His detailing of picking crops brought up several topics: the back pain he experienced, the strategies he used to avoid the smells of the pesticides, the purchases he made with his pay, and the meals he and his father had after returning from a day's work. Each time during these share-outs, another student who had picked tomatoes in Mexico would add to the class' understanding of the hardship of picking crops.

In considering how to ground Steinbeck's *Of Mice and Men*, which is set in Salinas a community within driving distance of Aloe, the student's experience picking grapes with his father and the use of student migrant workers in California's agricultural industry became a central lens of literary analysis. California has the highest number of student migrant workers (California Department of Education). Many student migrants work outside of federal guidelines—that are already less stringent for other sectors, most visibly the fact that children of any age can work on a small farm for up to 10 hours a day. According to the law student migrants are not allowed to work during school hours, however this law is routinely violated as students work long hours and travel from state-to-state following the harvest season (Human Rights Watch, 2010). The impact on student migrants' education is so debilitating that there are formalized structures such as the California Migrant Education Program (MEP), which provides tutoring and independent studies to migrant children with the hopes of mitigating the impact of high mobility. The documentary *The Harvest* (Romano, 2013) highlights the issues of both food security, economic instability, and lack of educational opportunities for student migrants. The lives of youth in the documentary resemble the lived experience of migrant youth in Salinas and some aspects of the lived experience of the students at PECA, most notably the need to work instead of attend school and the lack of consistent access to food.

Students at PECA experience cycles of food insecurity. Food insecurity is defined as "the state of being without reliable access to a sufficient quantity of affordable, nutritious food" (Feeding America, 2012). According to the data collected by Feeding America, a network of food banks, 41 percent of residents in the county PECA students reside have an income level 130 percent below the poverty level (Feeding America, 2012). The food insecurity in California is such a contradiction given that California produces nearly half of all the fruit, nuts and vegetables eaten in the nation and 14.7 percent of the overall United States agricultural exports. Despite the ability to produce an abundance of nutritional food commodities, high percentages of families in economically disadvantaged communities do not have access to consistent affordable, nutritious food. This contradiction and the lack of access to financial happiness grounded the unit and offered the opportunity to localize human rights. Students' investigated their personal and communal challenges to pursuing the declarations of Article 25*:*

1. *Everyone has the right to a standard of living adequate for the health and well-being of himself and of his family, including food, clothing, housing and medical care and necessary social services, and the right to security in the event of unemployment, sickness, disability, widowhood, old age or other lack of livelihood in circumstances beyond his control.*

and its intersection with Article 23:

2. *Everyone has the right to work, to free choice of employment, to just and favourable conditions of work and to protection against unemployment.*
3. *Everyone, without any discrimination, has the right to equal pay for equal work.*
4. *Everyone who works has the right to just and favourable remuneration ensuring for himself and his family an existence worthy of human dignity, and supplemented, if necessary, by other means of social protection.*
5. *Everyone has the right to form and to join trade unions for the protection of his interests.*

Students responded to journal prompts focused on their own access to certain foods and career choices, interviewed family members and friends about their food access and career opportunities, investigated the labor practices of their selected math exhibition business sector, and researched data—both quantitative and qualitative—to further inform their understanding of how aspects of Articles 23 and 25 were inadequately available to the marginalized populations within the United States and

Figure 6.1. Student examples of techniques perspective and bottom third

globally. To capture their research and findings, students leveraged technological tools: disposable cameras, cell phones, internet-search engines, camcorders, voice recording applications, and presentation software.

Localizing Human Rights through Technological Tools

Students identified a range of experiences that exemplified Articles 23 and 25 in the Universal Declaration of Human Rights: access to particular occupations, ability to make decisions about family planning, access to education, ability to practice religious freedom, and the impact of immigration to the United States. Students drew on their linguistic and cultural knowledge to design interview questions, identify community resources, and investigate the implications of the local experiences of their subject to the global experiences of marginalized people. Key to all aspects of the project was the leveraging of technological tools to document, obtain feedback, and distribute their work.

Use of Disposable and Digital Cameras

In order to support students' ability to capture the lived experience of community members and provide explicit instruction in how images communicate messages, students were given a disposable camera and a two-hour lesson in photography from Alejandro, a local photographer and videographer. The two-hour lesson included techniques such as the rule of two-thirds, creative cropping, and point of view—all techniques that highlight how perspective is generated through images. Once students practiced the various techniques on their own (Figure 6.1), Alejandro accompanied students on a school field trip to gather images with his professional digital camera of the community they identified as important to communicate the message of their pursuit of happiness project (see Figure 6.2).

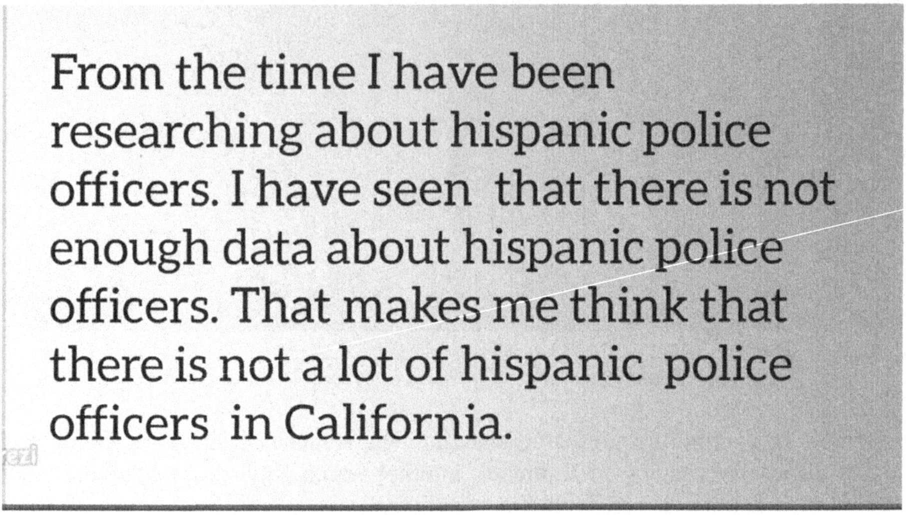

Figure 6.2. Screen shot of student's Prezi

Use of Smart Phones

Students translated their understanding of creating images and applied their out-of-school literacy of creating SnapChats and text messages to gather both visual and audio data of how community members strove for stable housing and employment. Students used their cell phones to record interviews, take photos, and arrange work times with peers. The use of cell phones created an opportunity for community voices to speak on everyday challenges to the pursuit of happiness such as the recording of a parent explaining the effects of housekeeping on her body or the narrative of a Tongan father immigrating to America. Use of cell phones also allowed students and their family to engage in conversation in the language most familiar—which ranged from dialects of English, Spanish and Tongan—as cell phones are an everyday tool of communication within the community and thus are not generally perceived as shifting the context in ways that more formal recording devices might.

Similar to the research findings of Schwartz et al. (2014) smart phones also mitigated the absences of students as communication and image sharing through text messaging allowed students to collaborate with both peers and the teacher without being physically present in the classroom. Redefining classroom participation is essential for students in low-socioeconomic households as students are more likely to miss school in order to support family either to babysit younger siblings or older relatives, earn an income, or recover from sicknesses caused by under or malnutrition.

Use of Digital Devices

As mentioned previously all students within the classroom were able to select and purchase a digital device for their academic and personal use. Students used their digital devices to create Prezis, iMovies, and PowerPoint presentations. Students also use the suite of Google tools to provide feedback, report progress, and store data. The personal devices allowed for smoother transitions between working on the projects at home and at school. Some students used the mobility of their digital devices to get support on their projects within after school programs. The flexibility of the personal devices removed the obstacle of students not being able to take school technology equipment home—an issue most prevalent in under-resourced schools. To support students use of digital devices at home, two portable Internet access squares were purchased and checked-out by students. The purchase of two squares was determined based on students' self-reported access to Internet outside of school.

A key use of the digital devices was researching and sharing information. Students researched and shared information on national and global data on issues of human rights, such as access to education and religious as well as freedom technical skills such as how to convert audio files from a smart phone to use within a platform such as Prezi or Garage Band. The inability to locate information through research, despite the current popular belief that anything can be found on the internet, led students to critically question the available information that framed particular issues. For example, one student began to develop a secondary inquiry question from his experience attempting to locate demographics of the local policing force.

In fact the student reported a tension when he asked local officers in his neighborhood about the demographics of the local police force. This example highlights the issue of information and human rights, as it is highly probable that the information the student sought is available, yet not in a format or repository that can be easily accessed by community members—despite the fact that the configuration of the police force impacts the students' real and perceived right to safety. On one level of the opportunity for Latino citizens to become police officers and a second level of the ability to engage in communication with the police force and ask about it representation or lack thereof of the community for which it serves.

The peer-to-peer teaching of data and technological skills made visible the capacity of students to become better informed and develop their technical skills. The engagement of both human rights knowledge and technological knowledge can potentially reposition students into pursuing careers within the technology fields. Technology fields are in need of diversity and in order to meet this need for diversity students of color require more opportunities to identify themselves as academically and technologically savvy (Schwartz et al., 2014). More importantly in order to ensure that the technology fields maintain an ethical component, as students of color pursue a career in technology it is important that they also take into the local and global impact, both positive and negative, that technology can have on a society.

THE HUMAN RIGHTS AND SOCIAL JUSTICE BLOG PROJECT: THE CASE OF TERRENCE

This example of HRE examines one San Francisco Bay Area teacher, and a transformative experience for one of her students. In March 2012, a class of middle school students at Rosewood School (pseudonym) in San Francisco started the day as they often did – with a discussion of current events. The classroom teacher, Kate (pseudonyms used for teachers and students), brought up the issue that gripped the country – the case of Trayvon Martin, an unarmed African-American teen in Florida killed by neighborhood watch coordinator, George Zimmerman (Blow, 2012). The students discussed the specifics of the case, the role of race, and their feelings about what had taken place. They talked about hooded sweatshirts, which had become one symbol of the injustice of the case. Terrence, a student who rarely spoke up during these discussions, uttered three words, "I wear hoodies."

Terrence was from San Francisco's Bayview Hunter's Point neighborhood and a product of a broken system. A 12-year-old African-American boy who had been in and out of foster care, Terrence reported to teachers that he was kicked out of six schools before he entered Rosewood in second grade. In January 2012, Terrence joined Kate's class at Rosewood. His reading was significantly below a sixth grade level and he refused to produce any writing. Terrence's behavior was more severe than the students already in the multi-age class. He attempted violence with students and teachers, and his resistance to academic work negatively impacted the classroom atmosphere. Kate described Terrence's frequent refrain, "I don't need school," as a defense mechanism to cover up a lack of academic confidence (Personal communication with one of the authors, October 27, 2012).

Kate was in her fifth year teaching at Rosewood, a non-public school, serving students with behavioral challenges and emotional disturbance. Kate taught a class of eight to nine students all of whom qualified as socio-economically disadvantaged, came from "some of the most marginalized communities in the Bay Area," and experienced the daily challenges of living with mental health disabilities (California Department of Education, 2011; Kate, Reflection, September 5, 2012). Kate taught all subjects to her students, created or adapted all of her curriculum and materials to meet the needs of individual students, and aimed to "present material that matters in their lives" (Kate, Reflection, September 5, 2012). She had a great deal of curricular freedom and access to classroom computers and high speed Internet.

Kate's classroom reflected a foundation of teaching *through* human rights. The walls contained charts with key concepts such as *social power* and *oppression*, maps, and rich vocabulary lists. She worked to support and challenge each student as an individual. In this special education setting, many students required help from the adult support staff and Kate and her assistants moved quietly from individual students to small groups aware of building independence. She kept expectations high, yet realistic and consciously asked a great deal of her students. Though many students came from special education classes filled with busy work and constant disruptions, she intentionally created an environment where each student was supported to produce high-level work.

Teaching social studies and language arts through current events within a curriculum based on human rights and social justice engaged Kate's students in a way that traditional methods did not. Using technology in conjunction with issues of social justice was a hook for students who struggled to connect with textbook content and were often more comfortable expressing themselves in through alternative forms of communication. In 2012, as part of a participatory action research (PAR) study, Kate was searching for a simple way for her students to draw upon their linguistic and cultural knowledge to consider issues that mattered to them, and to begin to take action on these issues.

Employing democratic classroom practice, the group made a list of ways to share their thoughts and feelings about issues important in the world and in their lives. Some students wanted to write a play, but others were not enthusiastic. "The thing they seemed most into was a blog" (Kate, February 9, 2012). As the blog discussion progressed, it became clear the students wanted it to be high quality. One student responded, "I don't want to do it again if it's crappy like last time" (February 9, 2012). She assured them that this attempt at blogging would be more polished and attractive than a limited classroom application they had tried the year before. Kate believed the motivation of using technology would rope in even reluctant students.

The header of the blog is shown below in Figure 6.3. Reflected in the descriptor is the importance of connecting students' lives at school and in their communities to broader issues of human rights and social justice.

The blog was simply laid out, consisting of student entries and comments posted in chronological order with the newest material at the top of the page. In addition, a *Free Speech* section allowed students to free associate about human rights and social justice topics, as well as engage in conversations with each other about these topics.

Until the day the class talked about Trayvon Martin, Terrence had been disengaged from classroom discussions and had not participated in the class blog. However, Trayvon Martin captured his attention. He signed up for the blog under the blogger

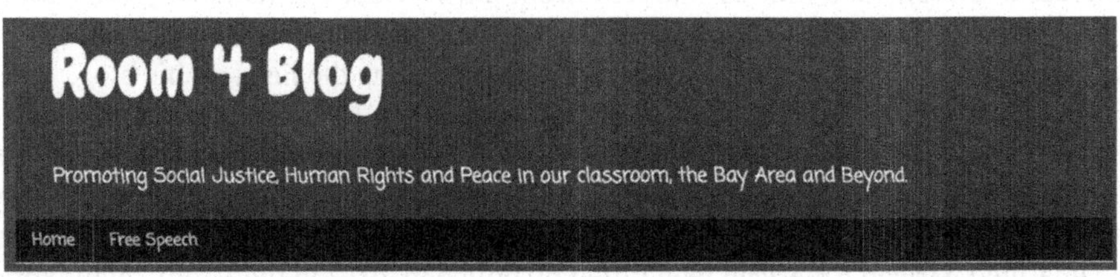

Figure 6.3. Header for Kate's Class Blog

name "im trayvon martin" and used the classroom computer to search for more information. Terrence immersed himself in articles and media related to the killing of Trayvon Martin. He created multiple postings on the class blog, including links to articles and music videos inspired by the case. In his first post, Terrence linked to a video entitled *Trayvon Martin PSA,* which portrayed 30 African-American males in hooded sweatshirts speaking out about how they were perceived in media and society. Below the video link, Terrence wrote: "hoodys up. trayvon martin rip" (Terrence, Blog post, April, 2012). Terrence also posted another music video, *Hoodiez,* by rapper Willie D., which was created as a tribute to Trayvon Martin.

Figure 6.4. Terrence's Blog post for the Willie D video, Hoodiez

He also posted a link to a National Public Radio piece recorded when George Zimmerman was arrested and charged in Trayvon's death. Though Terrence's posts on the class blog consisted mainly of links to videos and articles, and an occasional one-sentence reflection, he was engaged and participating in class in a way Kate had not seen before. Through the use of technology and access to videos, articles, and images posted on the Internet, Terrence engaged multiple literacies to research topics that interested him. He became more adept at navigating primary and secondary Internet sources and drew on a mix of sources—from music videos to nationally recognized news websites. Terrence also began to engage his classmates and blog readers in a conversation of solidarity through questions posted below the links on his blog.

Broadening his research about the Trayvon Martin case, Terrence investigated further issues of racial violence and injustice. He became interested in other victims of racial violence including Oscar Grant, an African American male killed in 2009 by a Bay Area Rapid Transit police officer in Oakland, California; Remarley Graham, an unarmed African American teen killed by police in 2012 in the Bronx, New York; and Kenneth Chamberlain, a 68 year-old African American man killed by police in

White Plains, NY. Terrence began to make connections and draw conclusions based on the multiple sources he found.

As Kate saw the positive effect blogging had on Terrence, she began to use it to help Terrence redirect his participation in class. When Kate could see negative behaviors beginning to escalate, she would send Terrence to the computer to blog. Terrence's research expanded even further to include investigation into the Black Panther Party[2]. He read and posted more videos and longer articles. With each post, he added a little more of his own writing.

About a month into his participation in the blog, Terrence posted a video clip from daily independent news program *Democracy Now* entitled, "They Can't Just Kill Us: Kenneth Chamberlain's Neighbors Speak Out as Police Avoid Charges." Under the video clip, he wrote the following commentary: "the police were using unauthorized force. the people in the neighborhood were mad. one teenage boy and his family moved because it was unsafe. but where is ''safe'' anyway? there is always some thing going down" (Terrence, Blog Post, May, 2012). Terrence's comments reflected his increasing knowledge of vocabulary related to police violence, language that questions and challenges the status quo, and his personal connection to living in a place that does not feel safe from police violence.

Terrence's engagement with media literacy increased his confidence and willingness to utilize more traditional forms of literacy, from writing to class discussion. Within weeks, Terrence was positively contributing to the class, was willing to participate in lessons Kate presented in language arts and social studies, and showed increased self confidence (Kate, Reflection, September 5, 2012). Connecting his cultural knowledge of living as a black male in the United States with the unfortunate death of Trayvon Martin, Terrence became engaged in issues that mattered to him – issues of social justice and human rights that impacted his community, the country, and the world. Through the use of technology, including the blog platform and primary and secondary Internet sources, Terrence employed multiple literacies including digital and media literacy, to bridge his informal linguistic practices—identifying videos and web articles—with the traditional literacy of schooling.

Kate's teaching exemplified the flexibility of HRE and created an environment where Terrence could follow his interest using digital tools. The classroom was organized around mutual respect and acknowledged the linguistic and cultural knowledge of each learner, thus promoting learning *through* human rights. As the class discussed current events issues, from immigration debates to the rights of gays and lesbians, Kate taught *about* human rights, linking to the study international UN documents including the Universal Declaration of Human Rights, which was posted on the wall of the classroom. Additionally, through the use of the blog platform, Kate's students made their voices heard and engaged in learning *for* human rights, educating others about the issues they felt needed to be discussed.

Human rights violations, such as the denial of Trayvon Martin's right to life and the racial bias within the U.S. justice system, provide rich content for classroom discussion and investigation, especially when connected to students' lived realities.

Unfortunately, countless teachers of urban youth find it difficult to address challenging topics within the classroom. Much worse, for many students, school is a place where they are subjected to surveillance, denied the right to speak their minds, forced to endure drawn-out standardized testing, and taught political indifference through 'objective' presentations of knowledge (Ayers & Ayers, 2011). The dominant culture of education disregards the role of power and justice in the curriculum and in educational policy (Kincheloe, 2008).

For students like Terrence, traditional forms of education leave little space for understanding power and oppression and how these forces impact their lives. "Every minute of every hour that teachers teach, they are faced with complex decisions concerning justice, democracy and competing ethical claims" (Kincheloe, 2008, p. 1). The field of human rights education holds the potential to begin a dialogue with students (Tibbitts, 2002).

DISCUSSION AND CONCLUSION

Literacy education that connects students to local, regional, and global struggles for justice through the use of technological tools can deepen students' ability to draw upon their linguistic and cultural knowledge to critique the world. In the two literacy-based examples provided, students evidenced the ability to engage the concepts of the Universal Declaration of Human Rights and evaluate the lived experiences of their community members. As Terrence shared when discussing the blogging project, his blog was a "world wide wall" that allowed him to make a specific cultural practice accessible globally. Furthermore Terrence's engagement with technology and his growing knowledge of media literacy increased his confidence and willingness to utilize more traditional forms of literacy. Not only did the blogging project provide an opportunity for Terrence to speak on the right to life, he was able to do so while enacting his right to education.

Students at PECA engaged in their right to education through accessing their right in education to draw upon their linguistic and cultural knowledge to explore the intersection between the local and global human rights issues. Significant to the students' investigations was the right to economic and social stability. The interviews, presentations, and research statistics highlighted the lack of access to particular careers, higher education, family planning, and stable housing. Students not only investigated issues important to them they also supported the learning of their peers by leveraging the technological tools communication capabilities such as SnapChat, voice recorder, text messaging, and data compression. These capabilities allowed students to expand the boundaries of traditional school learning and negated the notion of being present and active in class to the physical location of the classroom.

Human Rights Education is a necessary component of education. As the world becomes increasingly interconnected it is vital that educators teach both *through* and *for* human rights in local and global contexts. Technological tools support educators' goals of teaching *through* and *for* human rights as they afford students the capacity to draw upon their linguistic and cultural knowledge to read, write, and speak on the human rights issues—thus creating a visible thread between the local and global. We do not suggest that technological tools result in easy solutions to the human rights violations experienced by our students or global communities, however we do posit that technological tools support students experience of literacy as a political act and a tool for democratic engagement—vital for transformative education and the continued struggle for human rights.

NOTES

1 One of the co-authors met an educational technology entrepreneur at a local conference for educators interested in developing students' writing abilities. The entrepreneur asked to meet with the co-author in order to obtain feedback on an online lesson planning software and assessment tool he was creating. The meeting resulted in an agreement that the software would be piloted within the classroom and the entrepreneur would supply students with technical devices that would remain their possession after the pilot.

2 The Black Panther Party is a civil rights organization developed in response to the human rights abuses experienced by African Americans. Members of the organization were key in the 1960s protests to actualize human rights for African Americans afforded by the United States Constitution.

REFERENCES

Ayers, R., & Ayers, W. (2011). *Teaching the taboo.* New York, NY: Teachers College Press.

Babaci-Wilhite, Z. (2015). *Local languages as a human right in education: Comparative cases from Africa.* Rotterdam, The Netherlands: Sense Publishers.

Bajaj, M. (2011). Human rights education: Ideology, location, and approaches. *Human Rights Quarterly, 33,* 481–508.

Blow, C. (2012, March 16). The curious case of Trayvon Martin. *New York Times* [Online version]. Retrieved from http://www.nytimes.com/2012/03/17/

California Department of Education. (2011). *School accountability report card.* Retrieved from http://www.cde.ca.gov/ta/ac/sa/

Cardenas, S. (2005). Constructing rights? Human rights education and the state. *International Political Science Review, 26*(4), 363–379.

Coelho, P. (1993). *The alchemist.* San Francisco, CA: HarperOne.

Covell, K., Howe, R. B., & Polegato, J. (2011). Children's human rights education as a counter to social disadvantage: A case study from England. *Educational Research, 53*(2), 193–206.

Feeding America. (2012). *Map the meal gap.* Retrieved from http://map.feedingamerica.org/county/2013/overall

Flowers, N. (2000). *The human rights education handbook: Effective practices for learning, action, and change.* Minneapolis, MN: University of Minnesota.

Hornberg, S. (2002). Human rights education as an integral part of general education. *International Review of Education, 48*(3–4), 187–198.

Human Rights Watch. (2010). *US: Child farmworkers' dangerous: End legal double-standard that fails to protect children working in agriculture.* Retrieved from http://www.hrw.org/news/2010/05/05/us-child-farmworkers-dangerous-lives

Jennings, T. (2006). Reclaiming standards for a progressive agenda: Human rights education standards for teachers and teacher education. In J. Andrzejewski, M. Baltodano, & L. Symcox (Eds.), *Social justice, peace, and environmental education: Transformative standards.* New York, NY: Routlege.

Kincheloe, J. L. (2008). *Critical pedagogy.* New York, NY: Peter Lang Publishing.

Locke, J. (1905). *An essay concerning human understanding.* Amherst, NY: Prometheus Books.

Lorenscheit, C. (2002). International approaches in human rights education. *International Review of Education, 48*(3–4), 173–185.
Morrell, E. (2005). Critical English education. *English Education, 37*(4), 312–321.
Norlander, R. (2014). A digital approach to human rights education. *Peace Review: A Journal of Social Justice, 24*, 70–77.
Reardon, B. (1995). *Educating for human dignity.* Philadelphia, PA: University of Pennsylvania Press.
Romano, U. (Director). (2013). *The harvest* [documentary]. Burbank, CA: Cinema Libre.
Schwartz, L., Noguerón-Liu, S., & González, N. (2014). The compression of time and space in transnational social fields: Mobilizing the affordances of digital media with Latina students. In C. Compton-Lily & E. Halverson (Eds.), *Time and space in literacy research.* New York, NY: Taylor and Francis.
Steinbech, J. (1937). *Of mice and men.* New York, NY: Penguin Books.
Stone, A. (2002). Human rights education and public policy in the United States: Mapping the road ahead. *Human Rights Quarterly, 24*(2), 537–557.
Suarez, D. (2007). Education professionals and the construction of human rights education. *Comparative Education Review, 51*(1), 48–70.
Tarrow, N. (1990). Human rights education: A comparison of Canadian and U.S. approaches. *Educational Research Quarterly, 12*(4), 12–22.
Tibbitts, F. (2008). Human rights education. *Encyclopedia of Peace Education Online.* Retrieved from http://www.tc.edu/centers/epe/entries.html
United Nations. (2011). United Nations General Assembly Resolution 66/137. *United Nations Declaration on Human Rights Education and Training.*
Vasagar, J. (2010, November 2). Human rights teaching reduces bullying, study finds. *The Guardian* [Online version]. Retrieved from http://www.guardian.co.uk

Lanette Jimerson
Graduate School of Education
University of California, Berkeley
USA

Page Hersey
School of Education
University of San Francisco
USA

PART III

LANGUAGE OF INSTRUCTION IN SCIENCE AND TECHNOLOGY

BIRGIT BROCK-UTNE

7. ENGLISH AS THE LANGUAGE OF SCIENCE AND TECHNOLOGY

A PERSONAL INTRODUCTION

In 2005 I was for several weeks sitting hour after hour in the back of a class-room in a secondary school in Tanzania (Brock-Utne, 2005, 2013a). I observed students who did not understand what the teacher was saying when he spoke English, and often would ask the teacher to express himself in Kiswahili, a language they all knew very well. My eyes fell especially on one gentle looking boy who was completely passive and obviously did not understand anything of what was going on. Once I heard him ask one of his class-mates in Kiswahili what the teacher had said. When I spoke in Kiswahili to him during the break afterwards and mentioned that I had noticed that he did not understand the language of instruction, he admitted that my observation was correct. He had great difficulty following the teacher, especially if the teacher did not switch to Kiswahili during the lesson. I asked him if it would not have been much better for him had the lesson been given in Kiswahili throughout. He admitted that it certainly would have been much easier. Then he would be able to understand what the teacher was saying. When I then asked him whether he thought one should change the language of instruction, he said no he did not think so because English was the language of science and technology. English was the language of engineering, modernisation and all technological development. Without knowing English one could not get a good job. He had to learn English and could not see another way than having it as the language of instruction. The researcher Lakshman Weddirage (2009) in his work on the language of instruction in Sri Lanka finds it difficult to understand where the belief comes from that science is better learnt in English than in other languages and that "English is the language of science and technology."

SCIENCE EDUCATION AND ENGLISH MEDIUM: THE SRI LANKAN EXPERIENCE

The claim that "we need English as the language of technological development" is repeated several times in the report on the *Education System for the 21st Century in Tanzania* (URT, 1993). Yet the claim seems unfounded. As Rugemalira and colleagues (1990, p. 31) maintain:

Z. Babaci-Wilhite (Ed.), Human Rights in Language and STEM Education, 111–127.

> It should be demonstrated that countries such as Finland, Norway, China or Japan, which do not teach their children through the medium of an "international language" are isolated and have lost track of technological developments beyond their borders.

Lakshman K. Wedikkarage (2009 and further elaborated in Brock-Utne, 2013c) tells that Sri Lanka introduced her local languages, Sinhala and Tamil, as languages of instruction in education even before having obtained independence from Britain in 1948. Steps were taken to introduce these languages as media of instruction in all primary schools in 1945, secondary schools in 1953 and at universities in 1960. Wedikkarage argues that local educationists all argued that the change of medium of instruction from English to local languages enabled the majority of students to learn science subjects in their mother tongue, nullifying the previous belief that studying these subjects in English would be an advantage. Local educationists in Sri Lanka argue that the mother tongue medium policy in Sri Lanka has contributed remarkably to the development process of the country. Sri Lanka enjoys a literacy rate of 91 percent (the highest in South Asia and one of the best in developing countries). It can also boast of hundred percent participation rates in primary education. In addition to the increased life expectancy in general, Sri Lanka is the only country in South East Asia that is not considered a low income country. Mahinda Ranaweera, the former Director of Education at the Curriculum Development Centre, Ministry of Education, Sri Lanka wrote about the great advantages to the population of Sri Lanka of the introduction of Sinhala and Tamil as the languages of instruction to replace English—*especially* for the teaching of science and technology:

> The transition from English to the national languages as the medium of instruction in science helped to destroy the great barrier that existed between the privileged English educated classes; between the science educated elite and the non-science educated masses; between science itself and the people. It gave confidence to the common man that science is within his reach and to the teachers and pupils that a knowledge of English need not necessarily be a prerequisite for learning science. (Ranaweera, 1976, p. 423)

Ranaweera relates that the change of medium of instruction in science and mathematics lagged behind the other subjects because of special difficulties, like the absence of scientific and technical terms, textbooks, and proficient teachers. Yet he found the greatest need to switch over to the national languages in the science subjects. He gives two reasons for this claim.

- First, science education was considered the main instrument through which national development goals and improvements in the quality of life of the masses could be achieved. Thus, there was a need to expand science education. He says that English medium was a great constraint which hindered the expansion of science education.

- Secondly, he notes that in order to achieve the wider objectives of science education, such as inculcation of the methods and attitudes of science, the didactic teaching approach had to be replaced by an activity- and inquiry-based approach which requires greater dialogue, discussion, and interaction between the pupil and the teacher and among the pupils themselves. As Ranaweera (1976, p. 417) notes: "Such an approach makes a heavy demand on the language ability of the pupils and will be more successful if the medium of instruction is also the first language of the pupils."

However the educational authorities in Sri Lanka in 2001 reintroduced English as a medium of instruction for science classes at collegiate level (Grades 12 and 13) also known as General Certificate of Education Advanced Level (G.C.E.A.L) in certain selected government schools. Wedikkarage (2009) has analysed the discourses that led to a reverse in language of instruction for G.C.E. AL science classes in certain selected schools at a time when such subjects were comfortably being taught in local languages. Lakshman Wedikkarage found that the major objective of the reintroduction of English as a medium of instruction for G.C.E. (AL) science classes was in reality to improve English language competence of students. The idea that English medium will lead to improved English competence of the G.C.E. (AL) science students emerged as a central but totally unsubstantiated belief. In Sri Lanka, a country where science teaching has taken place for nearly 40 years in the local languages, Sinhala and Tamil, it was difficult to find teachers who were willing and competent to teach such subjects in English.

Generally, when admitting students to G.C.E. AL science classes in privileged government schools, a very strict selection procedure was adhered to. Since the demand for these schools is very high, only the very best students were selected. However, in order to fill up the English medium classes this strict selection mechanism was relaxed for students who promised to study AL science subjects in English medium. Many of these students soon experienced that they could not study these subjects in English and sought permission to go back to mother tongue medium, creating administrative problems in the schools.

According to the students Laksman interviewed for his Ph.D. (Wedikkarage, 2006) most of the teachers in these English medium classes resorted to either Sinhala or Tamil when they could not properly explain their lessons in English. According to the same students, the teachers were far more effective presenting their subject matter when they taught in their mother tongue.

The failure of the educational authorities to teach English effectively as a second language has been used as a pretext to reintroduce English medium in the public school system in Sri Lanka. The study by Wedikkarage of the difficulties both teachers and students face when using English as a medium of instruction indicates that what is required in the Sri Lankan context is not to go back to English medium, as the two local languages are effectively used in the provision of education, but for students to learn English well as a foreign, yet important language. Most private sector employers do not demand job seekers to have studied in English medium. What the private sector requires are people with a good proficiency in English.

THE CASE OF MALAYSIA

Some of us have argued that without mother-tongue education at all levels of education there is no future for African development (Prah & Brock-Utne, 2009). The experience of post-colonial Asia and Western Europe point irrefutably to the inherent value in mother-tongue education or at least education in popular, widely spoken, local languages. We have frequently argued that the prosperity and economic prowess of modern Asia is, in no small measure, attributable to the use of languages confidently understood, spoken and written by the overwhelming masses of the people.

In July 2009, Associated Press (*New York Times, Asia Pacific*) wrote that:

> Malaysia said Wednesday (8.7.2009) that it would abandon the use of English to teach mathematics and science, bowing to protesters who demanded more use of the national Malay language. Malay will be reinstated in state-financed schools starting in 2012 because teaching in English caused academic results in those subjects to slip, Education Minister Muhyiddin Yassin said. There have been months of high-profile demonstrations by politicians and linguists, especially from the ethnic Malay majority, who say a six-year-old policy of using English undermines their struggle to modernize their mother tongue. English was once the medium of instruction in most schools in Malaysia, a former British colony. Nationalist leaders switched to Malay less than two decades after independence in 1957. In 2003, Prime Minister Mahathir Mohammad started a programme to resume teaching math and science in English. Most other subjects are taught in Malay.[1]

By evidence of the record, Malaysia has covered a trying journey to where she stands today on the issue of language of instruction in education. In further elaboration and clarification of this in the news report, it was stated that:

> Deputy premier Muhyiddin Yassin said that from 2012 the subjects will be taught in Bahasa Malaysia in national schools, or in Chinese and Tamil in vernacular schools. Critics of the six-year policy of using English to teach the subjects argue that it has dragged down students' performance and is particularly unfair on children who are not proficient in the language. "I wouldn't say it's a complete failure but it has not achieved the desired objectives that it was supposed to achieve," Muhyiddin told a press conference. "The government is convinced that science and maths need to be taught in a language that will be easily understood by students, which is Bahasa Malaysia in national schools, Mandarin in Chinese schools and Tamil in Tamil schools.[2]

Interestingly, it would appear that whereas in Asia the broader masses demonstrate and protest in the streets in favour of the use of their local languages as languages of instruction for science and technological education, in Africa the elites lead the way in support of the use of the ex-colonial languages as languages of instruction. One can say that Africa, in this respect, is moving in the opposite direction to modern Asia.

Many Africans admire the visible success of contemporary Asia in all areas of the social and economic lives of Asians but are unable to easily see the connection between this scientific, technological and economic ascendancy of Asia and the use of local languages as languages of instruction in education. If language is understood to be the central feature of culture and development is seen as ultimately a cultural phenomenon, it is not difficult to see the interconnections between language and development. In our introduction to a book on multicultarism in Africa, Kwesi Kwaa Prah and I are not suggesting that the use of the multilingualism, the home language or the first language as the language of instruction automatically leads to social development (Prah & Brock-Utne, 2009). We are suggesting that there are other factors which contribute to development, but development cannot occur in the post-colonial circumstances of Africa and Asia without the centralization of the languages of the masses as languages of educational instruction.

Examples from Tanzania

One of my Tanzanian students in the introduction to her master's thesis recalls her own school days:

> I can recall from my school days about my Chemistry teacher who every ten minutes or so he would ask us: "Any question students?" Nobody answered and he would conclude: "If there are no questions, then you have understood everything!" We did not understand him at all, not only because he taught in English only, he spoke American English! – he was a Peace Corp. The issue was language, as it is in our contemporary schools. (Mwinsheikhe, 2001)

Halima Mwinsheikhe later got the opportunity to study the effects of using English or Kiswahili as the language of instruction in secondary schools in Tanzania. In their Ph.d. theses, which were undertaken under the umbrella of the LOITASA project, Halima Mwinsheikhe (2007) and Mwajuma Vuzo (2007) let the same teacher teach the same topic first in English or Code-Switching and then some days later in Kiswahili. Two different secondary schools were used and six weeks spent in each of them. The experiment was conducted in Form I and both quantitative and qualitative data were gathered. Halima concentrated on biology lessons while Mwajuma concentrated on lessons in geography. I spent three weeks with them in the first school and two weeks in the second school. This I did both to increase the reliability of the findings and also to get some first-hand field experience. My own data were of a qualitative kind (Brock-Utne, 2007a).

A high standard deviation suggests a large amount of variability of scores around the mean, whereas a small standard deviation indicates little variability. Mwajuma Vuzo (2007) found that the standard deviations from the two different tests showed that there was a much higher standard deviation in the group which was taught through the medium of English compared to the groups taught in code-switching or in Kiswahili. The smallest standard deviations were obtained in the group that had Kiswahili as language of instruction, then came the code-switching group and the largest standard deviation was found in the group with English as the language of instruction. This indicates that the use of Kiswahili as language of instruction facilitates a more equitable performance among students. Some of the teachers had the same impression. They claimed that students did not differ much in performance when Kiswahili was the language of instruction. The standard deviation for the group that had Kiswahili as the Language of Instruction (s=13.18) was the smallest compared to that of the group that had code-switching (s=15.83) and that of the group which had English as the language of instruction. (s=18.01). It is also implied from the mean scores and from graph I (see below – constructed by Vuzo, 2007) that there are immense differences in mean scores particularly between the group with Kiswahili as the language of instruction and the one with English as language of instruction in school Y.

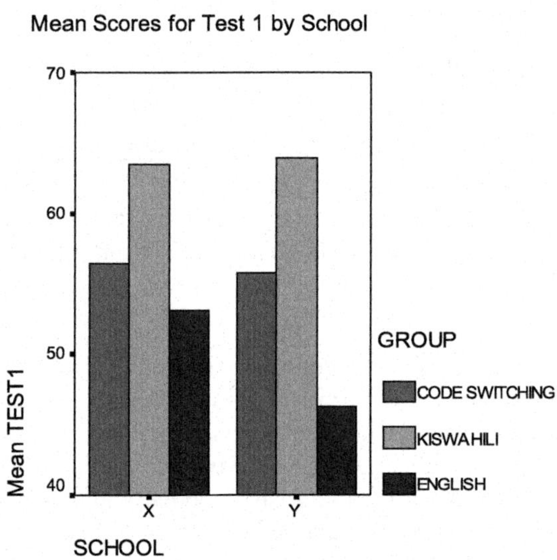

Figure 7.1. Mean score for test 1 by school

Both Mwinshehe (2007) and Vuzo (2007) found that when the same teacher taught students in Kiswahili and in either CS or English Medium of Instruction, students performed much better in Kiswahili. The large spread of scores when English was used actually shows how the use of the English language as language of instruction contributes to social segregation, the opposite of poverty alleviation. The examples given here show that students learn better if they understand what the teacher is saying. They learn better when they can use a familiar language as a language for acquiring new knowledge. The example from Vuzo's doctoral research also shows that by using English as the language of instruction, differences between students are increased.

The use of a language for instruction which is unfamiliar to most students is a recipe for increased inequality. It may benefit a very small group of students who have well to-do parents who take them to English speaking countries, have English speaking guests and a lot of English language books, videos and games. Frequently, these children are also given extra tuition in English. The use of an unfamiliar language as the language of instruction is, however, a strategy to keep the masses down, to stupidify them and make it difficult for them to rise from poverty. It is the opposite of poverty alleviation. As mentioned earlier I also spent some weeks together with Halima and Mwajuma in the schools where they made their experiments, observations, interviews and tests. I wrote an article summarizing my observations (Brock-Utne, 2007a). In the article I give excerpts from my field-notes describing two lessons being taught by the same female biology teacher, the first one in English, the second one in Kiswahili. The description is rather typical of the situations I experienced during my twenty hours of observation in August 2004. It shows the much greater interaction between teacher and students in the class taught in Kiswahili. It shows students challenging the teacher, asking questions, participating in a lively manner. It shows a smiling teacher who is at ease. When the same teacher teaches in English she hardly smiles, she punishes the students by having them stand by their desks if they cannot answer. The students are passive, afraid to say anything, hoping they are not called upon to answer questions.

UNDERSTANDING OF SCIENCE CONCEPTS

As part of the LOISA[3] project Keith Langenhoven (2006, 2010) looked at the understanding of science concepts among grade 6 learners who had been taught in English and grade 6 learners who had been taught in isiXhosa. The classes were found in the township school we have called Zama, one of the schools participating in our project. The following question was chosen for coding and analysis because it lent itself to expressions of learning competence through writing. The question was taken verbatim from the workbook that teachers used when teaching Grade 6: Module 1: Ecosystems and Environmental Balance: The Influence of Humans on the Ecosystem.

Here we look at the answers to the following question: Describe TWO natural resources that have been exploited (damaged) by humans. The results are reflected in the graph below.

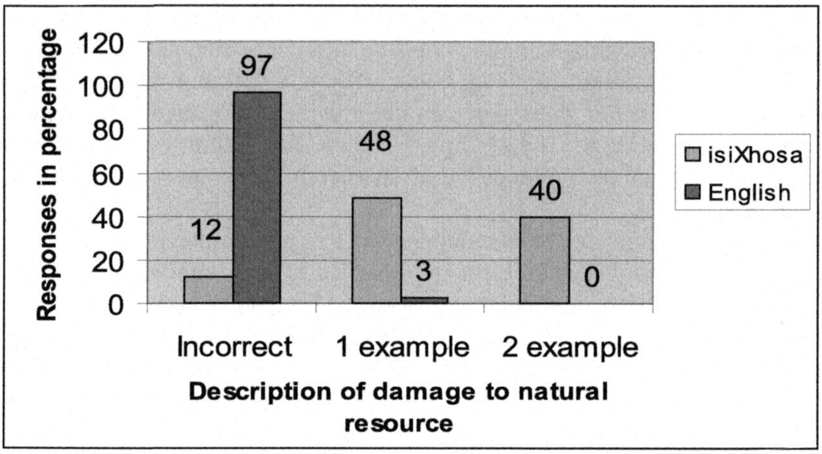

Figure 7.2. Respondents describing damaged natural resources, in percentage
Source: Langenhoven, 2006, p. 262

We see that 88% (22/25) of the isiXhosa taught pupils were able to write down one or two examples of natural resources that have been exploited (damaged) by humans. They wrote in isiXhosa, their mother tongue. They wrote in an understandable way. It was evident from their answers that the concept of damage to natural resources had been well understood. The English taught pupils were extremely weak in writing about this concept in English. Thirty out of thirty-one students (97%) gave incorrect answers. Langenhoven (2006, p. 262) comments: "This result is shocking and raises many questions about education policy and its outcomes." He gives some examples which illustrate some of the weak responses received when learners are forced to write in English:

Pupil 1: It say they bring death at our home but they don't.

Pupil 2: Sheep we damage them by killing them, after all we took them to the fire we make meat by them.

Pupil 3: Fungi and bacteria they stay in the bread. They damage in the bread, people do not eat the bread.

Pupil 4: The frog that we are the amphibian that are living organisms. This frog he live in a forest because is living thing. The frog that we have damages the forest. We have their damages of the frog that are.

Pupils are unable to understand the question asked because they do not understand what they read. They write about things that appear to have been rote-learned. The perception is that any writing about science is acceptable. This writing ranges from personal experiences to snatches of classroom teaching and learning experiences. The writing also reveals misunderstanding of definitive animal characteristics and definitive habitats.

A study undertaken by Vuyokazi Nomlomo at the Univeristy of Western Cape as part of her Ph.D. research (Nomlomo, 2007) showed a positive correlation between the use of the learner's mother tongue (isiXhosa) as a medium of instruction and learners' understanding and academic performance in science. That is, learning through the mother tongue (isiXhosa) enhances better understanding of science concepts than learning through English which in reality is a foreign language for children living in an informal settlement like Khyalitsha. Nomlomo (2008) is concerned with the fact that science teachers need both adequate subject matter knowledge and be able to get this subject knowledge across to learners. This is definitely easier in a language which both they and the learners have a good command of.

In a study from Botswana by Prophet and Dow (1994) a set of science concepts was taught to an experimental group in Setswana and to a control group in English. The researchers tested understanding of these concepts and found that Form I students taught in Setswana had developed a significantly better understanding of the concepts than those Form I students taught in English.

In the 2003 TIMSS (Trends in International Mathematics and Science Study) mathematics test for grade eight, it was reported that out of the 45 countries that participated Ghana finished as number 44. Ghanaian students scored 276 compared to the international average of 466. In two articles in *Ghana News* Y. Fredua-Kwarteng and Francis Ahia (2005a, 2005b) try to explain these low results. In the first article they discuss the results in mathematics, in the second the results in science. They start by explaining that a country, whose national mathematics pedagogy is compatible with the one undergirding the test, is more likely to do well than a country with different mathematics pedagogy. In Ghana, according to the authors:

> Mathematics teaching at the eighth grade is characterized by the transmission and the command models. Teachers merely transmit mathematical facts, principles and algorithms, and students are commanded to learn them in a passive and fearful manner. Students are not encouraged to pose questions or engage in problem-solving activities in order to attain both conceptual and procedural understanding of what they are being taught. Students simply memorize the algorithms and regurgitate them during tests or examinations.

They find the main reason for why the students do not learn problem-solving and problem-posing skills in the use of a foreign medium as the language of instruction:

> Since Ghanaian students took the test in English (the so-called official language of Ghana), those whose first language is non-English are at great disadvantage. We are not surprised that countries that top-performed in the mathematics test—Taiwan, Malaysia, Latvia, Russia- used their own language to teach and learn mathematics.

The two authors, who both are mathematics educators, argue that a Ghanaian student who is proficient in his or her mother tongue would be likely to answer most of the questions correctly if the questions were translated into the native language of the student. They are aware of the fact that the language of instruction in Ghanaian schools is a contentious issue. And they go on to note:

> Some Ghanaians theorize that a person becomes increasingly proficient in a foreign language after using it over and over for a long time. Applying this line of reasoning to the case under discussion, as our grade eight students go through the grade-ladder they would eventually attain English proficiency needed for mathematical problem-solving. Nevertheless, the unfortunate thing is that most of these students would psychologically drop out of mathematics before they attain English proficiency! ...Some Ghanaians also argue that using English for instruction makes it possible for Ghanaians to "transport" their education to any of the English-speaking countries. But as we have argued in one of our articles on mathematics education, when Ghanaian students at the secondary level enroll in schools in Canada they are confronted with two main tasks. They have to find the meaning of mathematical concepts and also the words to communicate the meaning of those concepts. Asian students, on the other hand, have to find the words to express their understanding of mathematical concepts. This is because they have already learnt the meanings of mathematics concepts in their own language. So whose education is more portable?

The authors further criticize the tests for being rooted in a western, especially American environment using concepts which are unfamiliar in Ghana, like a "parking lot." From their professional experience, students are more likely to solve mathematical problems if they can relate to the cultural context of the problem.

The Italian professor of chemistry Liliana Mammino (2010) presents interesting reflections based on over 20 years experience with teaching general chemistry and physical chemistry courses in Southern Africa – the last 14 years at the University of Venda (UNIVEN) in South Africa, where the teaching also included the process technology course (an introduction to chemical engineering, largely based on physical chemistry). Liliana Mammino holds a degree in Chemistry from the University of Pisa (Italy) and a Ph.D. in Chemistry from Moscow State University (Russia). Before joining UNIVEN (1997) she worked in other universities in Southern Africa (University of Zambia and National University of Lesotho). UNIVEN is a particularly disadvantaged institution, combining the *historical disadvantage* of being a *Historically Black University* (HBU) and the socio-economic disadvantages common to poor rural contexts. The HBUs were universities that were "for blacks only" during the apartheid period, which, according to the political criteria of those times, implied both poor resources and poor educational approaches, aimed at ensuring that black students would not have a chance to excel. Mammino notes that the country-wide general scarcity of qualified secondary school science teachers affects rural areas more extensively, resulting in the serious underpreparedness of most students entering UNIVEN. Mammino (2010) provides extensive documentation on the impact of using a language different from the mother tongue to teach chemistry. She analyzes the difficulties encountered by tertiary level chemistry students in second-language disadvantaged contexts. The results stress the importance of utilizing the mother tongue to teach chemistry. Mammino stresses that the *language of science* is not just terminology. Communication or thought generation require much more than technical terms because they depend on the words linking the technical terms to build a meaning, and these are the common words (verbs, adjectives, prepositions, logical connectives, etc.) pertaining to the language utilized: technical terms are thus *immersed* in a *sea* of common words that constitute the backbone of the communication.

> Knowing the meaning and roles of these common words, being able to understand what they communicate (on reading or listening) and to use them so as to communicate a wanted meaning (on speaking or writing) become essential instruments to ensure correctness and clarity in any form of communication. In particular, rigorous wording usage by the teacher enhances the quality of explanations and helps prevent confusion and misconceptions. The inevitable general inference is that language mastering is the key to science learning as well as to creativity in the sciences. This, together with the acknowledgment of the paramount internalization depth of word-concept and expression-concept correspondences within the mother tongue, points to the essential role of the mother tongue as the natural ground to develop language mastering up to the highest sophistication levels and, in particular, up to the levels that are needed for science communication and science learning. (Mammino, 2010, p. 8)

Mammino (2010) notes that Chemistry and chemistry education are particularly apt to highlight language-related aspects in science teaching/learning, for the same reasons for which chemistry can be viewed as an ideal area for *language-of-science* education. The simultaneous extensive presence of descriptions through language and through mathematics, with two intertwining description levels (macroscopic and microscopic), the use of a symbols system that is probably the most extensive and articulated in the sciences, and the continuous interplay between observation and interpretation, demand substantial language mastering at a sophisticated level. Mammino (2010) finds that the known difficulties students encounter at learning chemistry can often be traced to difficulties with understanding texts – with reference to both written and oral communication – and this, in turn, often depends on inadequacies in the language-mastering. Interventions aimed at enhancing language-mastering can thus be viewed as relevant components of chemistry education. In the text here referred to Mammino (2010) gives many examples showing how the poor mastery of English constitutes a barrier to learning of chemistry.

In an experiment Lilliana Mammino (1995) conducted at the National University of Lesotho, students who had written incorrect or meaningless statements in their chemistry papers were asked to explain their views on the given issues through their mother tongue to somebody who could then translate their answer into English. Mammino, who at that time taught in Lesotho, explains that she does not speak any African language. For those experiments, she asked a nonchemist to assist with the translation; to ensure that there would be no automatic improvement of the chemistry, as could occur spontaneously if a chemist translates chemistry. She notes that in several cases, the translated answer corresponded to reasonable chemistry. The students had understood the chemistry but not been able to express their insight in English. Further discussions highlighted the details of the language difficulties that had led to absurd or meaningless answers which often were related to grammar and sentence construction, but also to the selection of individual words, or how to combine them to express the desired meaning.

Mathematics and the sciences are normally difficult subjects for most children to learn. Yet they are important subjects for the development of any country. One would think that policy-makers would make a great effort to bring these important subjects to the people in a language they can easily understand. Strangely enough this is not happening. The recent change in the language policy of the primary schools in Zanzibar is a case in point (Babaci-Wilhite, 2013, 2014, 2015).

WHAT IS ZANZIBAR UP TO?

In connection with a World Bank loan to the education sector and a consultancy report (MEVT/University of Bristol, 2005) showing that the competence in English was very low among primary school pupils, even among those who had an extra year for preparing to use English as the language of instruction in secondary school, Zanzibar reintroduced a new curriculum where English will be the language of instruction from grade five in mathematics and science subjects in all government primary schools (MEVT, 2006). There is a strange discrepancy between the terms of reference given the researchers behind the consultancy report and the conclusions they come up with. They had *not* been asked by the Ministry to come up with suggestions for a new language of education policy for Zanzibar. They had been asked to evaluate the orientation secondary class (a bridging year between primary and secondary school for pupils who have not done well enough in English to use it as a language of instruction right after primary school) and to interview different stakeholders. They found that there is a low level of English language proficiency among pupils and teachers and that "primary English does not provide an adequate basis for the switch to English in the secondary phase" (MEVT/ University of Bristol, 2005, p. 4). Yet they recommended: Gradually introducing English medium teaching, starting first with one subject in, say std 4, increasing to core subjects, i.e. Maths, Science, Social Science by the end of Std.7 (MEVT/University of Bristol 2005, p. 5). The Ministry decided, partly built on this advice, to reintroduce English as the language of instruction from grade five in mathematics and science subjects in all government primary schools (MEVT, 2006).

On Tuesday 24th of November 2009 I had a meeting with the Deputy Principal Secretary in the Ministry of Education and Vocational Training Abdulla Mzee Abdulla, in his office in the Ministry in Stone Town, Zanzibar. He has a Ph.D. from the University of Bristol. He said that both he and Prof.Pauline Rea-Dickens, who was the lead author of the consultancy report, knew that children learn best in a language which is familiar to them. Yet his job as a politician was to listen to what his constituency wants and the parents want their children to be taught through the medium of English.

Changing a Belief System Which Has Become Common Knowledge

Having English as the language of instruction does not promote understanding of what is learnt in the majority schools in so-called anglophone Africa. As Ayo Bamgbose (2005, p. 255) correctly observes:

> Outside Africa, no one questions why the languages of countries with smaller populations in Europe should be used as a medium, even up to and including the university level. What seems to be lacking in many African countries is the political will to break away from the colonial policy and practice of limiting mother tongue education to lower primary classes. Where such a will exists, much can be done in a short period of time.

Foucault (1988) claims that belief systems gain momentum (and hence power) as more people come to accept the particular views associated with that belief system as *common knowledge*. Some ideas, being considered undeniable "truths," come to define a particular way of seeing the world. At the moment, those who stand the most to lose from having a foreign language used as the language of instruction consider it as an undeniable "truth" that having English as the language of instruction is the best way to learn English. This is a false belief as much research has shown. It is, however, a belief that both donors, the former colonial powers, the publishing industry in the west and the African elite have interest in promoting. These power groups are, however, like Gene Sharp (1980) points out, dependent for their positions and political powers upon the obedience, submission, and cooperation of their subjects. They have an interest in promoting and maintaining the misconception that having a language that teachers and students can barely communicate in will be of benefit for them. In reality this policy leads to Africans learning less than they could have, had they been taught in a language they know well. Misconceptions are possible to alter. When the masses of Africans understand how this misconception holds them down and works to the advantage of the powerful, the allegedly powerless may unite to do away with the misconception.

Avenstrup (2009) in an article on the values dimension in evaluation mentions that much development aid in education engages in the rhetoric of improving access and quality and measures this in terms of indicators such as enrolment, progression and completion rates, teacher/learner ratios, textbook/learner ratios, teacher formal qualifications:

> This sort of instrumentalism completely ignores the debate about the reproduction of cultural capital and power structures in educational systems. It assumes that literacy and numeracy are relevant in themselves, and ignores the discussion about what content and values are conveyed in the official and hidden curriculum, or contested in curriculum appropriation. (Avenstrup 2009, p. 257)

The curriculum which is hidden by having children study in a language they do not master is to have them look down on their own culture and to tell them that they are stupid when they do not understand what the teacher is trying to convey.

African academics like Bamgbose (2005), Mazrui (1997) Ouane (2009), Qorro (2009), Desai (2006), Rubagumya (2003), Prah (2005), Nomlomo (2007), Vuzo (2007), Bgoya (1992), Makelela (2005), Mwinsheikhe (2007) see how the formal school sector is an intellectual recolonisation of Africa through the curriculum. The views of these academics, because they are still few, may to-day seem marginal but the hope for Africa lies in such people gathering sufficient momentum, and hence power, which will lead to political will, to have their views accepted as *common knowledge*.

At a point in time it was common knowledge that men and whites were more intelligent than women and blacks. This is not common knowledge any more. It was common knowledge that neither women[4] nor blacks[5] could learn mathematics. Now the common knowledge is that women and blacks have the same ability for learning mathematics as men and whites have.

It may become common knowledge in Africa too that children learn better when they understand what the teacher is saying.

NOTES

1 Malaysia Drops English for 2 Subjects. The Associated Press (*New York Times, Asia Pacific*). 8th July 2009. A version of this article appeared in print on July 9, 2009, on page A5 of the New York edition. http://www.nytimes.com/2009/07/09/world/asia/09malaysia.html?_r=2

2 Malaysia Drops English for Math, Science Classes. Quoted from, Agence France Press (AFP). 9/07/2009. Mother Tongue Based Learning in the Philippines. A Blog for the MLE Consortium. http://mothertongue-based.blogspot.com/

3 The LOISA research project was funded through a joint grant from the National Research Foundation (NRF) in South Africa and NRC (Norwegian Research Council) in Norway. The project has been closely connected to the LOITASA project. The LOITASA project has been and is funded by NUFU (Norwegian Universities Committee for Development, Research and Education). The first phase of LOITASA ended with a conference in Norway called Languages and Education in Africa (Brock-Utne & Skattum (Eds) 2009), a co-operation between five NUFU funded projects at the University of Oslo, all dealing with languages and education in Africa.

4 Even at the time I went to elementary school in Norway the girls had fewer weekly lessons in mathematics than the boys had. There was a belief that girls could not learn mathematics and would not have any need for it. Instead we had home economics, a subject the boys did not have. In our

mathematics textbooks there were some tasks marked with an asterisk and after the asterisk it said: The girls do not need to do this task. The same was true at the final exam (Brock-Utne & Haukaa, 1980).

[5] Black students from South Africa tell me that during the apartheid there was a belief that Blacks could not learn mathematics and they consequently were not taught this subject.

REFERENCES

Avenstrup, R. (2009). Discoursing E-Value-ation: The values dimension. In B. Brock-Utne & G. Garbo (Eds.), *Language is power. The implications of language for peace and development* (pp. 249–259). East Lansing, MI: State University of Michigan Press.

Babaci-Wilhite, Z. (2013). The new education curriculum in Zanzibar. The rationale behind it. In Z. Desai, M. Qorro, & B. Brock-Utne (Eds.), *The role of language in teaching and learning science and mathematics* (pp. 127–151). Cape Town, South Africa: African Minds.

Babaci-Wilhite, Z. (2014) (Ed.) *Giving Space to African Voices. Rights to local languages and local curriculum*. Rotterdam, The Netherlands: Sense Publishers.

Babaci-Wilhite, Z. (2015). *Local languages as a human right in education. Comparative cases from Africa*. Rotterdam, The Netherlands: Sense Publishers.

Bamgbose, A. (2005). Mother tongue education: Lessons from the Yoruba experience. In B. Brock-Utne & R. K. Hopson (Eds.), *Languages of instruction for African emancipation: Focus on postcolonial contexts and considerations* (pp. 231–257). East Lansing, MI: Michigan State University Press.

Bgoya, W. (1992). The challenge of publishing in Tanzania. In P. Altbach (Ed.), *Publishing and development in the third world*. London, England: Hans Zell Publishers.

Brock-Utne, B. (2005). Language- in –education policies and practices in Africa with a special focus on Tanzania and South Africa insights from research in progress. In A. M. Y. Lin & P. W. M. Lin (Eds.), *Decolonisation, globalisation. Language-in -education policy and practice* (pp. 173–191). Clevedon, England: Multilingual Matters.

Brock-Utne, B. (2007a). Learning through a familiar language versus learning through a foreign language: A look into some secondary school classrooms in Tanzania. *International Journal of Educational Development, 27*(5), 487–498.

Brock-Utne, B. (2007b). Language of instruction and student performance: New insights from research in Tanzania and South Africa. *International Review of Education, 53*(5/6), 509–530.

Brock-Utne, B. (2013a). Teaching science for development – in whose language? In Z. Desai, M. Qorro, & B. Brock-Utne (Eds.), *The role of language in teaching and learning science and mathematics* (pp. 11–33). Cape Town, South Africa: African Minds.

Brock-Utne, B. (2013b). Language of Instruction and learning in mathematics and science in some African countries. *African and Asian Studies, 12*, 83–99.

Brock-Utne, B. (2013c). Language and liberation. Language of instruction for mathematics and science: A comparison and contrast of practices focusing on Tanzania. In C. Benson & K. Kosonen (Eds.), *Language issues in comparative education. Inclusive teaching and learning in non-dominant languages and cultures* (pp. 77–97). Rotterdam, The Netherlands: Sense Publishers.

Brock-Utne, B. (2014). Language of instruction in Africa – The most important and least appreciated issue *International Journal of Educational Development in Africa (IJEDA), 1*(1), 4–19.

Brock-Utne, B., & Haukaa, R. (1980). *Kunnskap uten Makt. Kvinner som Lærere og Elever (Knowledge without Power. Women as Teachers and Pupils)*. Oslo, Norway: Universitets-forlaget. (Reprinted 1981 and 1984. German edition in 1986. *Wissen ohne Macht Frauen als Lehrerinnen und Schülerinnen.* Gießen: Focus Verlag. 195. pp)

Desai, Z. (2006). Reflections on the LOITASA project in South Africa – Three years later. In B. Brock-Utne, Z. Desai, & M. Qorro (Eds.), *Focus on fresh data on the language of instruction debate in Tanzania and South Africa* (pp. 102–112). Cape Town, South Africa: African Minds.

Fredua-Kwarteng, Y., & Ahia, F. (2005a, January 8). Ghana flunks at math and science: Analysis (1). Feature article. *Ghana News*.

Fredua-Kwarteng, Y., & Ahia, F. (2005b, February 23). Ghana flunks at math and science: Analysis (2). Feature article. *Ghana News.*

Langenhoven, K. R. (2006). Teachers speak about the language of instruction in natural science classrooms in the Western Cape in South Africa. In B. Brock-Utne, Z. Desai, & M. Qorro (Eds.), *Focus on fresh data on the language of instruction debate in Tanzania and South Africa* (pp. 251–265). Cape Town, South Africa: African Minds.

Langenhoven, K. R. (2010). Teachers speak about the language of instruction in natural science classrooms in the Western Cape in South Africa. In B. Brock-Utne, Z. Desai, M. Qorro, & A. Pitman (Eds.), *Language of instruction in Tanzania and South Africa – Highlights from a project* (pp. 133–143). Boston, MA: Sense Publishers.

Langenhoven, K. R. (2013). Scientific literacy: Experiences in using a workshop environment to understand and develop natural science terminology and concepts in isiXhosa. In Z. Desai, M. Qorro, & B. Brock-Utne (Eds.), *The role of language in teaching and learning science and mathematics* (pp. 57–69). Cape Town, South Africa: African Minds.

Makalela, L. (2005). We speak eleven tongues. Reconstructing multilingualism in South Africa. In B. Brock-Utne & R. K. Hopson (Eds.), *Languages of instruction for African emancipation: Focus on postcolonial contexts and considerations* (pp. 147–175). East Lansing, MI: State University of Michigan Press

Mammino, L. (1995). Teaching/learning theoretical chemistry at undergraduate level. *Southern Africa Journal of Mathematics and Science Education, 2*(1,2), 69–88.

Mammino, L. (2010). The mother tongue as a fundamental key to the mastering of chemistry language. In C. Flener & P. Kelter (Eds.), *Chemistry as a second language: Chemical education in a globalized society* (pp. 7–42). Washington, DC: American Chemical Society Syposium Series.

Mammino, L. (2013). Importance of language mastery and mother-tongue instruction in chemistry learning. In Z. Desai, M. Qorro, & B. Brock-Utne (Eds.), *The role of language in teaching and learning science and mathematics* (pp. 33–57). Cape Town, South Africa: African Minds.

Mazrui, A. (1997). The world bank, the language question, and the future of African education. *Race & Class, 38*(3), 25–49.

MEVT (Ministry of Education and Vocational Training). (2006). *Education policy.* Stone Town, Zanzibar: Author.

Mwinsheikhe, H. (2007). *Revisiting the language of instruction policy in Tanzanian secondary schools: A comparative study of biology classes taught in Kiswahili and English* (Ph.D. dissertation). Unit for Comparative and International Education, Oslo, Norway.

Mwinsheikhe, H. M. (2001). *Science and the language barrier: Using Kiswahili as a medium of instruction in Tanzania secondary schools as a strategy of improving student participation and performance in science* (Master thesis for the M.Phil. in Comparative and International Education). University of Oslo, Institute for Educational Research, Oslo, Norway.

Nomlomo, V. S. (2007). *Science teaching and learning through the medium of English and isiXhosa: A comparative study at two primary schools in the Western Cape* (Unpublished doctoral thesis). University of the Western Cape, Bellville, South Africa.

Nomlomo, V. S. (2008). IsiXhosa as a medium of instruction in science teaching in primary education in South Africa: Challenges and prospects. In M. Qorro, Z. Desai, & B. Brock-Utne (Eds.), *LOITASA reflecting on Phase I and entering Phase II* (pp. 81–102). East Lansing, MI: Michigan State University Press.

Ouane, A. (2009). My journey to and through a multilingual landscape. In K. K. Prah & B. Brock-Utne (Eds.), *Multilingualism: A paradigm shift in African language of instruction polices* (pp. 52–62). Cape Town, South Africa: CASAS.

Prah, K. K. (2005). Languages of instruction for education, development and African emancipation. In B. Brock-Utne & R. K. Hopson (Eds.), *Languages of instruction for African emancipation: Focus on postcolonial contexts and considerations* (pp. 23–51). East Lansing, MI: State University of Michigan Press.

Prah, K. K., & Brock-Utne, B. (Eds.). (2009). *Multilingualism: An African advantage. A paradigm shift in African languages of instruction policies* (pp. 1–17). Cape Town, South Africa: CASAS.

Prophet, R., & Dow, J. (1994). Mother tongue language and concept development in science. A Botswana case study. *Language, Culture and Curriculum, 7*(3), 205–217.

Qorro, M. (2009). Parents' and policy-makers' insistence on foreign languages as media of education in Africa: Restricting access to quality education. In B. Brock-Utne & I. Skattum (Eds.), *Languages and education in Africa: A comparative and transdisciplinary discussion* (pp. 57–82). Oxford, England: Symposium Books.

Ranaweera, A. M. (1976). Sri Lanka: Science teaching in the national languages. *Prospects, 6*(3), 416–423.

Rubagumya, C. (2003). English medium primary schools in Tanzania: A new "linguistic market" in education? In B. Brock-Utne, Z. Desai, & M. Qorro (Eds.), *Language of instruction in Tanzania and South Africa* (pp. 149–170). Dar es Salaam, Tanzania: E & D Limited.

Rugemalira, J. M., Rubagumya, C. M., Kapinga, M. K., Lwaitama, A. F., & Tetlow, J. G. (1990). Reflections on recent developments in language policy in Tanzania. In C. M. Rubagumya (Ed.), *Language in education in Africa: A Tanzanian perspective* (pp. 25–36). Philadelphia, PA: Multilingual Matters.

URT (United Republic of Tanzania). (1993). *The Tanzania education system for the 21st century: Report of the task force.* Dar es Salaam, Tanzania: United Republic of Tanzania, Ministry of Education and Culture and Ministry of Science, Technology, and Higher Education. (Printed by the Print Service, a division of University of Leeds Media Services)

Vuzo, M. (2007). *Revisiting the language of instruction policy in Tanzanian secondary schools: A comparative study of geography classes taught in Kiswahili and English* (Ph.D.dissertation). Unit for Comparative and International Education, Oslo, Norway.

Wedikkarage, L. (2006). *English as medium of instruction for collegiate level science classes in Sri Lanka: Theory, policy and practice* (Ph.D. dissertation). Unipub AS, Oslo, Norway.

Wedikkarage, L. K. (2009). Science education and English medium: The Sri Lankan experience. In B. Brock-Utne & G. Garbo (Eds.), *Language is power: The implications of language for peace and development.* Dar es Salaam, Tanzania: Mkuki na Nyota.

Birgit Brock-Utne
Department of Education
University of Oslo
Norway

SAM MCHOMBO

8. LANGUAGE, SCIENTIFIC KNOWLEDGE, AND THE "CONTEXT OF LEARNING" IN AFRICAN EDUCATION[1]

LANGUAGE AND EDUCATION

Every human society has an educational program, i.e., a system of training the young to become responsible citizens. The education includes the transmission of cultural values, morals, kinship systems, inheritance procedures, specialized skills needed for survival, religious beliefs, legends and history of the community, customs and traditions pertaining to various aspects of life and living (including dying), rites of passage, duties and responsibilities associated with age as well as sex and gender etc. Language, central to the transmission of knowledge, facilitates the learning process and contributes to cultural identity.

Language is primarily expressed through the medium of the spoken word. As such, every society makes use of human sounds for the transmission of knowledge. Such oral performance remains central to education of citizenship training, the cultivation of moral values, training the young in practical skills, socialization into the cultural practices and religious beliefs, cognitive development, etc. In brief, the oral tradition has served as the main vehicle of education and of knowledge representation in many societies. African indigenous knowledge systems have been in the form of oral performance and participation in overall cultural life.

Indigenous knowledge has, as a consequence, been rightly regarded as holistic rather than compartmentalized into discrete and disparate disciplines, lacking individual authorship or acclaimed intellectual property rights. Further, oral transmission of knowledge lacks permanence in knowledge preservation except insofar as it resides in the memory banks of the agents.

Besides the determination of the boundaries of the nation-states of Africa, colonialism introduced western type education. The former resulted in the creation of multi-ethnic and multi-lingual states. The latter, on the other hand, was responsible for the introduction of Western knowledge systems. These included Europeans religious beliefs and practices, natural and social science, cultural values, introduced in European style education system, dutifully imparted through, and served by, European languages.

The work of missionaries who "strove for the complete destruction of African worldview, thus turning Africans into little Englishmen or into subjects of another

Z. Babaci-Wilhite (Ed.), Human Rights in Language and STEM Education, 129–150.

European nation" (Jagusah, 2001, p. 116) contributed to the replacement of African social and cultural systems with those of the West. Thus, the asymmetric power relations that obtained between the Europeans and the Africans that undergirded colonialism with its purportedly civilizing mission was codified through Eurocentric education and western religion.

CONCEPTUAL/LINGUISTIC INCARCERATION OF AFRICAN EDUCATION

The crucial component of the education brought into Africa by the colonialists was the use of the written medium. In Africa, knowledge preservation had depended on the agents, with its transmission highly localized, requiring direct interaction or participation. The advent of writing systems transformed that. The invention of writing systems facilitated the representation of language in a medium other than that of speech, affecting aspects of culture and communication, as well as modes of knowledge representation, preservation, transmission and, diffusion.

The centrality of the written word to formal schooling put a focus on the acquisition of knowledge encoded in written form. Inevitably, this was in the languages of the colonialists, representing Western knowledge systems that got identified with intelligence and civilization. In the colonial era education served to enhance and maintain power relations in that it became the most powerful weapon for suppressing African values. Gary Urton, cited in Greer, Mukhopadhay and Roth, claims that "European colonizers did not impose political and economic systems – and, most generally, worldviews – on blank slates, but rather strove to suppress and replace existing systems" (Greer et al., 2012, p. 12).

Quickly, education became a program of knowledge acquisition in diverse disciplines that was codified in written form (cf. Mchombo, 2014a). It, also, unwittingly, contributed towards a negative evaluation of African indigenous knowledge systems (AIKS) or African languages. These were viewed as deficient in the grammatical complexity and/or lexis requisite for the codification of, especially, scientific knowledge. They were not even regarded as languages at all, a view exacerbated by the characterization of the notion 'language' that reserved it for "the standard, written form" while 'dialect' was reserved for "the nonstandard, substandard, or unwritten form" (Nurse & Hinnebusch, 1993, p. 37). With regard to the term "literature" *The American Heritage Dictionary of the English Language 5th edition* defines it as "the body of written works of a language, period, or culture." Thus, the written form is a crucial criterion of the designation of a communication system as a language or of a body of knowledge as literature. The absence of African languages in written form contributed to their inevitable racialized conception that they did not qualify as languages. Africans got projected as lacking history or knowledge, the latter even getting defined as "…what can be extracted from a written page" (Feyerabend, 1987, p. 110). Naturally, this undercut any claims to knowledge that oral communities might make.

With regard to the language situation the reality about African countries is that most of them are multilingual, with a presence of at least two languages spoken within their borders. These languages are "…of four types, namely, ex-colonial languages, nationally dominant languages, provincially dominant languages, and localized languages" (Batibo, 2013, p. 109). To achieve national unity many countries "adopted the ex-colonial language (English, French or Portuguese) as the official language due to its being advanced technologically, giving access to international relations and foreign dealings and being, above all, the elitist medium and considered neutral, since it is not a mother tongue to most people" (ibid, p. 110).

This Eurocentric education, delivered in ex-colonial languages as the official languages, has remained the order of the day in many African countries, creating what Ayo Bamgbose has termed the *"inheritance problem."* This has to do with "… how the colonial experience continues to shape and define post-colonial problems and practices" (Bamgbose, 1991, p. 69). This is evident in the formulation of policies in education that merely carry out the logic and practices of the past that include the retention of Eurocentric curricula on the one hand, and resistance to the use of African languages as LoIs on the other. This has made for a "decontextualized" education program. The local context is eliminated and the children have to learn foreign values in foreign languages. This undermines Nyerere's program of altering Eurocentricism in education (cf. Nyerere, 1968, Warren et al., 1995). Francis Moto commented about the literature curriculum at the University of Malawi during the 1970s that:

> There was an obvious leaning toward the continued maintenance of the colonial legacy, seen in the promotion of foreign literature by such novelists, poets and playwrights as Jane Austen, T.S. Eliot and William Shakespeare. This is not to say that there is anything wrong in teaching and learning European literature, but in my view it is not acceptable to teach and learn a foreign literature to the total exclusion of one's own literature in one's own university, and one's own country of birth. (Moto, 2009, p. 146)

This does not serve the African child meaningfully in that African education, unlike its counterparts elsewhere in the world, merely succeeds in depriving the children of the ability to "…see the world from the perspective of African cultures" (Spring, 2000, p. 191, citing Asante, 1994). To redress such imbalances there has been political activism to review educational policy and revise the content of learning so that the curricula reflect local knowledge and cultural values, as well as literary heritage. Wade Nobles termed the emphasis on European knowledge systems and values in the curricula of Africans, both in Africa and the diaspora, as "conceptual incarceration." The term refers to "the state of intellectual imprisonment in European value and belief systems occasioned by ignorance of African and Native American philosophical, cultural and historical truths" (Hotep, 2003, p. 6).

The proscription of African languages as LoIs has been another contentious issue. It constitutes a violation of rights in education in that it requires children with different linguistic heritage to go to school and, immediately, adjust to a foreign "body of disciplinary content, a set of literacy (reading and writing) process, and a new official (probably European) language. Three unknowns is a lot for a 6 years old" (Pearson, 2015, p. vii). Further, the use of a foreign language as the LoI, runs contrary to accepted research results that have concluded that "primary education, particularly early primary education and early literacy, is most effectively conducted in a language familiar to the pupil. The arguments, which need not be rehearsed at length here, are familiar: instruction through a home or local/regional language improves the quality and quantity of interaction between pupil and teacher [...]; cognitive development and literacy is best fostered in a familiar language [...]; instruction in a home language eases the transition between home and school, and so on" (Ferguson, 2013, p. 17). The removal of education from the socio-cultural context denies the child the opportunity to understand or appreciate African cultural values or knowledge systems. The social and cultural contexts and processes are central to learning (cf. Nasir & Hand, 2006).

Local language and local curriculum are crucial variables at the core of the achievement of quality education and children's rights in education (cf. Babaci-Wilhite, 2015). UNESCO noted in its document of vernacular languages:

> We take it as axiomatic that every child of school age should attend school and that every illiterate should be made literate. We take it as axiomatic, too, that the best medium for teaching is the mother tongue of the pupil. (UNESCO, 1953, p. 6)

The proscription of African languages as LoIs also constitutes what Mugane has referred to as "linguistic incarceration." Mugane advances the thesis that the "first language of the child is incarcerated, reducing education to the pursuit of fluency in English mediated by markedly non-proficient instructors. Whenever the switch is made from the child's first language to the language of the school there is always an instructional blackout. For the vast majority of children, the blackout is total and final. Learning is then reduced to verbatim memorization (and in numerous cases good hand writing)! To arrest the use of indigenous languages where they are most needed begins the process of necro-linguistics, the erasure or non-mastery of the vernacular" (Mugane, 2006, p. 14).

There is need for schooling in Africa to rid itself of the twin "evils" of conceptual and linguistic incarceration. Indeed, while some accommodation has been made for "African studies" in the curricula of, especially, higher education, itself a result of political activism and agitation (cf. Ngugi wa Thing'o, 2014), it remains a spirited battle to convince the governments to adopt measures to free the education from "linguistic incarceration." The problems here emanate from a mosaic of reasons.

For a start, as Batibo has observed, African countries retained the ex-colonial languages as official languages because they were viewed as advanced technologically, giving access to international relations and foreign dealings. Further, and this is quite significant, they were the "elitist medium," that were also considered "neutral" among the various ethnicities within African states. The neutrality of the colonial languages is, to a large extent, merely a convenient subterfuge. It is the access to power or, rather, the elimination of the masses from the arena of power that accords them the preferential treatment from African leaders.[2] European languages remain valorized as languages of power that provide access to the reins of government, employment in urban areas, provide a route out of rural or provincial life, and open up prospects for international travel for employment or business. The valorization of European languages has, traditionally, gone in tandem with negative attitudes towards African languages. For a start most African languages rarely promise the socio-economic mobility that is attributed to English "largely because of their more restricted socio-economic functions. And these same languages, particularly the 'pure' standardized versions of them, are often seen by younger generations as indexing undesirable identities, ones that are more rural, static and narrowly ethnic" (Ferguson, op. cit. p. 17).

The negative attitudes towards African languages are widespread and, "surprisingly, shared by many African people and expatriate government advisors" (Wolffe, 2006, p. 32). According to Obanya (1999) the negative attitudes towards African languages are deeply rooted in the fear of social change among members of the post-colonial elites as well as on the part of their expatriate advisors from donor countries and agencies, who baulk at the potential for marginalized sections of the population, illiterates, women, and even children, to become empowered through official recognition of their languages, a fact which would detrimentally affect the balance of power and threaten the privileges of the dominant elite.[3] The need to appease the "expatriate advisors from the donor countries and agencies" lends credence to the view that "the colonial mind in African education has been perpetuated under the guise of international development" that, through the intervention of international aid from donor countries, has reduced education in Africa to "an export commodity from the center and is accepted in Africa as a result of Western-based ideas of what it means to be developed" (Babaci-Wilhite, 2015, p. 18). Such economic dependency, giving access to the English media from the developed countries, the Center, to penetrate the media of the developing nations, the Periphery, has comprised a major aspect of globalization (cf. Pennycook, 1994; Zeleza, 1997).

Admittedly, political leaders in Africa would not readily endorse such an assessment of their commitment to the promotion of African values or linguistic heritage. It would be imprudent. Rather, the arguments for the retention of European languages in African education reside in pragmatism in the quest for economic progress or national development. This progress is intimately connected to incorporation of, and adjustments to, modern technology with corresponding improvements in science and math education. Routinely, scientific knowledge is taken as central to development and technological advancement. As such, there is need for massive investment in science education. The centrality of science and technology to national development has normally gone together with the belief that scientific knowledge is best expressed in English (or other European languages). African languages simply lack the technical resources to properly represent scientific knowledge. We will examine this view closely here.

LANGUAGE POLITICS AND THE CULTURE OF SCIENCE

In 2006, the island of Zanzibar "initiated a review of its educational strategy, which resulted in a reform of its curriculum. In addition to the change in LoI, from Kiswahili to English for Mathematics and Science subjects from Standard 5, it introduced a new subject, Information, Communication and Technology (ICT) from Standard 5 (in English), as well as Arabic from Standard 1" (Babaci-Wilhite, 2015, p. 42). Zanzibar, part of the United Republic of Tanzania, is an island where 100% of its population is monolingual in Kiswahili. It has a school system that is autonomous from that of the Tanzanian mainland. The educational review that Zanzibar has implemented clearly departs from the tenets of *Elimu ya Kujitegemea* (Education for Self Reliance (ESR)) program in which Nyerere tried to democratize education through, *inter alia*, promotion of the use of Kiswahili as LoI while giving emphasis to the learning of English as a foreign or international language.[4] However, the deviation of Zanzibar from this policy is neither unexpected nor isolated. To improve educational performance especially in math and science, African countries believe that children must improve their proficiency in English. The belief is that this can be achieved if English remains the LoI from the very beginning. Thus, in Malawi the Ministry of Education announced in March 2014 that all subjects would be taught in English, with the exception of, predictably, the local (national) language, Chichewa. Making the announcement, the then Minister of Education, Dr. Lucius Kanyumba, claimed that this would be in line with "a new Education Act which will see pupils being taught in English right away from Standard 1."[5] The high failure rate in education is blamed on a dismal lack of proficiency in English among the teachers and among the school children.[6] The presumed remedy is in the use of English as LoI.

The statements made by, *inter alia*, Batibo, Obanya, about the connection between ex-colonial languages and advancement in science and technology are recycled in these other cases. Mtenje (2013) points out that in 1996 the ministry responsible for education in Malawi produced a circular that permitted primary school teachers to use local languages to explain difficult concepts wherever necessary. The circular, construed as a policy position, attracted strong reaction from the public, the most significant objection was that "local languages cannot be used for teaching science and technology since they are under-developed and do not have appropriate technical terminologies" (Mtenje, 2013, p. 97). Shidza makes comparable observations about the situation in Zimbabwe, again typifying the general view of language in education in Africa where "… foreign languages, such as English, French and Portuguese are viewed as scientific languages, while indigenous languages are perceived as shallow and inadequate" (Shidza, 2012, p. 788). There were comparable reactions in Kenya when, in 2014, there was proposal to implement mother-tongue education in lower primary grades in compliance with an educational policy that had already been officially enunciated almost thirty years previously.[7]

The appeal to improvements in math and science education, coupled with the centrality of those subjects to technological advancement, taken as a necessary condition for economic development, does camouflage the political aversion to the implementation of mother tongue or multilingual education in Africa (see Walibora, 2015).

Thus, it is pragmatic to use ex-colonial languages, given their instrumentality to the goal of national unity and economic development. Scientific knowledge, central to economic development is, equally, better represented in those languages. Therefore, with seemingly perfect logical deduction, they should get entrenched in education in Africa. This view is spurious, with characteristics of an "urban legend." It is not based on valid cross-linguistic examination of structural patterns of different languages and their suitability, or lack thereof, to expression of any specific ideas. The belief owes as much to racist assumptions about the inferiority of African languages and cultural values or the non-existence of AIKS as it does to the Eurocentric orientation of the educational curricula. It is a myth rooted in the legacy of colonialism. It survives because, as is common with myths, "once the public has decided to accept something as an interesting fact, it becomes almost impossible to get the acceptance rescinded" (Pullum, 1991, p. 159). Its adoption by African leaders and relevance to the political and economic dependency of the periphery nations on the center countries makes it practically difficult to jettison. However, it is a myth without coherent or substantial basis. Brock-Utne (2012c, p. 9) states bluntly, "it is difficult to understand where the belief that science is better learnt in English than in other languages originates. While it is a belief one often comes across in Africa, the claim seems so unsubstantiated."[8]

Scientific knowledge is prized because of the perceived "goods" that society enjoys that are attributed to scientific research. They include the many diseases that have received a cure, consumer goods that have been manufactured, the reclamation of living and farming space, etc. These exemplify the instrumental good of the business of science. Of course science is also credited with the increase in, and the sophistication of, the destructive power of weapons of mass destruction that have come to characterize modern warfare. The enumeration of such benefits of science does not, in and of itself, lend support to any view about the language in which scientific knowledge resides. All societies have had to respond to local problems through the application of local knowledge and use of local resources. To deny such knowledge the status of being scientific would require resort to specific stipulations about the nature of scientific knowledge. Crucially, in science there is "…systematization of the facts that it encompasses. The attempt by the scientists to place facts, usually derived from observation, into some order is directed at arriving at more than a mere record of data. In general the scientist hopes to understand what is happening and why it is happening" (Miller, 1990, p. 220). Thus, a crucial feature of scientific knowledge has less to do with what it consists in but more with how the knowledge is arrived at. It is the methodological aspect that is central to science. The formulation of falsifiable theories that are based on empirical observations of data means that scientists avoid dogma.

In advanced work scientists also contend with more intricate issues as well as evaluations of competing theories of the observed facts. The step of formulating the hypothesis may be pivotal in the scientific enterprise. However, the "…developing of hypotheses is an act of creation no less mysterious (and wonderful) than the painting of a picture or the composition of a symphony…A well-done hypothesis takes into account all the information available, ties it into a coherent package and suggests avenues for further investigation" (ibid., p. 237). Different hypotheses can and, usually, get proposed and determination of the best theory is another component of scientific practice. Still, even with such an elementary characterization of the scientific enterprise, it is evident that nothing in the content or practice of science ties it to any one set of languages as appropriate for its expression or representation. Hence, the alleged inadequacy or unsuitability of African languages as vehicles of expression or representation of scientific knowledge has to seek justification from other considerations.

ON MATHEMATICS AND SCIENCE EDUCATION

The preceding argument to debunk the myth of the inadequacy of African languages for representation of scientific knowledge has been at a conceptual level. No consideration has been given to the widely held views about science, math, as well as logic, that they are disciplines that are intertwined and share the properties of being culture-free, objective, and universal. Science traverses cultural and national borders, seeks universal truths, using unique methods that, ostensibly, are not shackled to any culture-specific beliefs or practices. Advances in technology and other benefits to society that have been attributed to scientific practice and investigation have contributed tremendously to its respectability in modern society and its dominance in intellectual activity. In fact, scientific reasoning is so highly valued and so central to investigation of phenomena that for beliefs to be labeled unscientific or of reasoning to be labeled illogical, amounts to a condemnation of the work or the inferential engine, with the results rapidly relegated to quackery, guesswork, lack of theoretical underpinning, un-falsifiable, non-distinct from superstition, non-empirical, pure dogma, etc. The scientific enterprise is concerned with "a search for explanatory laws or principles" that possesses the unique property of being "testable (or 'refutable' or 'falsifiable') predictions" (Smith, 2005, p. 31). Thus, science thrives on formulation of claims, hypotheses, theories, etc. that are supported by reasons and/or empirical data, are susceptible to refutation, have logical coherence, and are not tied to the cultural beliefs of any one community.

The claims of objectivity and universality of science, math, and logic, together with their being culture-free, as well as being instrumental to technological innovations, have done much to obscure their role in the dominance of Western values and knowledge systems. The practice of science and math, and the precision in expression of content that it has demanded, has made for the development of highly specialized notation. In other words, in science and math, special symbolization got developed designed to guarantee precision in meaning. The special advantage of mathematical messages is that “they may be symbolized, manipulated and transformed according to precise rules, without loss of meaning” (Ziman, 1978, p. 17). The technical vocabulary abstracts away from aspects of ordinary language. Admittedly, most of it is based on English lexis or English derivational morphology. This has given math and science the image of resisting expression in non-western languages. The issue here is quite straightforward: science and math are not culture-independent. They have received expression, even when highly stylized, in the languages and culture of those engaged in the practice. The specialized jargon abstracts away from ordinary language usage, posing difficulties even to “scientifically uninitiated” speakers of those languages.[9] In fact, the technical language of science and math abstracts away from ordinary language so much that Noam Chomsky provides the following observation about these subjects:

> In the enterprise of science and mathematics, one might argue, the goal is to develop a ‘perfect language’ to express a ‘common treasure of thoughts’ with terms that refer to actual things in the world, including natural kinds, understood as kinds of nature, a concept foreign to natural language. (Chomsky, 1993, pp. 27–28)

The use of technical jargon or specialized use of ordinary vocabulary in math and science does not undermine the use of ordinary language to communicate the ideas or concepts. While “English has acquired the summit position as the language of choice for communication…in Science” (Mishra, 2009), it neither constitutes nor provides valid grounds for belief in the incapability or inferiority of any other languages to represent scientific knowledge.

The unsuitability of African languages to math and science education normally revolves around availability of equivalents of the scientific jargon. In order to teach math and science in languages other than English, the languages must have terminology that captures the relevant concepts. The reality is that those languages seem to lack the technical terminology. However, Rugemalira et al. (1990) make a relevant comment with regard to Kiswahili, that “a major objection to Kiswahili has traditionally been the supposed inadequacy of the language with regard to technical terminology…and, further, that the language does not have the same international role as English…” (Rugemalira et al., 1990, pp. 30–31). English’s international role is, technically, independent of its use as LoI for science or education in general. It is the “supposed inadequacy of Kiswahili with regard to technical terminology” that constitutes the basis for its disqualification or its inferiority. As such, efforts to examine the extent to which African languages could be adapted to math and science education have normally focused on the extent to which such technical terminology could lend itself to adaptation or translation (cf. Kaphesi, 2003; Kazima, 2008).

The alleged inadequacy of African languages for science education is, largely, an axiom that derives from the history of formal education in Africa, itself imbued with the colonialists' prejudicial view about Africans. In fact, Rugemalira et al. proceed to dispute the claim about the inadequacy of Kiswahili, appealing to views about inter-translatability between languages and the expressiveness of language that allows for the expression in any language of any conceivable human thought (cf. Katz, 1972). The question of adaptability or translatability of technical terminology was to show that the alleged inadequacy of African languages is no more than a 'red herring,' that it is surmountable. However, others have imputed logical deficiency in African linguistic structure as the basis for their unsuitability. Dalvit, Murray and Terzoli cite the following remarks, dutifully attributed to the original source, that "[T]he different structure of African languages compared to English might cause problems in scientific discourse. For example, the African languages make little use of logical connectives, which are a common feature of scientific writing. African languages do not use the English articles 'the' and 'a', hence 'copper is a metal which conducts electricity' and 'copper is the metal which conducts electricity' could cause confusion when written in an African language (Grayson, cited in Finlayson & Madiba, 2002, p. 48)" (Davit et al., 2009, p. 41). This statement is pernicious in that it actually suggests that African languages lack crucial linguistic resources or logical devices needed for scientific exposition. What better proof could one seek to eliminate African languages from math and science education?

Unfortunately, this claim is false. One view about natural languages and logic is that "All mankind, through its natural languages, exhibits adherence to the elementary principles of logic and through its senses discovers mentally coherent world of invariant objects and spatial relations, patterns of sound and color, permanence and movement, time and change. This *matter-of-fact* world, projected into the noetic domain by language, is nearly the same for us all" (Ziman, op. cit. p. 120). This undercuts the claim made by Grayson, a claim that dwells on linguistic expression of quantification that, like relative clause formation strategies, exploits morpho-syntactic resources in languages (cf. Bach et al., 1995; Bresnan & Mchombo, 1987; Mchombo, 1999). In Kiswahili there are three different ways of constructing relative clauses one of which intersects with quantificational structure in relevant ways (cf. Ngonyani, 1999).[10]

SCIENTIFIC KNOWLEDGE AND THE CONTEXT OF LEARNING

The sciences, and mathematics in particular, require the acquisition and coordination of three kinds of knowledge: *Conceptual Knowledge*, *Procedural Knowledge* and *Utilization Knowledge*. These comprise, respectively, "the ability to understand the principles that underpin the problem; the ability to carry out a sequence of actions to solve a problem; and, the ability to know when to apply particular procedures" (Cole & Cole, 1993, p. 482). Research has shown that "...most children arrive at school with some of each kind of knowledge, and cross-cultural research reveals that even societies with no tradition of schooling and literacy use methods of counting and solving arithmetic problems…" (ibid). Such early acquisition of knowledge is totally dependent upon the child's exposure to the mother tongue, knowledge acquired within a socio-cultural context (cf. Kyeu, 2014; Vygotsky, 1978). Knowledge is "…a local commodity designed to satisfy local needs and to solve local problems" (Feyerabend, 1987, p. 28). The knowledge includes, *inter alia*, introducing the young into life of "the society where they are born and into the physical universe that surrounds the society" (Feyerabend, 1981, p. 163). Education equips the individual with that socio-culturally determined knowledge shaped by local contextual parameters. Such indigenous or traditional knowledge includes that commonly labeled "scientific," acquired in the language of the society. The "science" that is associated with formal schooling is, itself, the product of specific methodological aspects of inquiry fostered by, and supported within, specific socio-cultural-politico-economic contexts. It is not any more culture-free or objective or universal than other forms of knowledge. The methodological approach may give it a degree of uniqueness but it is not value or culture free. The claims of objectivity, universality, or culture independence derive from power relations or political dominance of the agents of knowledge production. While those notions get invoked to suggest that the knowledge is "…valid irrespective of human expectations, ideas, attitudes, wishes" (Feyerabend, 1987, p. 5) or as "…a measure by which theoretical suggestions and practical achievements must be judged" (ibid., p. 99), they are "older than science and independent of it" arising "when a nation or a tribe or a civilization identified its ways of life with the laws of the (physical and moral) universe and it became apparent when different cultures with different objective views confronted each other" (ibid). In other words, they are notions that have to do with political dominance or power relations.

Science in formal education is imbued with those values. Agents of knowledge production in science have rarely ever consisted of the general public. On the contrary, "scientific knowledge is generated and validated by a *scientific community*, which is as far as could be from a random sample of unregenerate mankind. By social division of labor, modern society entrusts the cultivation of science to a highly specialized professional group, characterized both by expertise and extreme commitment to science as a social institution. What should, in philosophical principle, be done by all men, is given into the hands of proxies, who bear collectively the powers and the responsibilities of science within society at large" (Ziman, op. cit. p. 125).

This "scientific community" has, traditionally, consisted of white males, excluding people of color and, for the great part, women. This has, effectively, made science a weapon of white male dominance, according them the power to determine the nature and mode of production of "valid" knowledge. Thus the *scientific community*, empowered to formulate theories that make basic claims about the nature of knowledge, acquires the mandate to determine who can know, how we can know, and what counts as valid evidence. This is more of cultural domination and exercise of political power than mere production of scientific knowledge. The dominance of white male culture in research and knowledge production, and other aspects of life in general, has been a central sociological feature of politics, education, and much else in society. Feminist scholars, critical race theorists, advocates of Afrocentric education, and many others committed to social justice, democratization of knowledge production, freedom from mental or physical slavery and colonialist-inspired domination have increasingly questioned the legitimacy of such a paradigm of cultural dominance of white males who have had the power and ability to, *inter alia*, define reality and to convince other people that it is their definition (cf. Nobles, 1986). It is a paradigm that diminishes liberty or social justice.

Science and math education is influenced by the worldview of, predominantly, white males. Further, the validity and universality accorded to that knowledge goes together with the estimation of their (variety of) language as the "standard" language of knowledge and scholarship. Certainly in African education this has been the operative principle, where ex-colonial languages are languages of power.

The connection between language and scientific knowledge cannot be independent of the socio-cultural "context of learning." Western scientific knowledge receives articulation in the context of western culture, values, and educational system and is represented in western languages. The introduction of formal education in Africa reflected the usual focus "...on mastery of two basic symbol systems, written language and mathematics" (Cole & Cole, op. cit. p. 476). Both the written languages and the knowledge were those of Europeans. The education for African children got steeped into curricula of European cultural values and knowledge systems, delivered in European languages. The incorporation of African cultures and knowledge systems could not be envisioned in light of their unavailability in written form hence deemed non-existent. African languages were relegated to dialects, not sufficiently developed to represent scientific knowledge with all its technical vocabulary.

The retention of the colonial educational system, with the accompanying ideology of the inferiority of AIKS and African languages normalized the "conceptual and linguistic incarceration" that has characterized African education. The Eurocentric curriculum remains a major hurdle to educational reforms that accommodate African languages or AIKS. It is an education that marginalizes "... African languages in favor of Euro-languages, thus creating a linguistic configuration that served to legitimize and reproduce the unequal division of power and resources between the speakers of Euro-languages and the speakers of African languages. The overwhelming majority of post-colonial African governments thus inherited educational systems with imperial languages as the predominant media of instruction" (Mazrui, 2004, p. 40).

Science education has contributed to solidification of this "unequal division of power and resources" through its alleged objectivity, universality, and culture independence. The inadequacy of African languages lay, ostensibly, in their lack of logical or grammatical structure and lexis requisite for representation of scientific knowledge. The failure of science and math education to take into account the science and technology of local people, the knowledge and skills they have and what problems they feel are important to consider, contributed to the alien nature of school science. African scientific knowledge based on different conceptual and cognitive models from those of the colonialists who first introduced Western education to Africa did not receive any recognition; thus, the colonialists could assume that "... African peoples had no science" (Jegede, 1995, p. 121). This is unfortunate since it obfuscates science education. Like all knowledge that depends for its acquisition on socio-cultural context, the learning of science is "a process of culture-acquisition, where students get to know and practice the 'ways of seeing', 'ways of talking', and 'ways of doing'—characteristics for the science class specific subculture" (Akpanglo-Nartey et al., 2012, p. 70).

In science education the socio-cultural environment of the learners is significant because the concepts, explanations and interpretations that learners get from the personal experiences at home and in their community constitute personal and cultural knowledge that gets used as backdrop against which to evaluate or interpret the knowledge and expertise that they encounter at school. The exclusion of such local knowledge poses difficulties for the students. In addition, the elimination of the native language in favor of the exclusive use of foreign languages for instruction does result in non-participation and frustration, eventually leading the children to making choices against science and science-related disciplines. Further, the fact that science and mathematical knowledge is not culture free, that it results from social interactions in which relevant ideas, facts, concepts, principles, and skills are acquired as a result of cultural context (cf. Dossey, 1992; Orey, 2000) means that "mathematical skills that students learn in schools are not logically constructed based on abstract cognitive structures but rather forged out of a combination of previously acquired knowledge and skills and new cultural inputs" (Rosa & Orey, 2011, p. 41). The suppression of local knowledge and languages in science and math education rests on power relation, effectively offering perfect demonstration of the claim that "the enterprise of mathematics education is no different than other societal contexts characterized by power relations" (Martin et al., 2010, p. 21).

LANGUAGE AND THE TEACHING OF SCIENCE

Knowledge acquisition happens within a socio-cultural context just as agents of knowledge production develop within specific socio-cultural environments that influence their cognitive development. The socio-politico-economic factors are not irrelevant to the conduct of knowledge acquisition or promotion of innovation. Literacy contributed immensely to progress in research. People in oral cultures tend to be traditionalist and conservative. They acquire knowledge and skill through personal participation and practice; their conceptual categories are invariably concrete and are interiorized as communal knowledge. In contrast the literate world provides a different outlook. According to Walter Ong (1986), the advent of the written medium made for profound transformation of the mode of codification and structuration of knowledge, leading to a cultural regimen that placed greater premium on innovativeness, inventiveness, and 'objectivity'. The discourses from such a milieu tend to be abstract, analytic, syllogistic, and definitional, and their immediate context of production is generally privatist. Literacy is relevant to modes of logical explanation in that "… all philosophy depends on writing because all elaborate, linear, so-called 'logical' explanation depends on writing. Oral persons can be wise, as wise as anyone, and they can of course give some explanation for things. But the elaborate, intricate, seemingly endless but exact cause-effect sequences required by what we call philosophy and by extended scientific thinking are unknown among oral peoples, including the early Greeks before their development of the first vocalic alphabet" (Ong, 1986, p. 42).

Literacy gave Eurocentric education the superiority associated with Western civilization and knowledge. The abstract, analytic, syllogistic, and definitional discourse that went with the individualistic approach to knowledge production made for a paradigm shift in conception of education. This educational practice did not fit in with the African context. That node of knowledge production and dissemination did not make it culture free or ahistorical. The ahistoricity of scientific knowledge derives from the distinction made in science between the "context of discovery" and the "context of justification." The (dis)-confirmation of scientific hypotheses does not ordinarily include the context of formulation of a theory. Scientific methodology, aided by the complexity of reasoning that literacy affords, contributed greatly to the image of science and mathematics as culture free, universal, and objective. Indeed scientific methodology does train people to think in a systematic and logical manner. However, "scientists, like all human beings, work in cultural and social contexts, they have passions and dreams and they accomplish their goals within certain human limitations" (Gitari, 2012, p. 33).

Literacy and its transformative effects on society proved instrumental in the formation of the emergent practices in science and the "rationality" associated with it. That provides no grounds for relegation of African languages or AIKS to the inferior, illogical, irrational, etc. Such evaluation derived from racist ideology that de-humanized African people to provide justification for their enslavement and oppression. Ochieng-Odhiambo dutifully notes that Western philosophers "… had been skeptical of the existence, if not certain of the absence, of rationality and reflective thought in African minds" (Ochieng-Odhimabo, 2010, p. 9). To this Ani adds the following observation:

> Besides the role of slave trade and colonialism in institutionalizing the inferiority of African, the works of renowned European academic thinkers such as that of Immanuel Kant and David Hume, among others, reinforced the idea of the inferiority of African minds and serviced Western expansionism. In his *Essay 'Of National Characters'* (1776: 152), Hume (1997: 33) in a footnote states, 'I am apt to suspect the Negros and in general all other species of men (for there are four or five different kinds) to be naturally inferior to the whites. (Ani, 2013, p. 296)

The racist ideology that has been the centerpiece of relations between African peoples and the white races contributed to the dehumanization of the colonized and enslaved Africans. The colonial education and the Christian missionary enterprise effectively became "...the main agents of perpetrating these racist images of the 'native.' Both the exercises were intended to elevate the culture of the colonizer and debase the culture of the colonized" (Mazrui & Mazrui, 1998, p. 58). Science and mathematics, with their rationality, logic, objectivity, analyticity, and culture-independence, too advanced for the allegedly culture-bound conceptual systems of the Africans and their purportedly logic deficient "rudimentary" systems of communication. This was unfortunate because scientific knowledge is part of general knowledge, its specialization notwithstanding, and it is culture dependent. Iaccarino states that "science is part of culture, and how ... science is done largely depends on the culture in which it is practiced" (Iaccarino, 2003, p. 1).

For scientific concepts to be comprehensible and for the knowledge, including scientific, to be appreciated, it must derive from the socio-cultural context, with a focus on the local as the basis for dealing with the 'global.' The use of the local languages for delivery of such knowledge is essential. In other words, education should indigenize the curriculum and LoI in order to strengthen young learners' acquisition of scientific knowledge and skills. Gitari has termed such indigenized knowledge "Endogenous science" and defined it as "the scientific literacy that is inspired by local needs, but crafted using knowledge from all domains of life" (Gitari, op. cit. p. 31). Gitari refers to the practice of science that is dominant in the curricula as the "narrow view of scientific literacy" that, he believes, "... has contributed to the devaluing of the role of Indigenous local knowledge in the development of endogenous science. Such devaluing emanates from the mode of knowledge production in the academy that privileges certain forms of human experiences over other forms. For instance, analytical methodologies are advanced over holistic methodologies, written communication is favored over oral communication, performance is given prominence over participation, competition is nurtured whereas cooperation is marginalized and the sense of sight is elevated over other human senses" (ibid.). The indigenization of the curriculum with the use of the most ignored of the cultural heritage, namely, local languages as LoIs would make for an appreciable "contextualization of education" (cf. Mchombo, 2014b). It would also free African education from conceptual and linguistic incarceration.

CONCLUSION

Literacy has played a major role in knowledge production and preservation, and is central to modern education. The profound transformative effects of the written word included a major alteration "... in the way the consciousness of Western men and women is organized. There was a paradigmatic shift from a time-oriented focus of communicative consciousness to a space-oriented one. Even more importantly perhaps, there was a change in the style of knowledge presentation resulting in a dominance of discourses that were more and more definitional, descriptive, and analytical [...]. Here was the origin of Western science and philosophy" (Biakolo, 2003, p. 14).

The need for Africa education to be responsive to the African context necessitates the development and harmonization of orthographic conventions for African languages. The lack of literacy in pre-colonial Africa contributed to the perceived inferiority of AIKS and of African languages. Problems of education in pre-colonial Africa had roots in "the inability of the people to write and keep records, which would have enabled local teachers and master-craftsmen and herbalists to preserve their wisdom and knowledge for the use of the younger generation. Since the knowledge concerning many professions remains undocumented, it is high time for Africans to document the cognitive aspect of their professions in black and white. A situation where a master in a profession dies with his knowledge should not be allowed to continue" (Adeyemi & Adeyinka, 2002, p. 237).[11]

The fundamental goals of early primary education include aiding the children with the acquisition of the social skills of reading and writing. Education conducted in their mother tongues or familiar local/regional languages facilitates the attainment of those goals. Alphabetic writing, as the most common orthographic system, is independent of any particular language. Once acquired, it is transferrable to other languages with minimal difficulty. Therefore, it makes sense "...to teach this principle to a young child in his own language. And it makes no sense whatsoever to teach it to him first in a foreign language, thus placing on him a double learning task" (Hale, 1974, p. 2). Further, the principle of alphabetic writing can be delivered more effectively to a learner who is in full control of the phonology of the language for which the writing system is designed, since the learner can grasp the relationship between the alphabetic symbols and the phonological segments of the language. Hale makes the observation that "children of school age are in control of the phonology of their native language—in other words, they come to school with precisely those attainments which are necessary to learn the principle which is basic to literacy; once they learn that principle...they will be able to apply the principle to any new language which they learn at a later date. The sooner they learn the principle the better—they can learn it almost immediately if it is taught to them in their own language" (ibid). Such arguments, supported by, yes, "scientific" research, call for a re-examination of the language in education policies that decry the use of the mother tongue especially in the early formative years of the child's growth and development. This would address the issue of "rights in education" and would integrate the linguistic and cultural wealth of the society into the formal education.

A focus on culturally relevant knowledge does not constitute an argument for the marginalization of "global" knowledge. Knowledge of the languages, cultural values, and systems of knowledge of other societies is essential for a globalized world. The argument is for local knowledge to be foundational to basic education. African education cannot get insulated from world systems of knowledge. That knowledge should be sought and get integrated to the foundational local knowledge. There is no need to insulate indigenous knowledge or protect it from interaction with other systems of knowledge.

Ngara provides a pertinent proverb from the Shona of Zimbabwe. It says that *Kakova kanozara nemadirirano* (The river is flooded by tributaries). Ngara's comment on this is that it "...implies that, to get a more complete and realistic understanding of the world, we need to value all the available alternative ways of knowing as much as our own" (Ngara, 2007, p. 10). The diffusion of ideas has been crucial to advances in knowledge. Western science and math gained much from the knowledge systems of other societies. Bishop remarked about Western math that "one of the greatest ironies in this whole field is that several different cultures and societies have contributed to the development of what is called western mathematics: the Egyptians, the Chinese, the Indians, the Arabs, the Greeks, as well as the western Europeans. Yet when western cultural imperialism imposed its version of mathematics on the colonized societies, it was scarcely recognizable as anything to which these societies might have contributed..." (Bishop, 1990, p. 61).

The suggested review of curricula and LoI would address the perceived shortcomings of education in Africa. According to a 2014 report of the cultural agency of the UN, the education has failed so miserably in its mission that "while worldwide a quarter of a billion children are failing to learn basic reading and math skills in an education crisis that costs governments $129 billion annually" in Africa the situation is so dismal that "four in 10 African children 'cannot read a sentence.'" The report did not highlight in what languages the children cannot read sentences. Clearly, the sentences were not in the children's languages. Equally missing was any statement on the content of curricula or the LoIs of the education that the children got. The contributory factors mentioned included the fact that "fewer than three-quarters of existing primary school teachers were trained to national standards, while 120 million primary age children across the world had little or no experience of school." These definitely deserve attention and academic performance might, plausibly, improve with corrective measures to the noted shortcomings. However, these need to get addressed together with the implementation of culturally relevant curricula delivered in languages that do not place both teachers and students at a disadvantage with respect to proficiency or comprehension. Such an education would preserve social justice, promote rights in education, and create the "context of learning" in which cultural knowledge is brought to bear on the acquisition of other knowledge systems.

NOTES

1 This chapter owes much to regular discussions and constant interaction with Dennis Banda, Herman Batibo, Adams Bodomo, Samuel Bruce, Jarvis Givens, Monica Kahumburu, William Kamkwamba, Langa Khumalo, Pascal Kishindo, Andrew Tilimbe Kulemeka, Patricia Kwok, David Kyeu, Francis Moto, Al Mtenje, John Mugane, Kennedy Mutothori Mugo, Thoko Mtchona, Na'ilah Suad Nasir, Deo Ngonyani, Betty Nyasulu, Samuel Obeng, Tryphenia Peele-Eady and Mathews Sapemba Tisatayane. I thank them for their input. They do not necessarily agree with the way that their ideas have been used nor do they necessarily endorse the views that are expressed in this work. For that I take full responsibility. Zehlia Babaci-Wilhite deserves special recognition for stimulating discussions of issues

of language in education in Africa. This work sprung from those discussions. I am grateful to Grace Deniece Gipson for assistance with formatting. Thanks to Janiya Banez for constantly reminding me, cheerfully but resolutely, that a fit and healthy body aids intellectual productivity; and, to Teri Urrutia for the advice that fun in life extends beyond the confines of research or intellectual activity.

2 In Malawi individuals seeking to become members of parliament needed to provide evidence of proficiency in English. Parliamentary proceedings are conducted entirely in English.

3 In personal (email) communication Samuel Bruce pointed out that "it is indeed in the interests of many current government elites to preserve an elite language as an official language of a country. If serious business (science, trade, politics all included) is conducted in English, French, or Portuguese, and if they have a solid mastery of those languages, then it's a convenient way to reduce the possible impact of political, economic and even academic competitors. Sad, but possibly true

4 On February 15, 2015, the government of Tanzania announced a policy to extend basic education to the end of high school, rather than terminating basic education after middle school. This would apply to state-run schools. With the new policy the language of instruction would be Kiswahili, not English. Zanzibar has an autonomous school system hence it is not clear whether it would adopt and implement the policy as well.

5 http://nysasatimes.com: Malawi Std. 1 pupils to start learning in English all subjects, March 5, 2014.

6 The problems that African teachers face with teaching or examining in English are real. In private correspondence Samuel Bruce, based at Oxford University, remarked, "speaking as a former English teacher (I taught in Rwanda for a few weeks), one thing I noticed was that there is a dire lack of understanding among not only teachers, but also the *examiners* of national exams. I know because I taught them English. It's clear that many weren't really in a competent position to assess examination scripts beyond a primitive level of English difficulty. That might be a worthwhile point which adds strength to the thesis – there will be better testing in science in Africa if it is conducted in local languages."

7 http://www.nation.co.ke/news/Use-local-languages-insists-minister/-/1056/2229690/-/7sq64b/-/index.html.

8 William Kamkwamba dropped out of school in Malawi during his early teens because of lack of funding. Yet, using local resources, he proceeded to build a windmill in his village in Kasungu district that generated power for electricity to his house. His accomplishments got recorded in the *New York Times* best seller of 2010, *The Boy Who Harnessed the Wind. Creating Currents of Electricity and Hope*, NY: Harper Perennial. Significantly, the scientific knowledge and the work did not depend on proficiency in English. At the time he hardly spoke the language.

9 Kazima (2008) provides interesting examples of how common linguistic expressions such as 'difference' 'product' 'set' etc. receive specialized interpretations in mathematics. This goes together with other words, such as 'isosceles,' that are peculiar to math, not common in ordinary discourse.

10 Swahili has three strategies of relative clause formation as noted in the work of Deo Ngonyani (1999). One of the relative clause constructions, ordinarily lacking tense morphology, as in the expression kipandacho hushuka "what goes up comes down" usually gets a generic or universal quantificational interpretation. I am grateful to David Kyeu for corroborating this observation.

11 In Chichewa/Chinyanja, there is a proverb with comparable import. The proverb is *Kumbile adamka nawo* "literally, "'dig–for-me' went with it (knowledge)". The admonition is that depending on someone to "dig" the roots (herbs) for one without learning from that person results in the knowledge "going" with that her/him.

REFERENCES

Adeyemi, M. B., & Adeyinka, A. (2002). Some key issues in African tradition education. *McGill Journal of Education, 37*(2), 223–239.

Akpanglo-Nartey, R. K., Asabere-Ameyaw, A., Dei, G. J. S., & Taale, K. D. (2012). Children's indigenous ideas and the learning of conventional science. In A. Asabere-Ameyaw, G. J. S. Dei, K. Raheem, & J. Anamuah-Mensah (Eds.), *Contemporary issues in African sciences and science education* (pp. 63–72). Rotterdam, The Netherlands: Sense Publishers.

Ani, N. C. (2013). Appraisal of African epistemology in the global system. *Alternation, 20*(1), 295–320.

Asabere-Ameyaw, A., & Ayelsoma, S. J. (2012). Language proficiency and science learning. In A. Asabere-Ameyaw, G. J. Sefa Dei, K. Raheem, & J. Anamuah-Mensah (Eds.), *Contemporary issues in African sciences and science education* (pp. 55–62). Rotterdam, The Netherlands: Sense Publishers.

Asante, M. K. (1991). The Afrocentric idea in education. *The Journal of Negro Education, 60*(2), 170–180.

Babaci-Wilhite, Z. (2012). A human rights-based approach to Zanzibar's language-in-education policy. *World Studies in Education, 13*(2), 17–33.

Babaci-Wilhite, Z. (2013). Local languages of instruction as a right in education for sustainable development in Africa. *Sustainability, 5*, 1994–2017.

Babaci-Wilhite, Z. (2014). E-Learning science literacy as human right in education in Africa. *Current Studies in Comparative Education, Science and Technology, ISCEST Conference Journal, 1*, 115–130.

Babaci-Wilhite, Z. (2015). *Local languages as a human right in education*. Rotterdam, The Netherlands: Sense Publishers.

Babaci-Wilhite, Z., & Geo-JaJa, M. A. (2013). Localization of instruction as a right in education: Tanzania and Nigeria language-in-education's policies. In Z. Babaci-Wilhite (Ed.), *The voices of the South: Right in local language and local curriculum*. Rotterdam, The Netherlands: Sense Publishers.

Babaci-Wilhite, Z., Geo-JaJa, M. A., & Lou, S. (2012). Education and language: A human right for sustainable development in Africa. *International Review of Education, 58*(5), 619–647. doi:10.1007/s11159-012-9311-7

Bach, E., Jelinek, E., Kratzer, A., & Partee, B. H. (Eds.). (1995). *Quantification in natural languages*. Dordrecht, The Netherlands: Kluwer Academic Publishers.

Bamgbose, A. (1991). *Language and the nation. The language question in Sub-Saharan Africa*. Edinburgh, Scotland: Edinburgh University Press.

Batibo, H. (2013). Maximizing people's participation through optimal language policy: Lessons from the SADC region. In H. McIlwraith (Ed.), *Multilingual education in Africa: Lessons from the Juba language-in-education conference* (pp. 109–116). London, England: British Council.

Biakolo, E. (2003). Categories of cross-cultural cognition and the African condition. In P. H. Coetzee & A. P. J. Roux (Eds.), *The African philosophy reader: A text with readings* (2nd ed., pp. 9–21). New York, NY: Routledge.

Bishop, A. J. (1990). Western mathematics: The secret weapon of cultural imperialism, *Race & Class, 32*(2), 51–65.

Bresnan, J., & Mchombo, S. (1987). Topic, pronoun, and agreement in Chichewa. *Language, 63*(4), 741–782.

Brock-Utne, B. (2000). *Whose education for all? The recolonizatioin of the African mind.* New York, NY: Falmer Press.

Brock-Utne, B. (2007, November 28–30). *Democracy in a multilingual and multicultural society*. Keynote lecture at conference on mother tongue, bilingualism and democracy: Myths, realities and consensus, Copenhagen, Denmark.

Brock-Utne, B. (2012a). Learning for all of Africa's children—but in whose language? *Commonwealth Education Partnerships, 2012*(13), 147–150.

Brock-Utne, B. (2012b). Language and inequality: Global challenges to education. *Compare: A Journal of Comparative and International Education, 2012*, 1–21.

Brock-Utne, B. (2012c). Language policy and science: Could African countries learn from some Asian countries? *International Review of Education, 58*(4), 481–503.

Bwenge, C. (2012). English in Tanzania: A linguistic cultural perspective. *International Journal of Language, Translation and Intercultural Communication, 1*(1), 167–182.

Chomsky, N. (1993). *Language and thought*. Wakefield, RI: Moyer Bell.

Cole, M., & Cole, S. R. (1993). *The development of children* (2nd ed.). New York, NY: Scientific American Books.

Dalvit, L., Murray, S., & Terzoli, A. (2009). Deconstructing language myths: Which languages of learning and teaching in South Africa? *Journal of Education, 46*, 33–55.

Dossey, J. A. (1992). The nature of mathematics: Its role and its influence. In D. A. Grouws (Ed.), *Handbook of research on mathematics teaching and learning: A project of the national council of teachers of mathematics* (pp. 39–48). New York, NY: Macmillan.

Ferguson, G. (2013). The language of instruction issue: Reality, aspiration and the wider context. In H. McIlwraith (Ed.), *Multilingual education in Africa: Lessons from the Juba language-in-education conference* (pp. 17–22). London, England: British Council.

Feyerabend, P. (1981). How to defend society against science. In I. Hacking (Ed.), *Scientific revolutions: Oxford readings in philosophy* (pp. 156–167). Oxford, England: Open University Press.

Feyerabend, P. (1987). *Farewell to reason.* New York, NY: Verso.

Gitari, W. (2012). Engaging scientific activities to build endogenous science and lay foundation for improvement of living conditions in Africa. In A. Asabere-Ameyaw, G. J. Sefa Dei, & K. Raheem (Eds.), *Contemporary issues in African sciences and science education* (pp. 29–54). Rotterdam, The Netherlands: Sense Publishers.

Greer, B., Swapna, M., & Wolff-Michael, R. (2012). Mathematics and politics of knowledge. In M. Swapna & R. Wolff-Michael (Eds.), *Alternative forms of knowing (in) mathematics: Celebrations of diversity of mathematical practices* (pp. 9–16). Rotterdam, The Netherlands: Sense Publishers.

Hale, K. L. (1974). Navajo Linguistics II, Special Lecture series

Hotep, U. (2003). *Decolonizing the African mind: Further analysis and strategy.* Paper presented at the Kwame Ture Youth Leadership Institute, Pittsburgh, PA. Retrieved from http://whgbetc.com/ifbm/decolonizing.html

Iaccarino, M. (2003). Science and culture. *Science and Society Viewpoint,* 1–6.

Jagusah, O. I. W. (2001). Educational policy in Africa and the issue(s) of context: The case of Nigeria and South Africa. *International Education Journal, 2*(5), 113–125. Retrieved from http://iej.cjb.net 113

Jegede, O. J. (1995). Collateral learning and the eco-cultural paradigm in science and mathematics education in Africa. *Studies in Science Education, 25,* 97–137.

Kamkwamba, W., & Mealer, B. (2009). *The boy who harnessed the wind creating currents of electricity and hope.* New York, NY: Harper Perennial.

Kamwendo, G. (2013). Language in education in South Sudan: Any lessons from Botswana and Malawi? In H. McIlwraith (Ed.), *Multilingual education in Africa: Lessons from the Juba language-in-education conference* (pp. 103–108). London, England: British Council.

Kaphesi, E. S. (2003). The influence of language policy in education on mathematics classroom discourse in Malawi: The teachers' perspective. *Teacher Development, 7*(2), 265–285.

Katz, J. J. (1972). *Semantic theory.* New York, NY: Harper & Row.

Kazima, M. (2008). Mother tongue policies and mathematical terminology in the teaching of mathematics. *Pythagoras, 67,* 56–63.

Kyeu, D. (2104). The impact of Facebook chat on Swahili essay writing. *Current Studies in Comparative Education, Science and Technology, 1,* 34–64.

Martin, D. B, Ghloson, M. L., & Leonard J. (2010). Mathematics as gatekeeper: Power and privilege in the production of knowledge. *Journal of Urban Mathematics Education, 3*(2), 12–24.

Matiki, A. (2002, April 16–20). Language planning and linguistic exclusion in the legislative process in Malawi. *World Congress on Language Policies* (Barcelona), 1–17.

Mazrui, A. M. (2004). *English in Africa. After the Cold War.* Clevedon, England: Multilingual Matters Ltd.

Mazrui, A. A., & Mazrui, A. M. (1998). *The power of Babel: Language & governance in the African experience.* Chicago, IL: The University of Chicago Press.

Mchombo, S. (1999). Quantification and verb morphology: The case of reciprocals in African languages. *Linguistic Analysis, 29*(1–2), 182–213.

Mchombo, S. (2014a). Language, learning and educational for all in Africa. In Z. Babaci-Wilhite (Ed.), *Giving space to African voices. Rights in local languages and local curriculum* (pp. 21–47). Rotterdam, The Netherlands: Sense Publishers.

Mchombo, S. (2014b). Language and the contextualization of education in Africa. *Russian-American Education Forum: An Online Journal, 6*(3).

Miller, R. W. (1990). *Study guide, Introduction of logic (8th edition) Irving M Copi & Carl Cohen.* New York, NY: Macmillan Publishing Company.

Mishra, R. (2009). *Science in vernacular language: A boon or a bane.* Retrieved from IndiaBioscience.org

Moto, F. (2009). *Language, power & society.* Pretoria, South Africa: University of South Africa Press.

Mtenje, A. (2013). Developing a language policy in an African country: Lessons from the Malawi experience. In H. McIlwraith (Ed.), *Multilingual education in Africa: Lessons from the juba language-in-education conference* (pp. 95–102). London, England: British Council.

Mugane, J. (2006). Necrolinguistics: The linguistically stranded. In J. Mugane, J. P. Hutchison, & D. A. Worman (Eds.), *Selected Proceedings of the 35th Annual Conference on African Linguistics* (pp. 10–21). Somerville, MA: Cascadilla Proceedings Project.

Nasir, N. S., & Hand, V. M. (2006). Exploring sociocultural perspectives on race, culture, and learning. *Review of Educational Research, 76*(4), 449–475.

Neke, S. M. (2003). *English in Tanzania: An anatomy of hegemony* (PhD dissertation). Universiteit Gent, Belgium.

Ngara, C. (2007). African ways of knowing and pedagogy revisited. *Journal of Contemporary Issues in Education, 2*(2), 7–20.

Ngonyani, D. (1999). X^0-Movement in Kiswahili relative clause verbs. *Linguistic Analysis, 29*(1–2), 137–159.

Ngugi wa Thong'o. (2014). Challenging colonialism and neo-colonialism. *Africa Speaks Reasoning Forum*. Retrieved from http://www.mg.co.za/article/2007-02-07-ngugi-wa-thiongo-still-decolonising-themind

Nobles, W. (1986). *African psychology: Toward its reclamation, revitalization and re-ascension*. Oakland, CA: Black Family Institute.

Nurse, D., & Hinnebusch, T. (1993). *Swahili and Sabaki: A linguistic history* (Vol. 121). Berkeley, CA: University of California Press Linguistics.

Nyerere, J. (1968). *Education for self- reliance*. Dar-es-Salaam, Isreal: Government Printers.

Obanya, P. (1999). Popular fallacies on the use of African languages in education. *Social Dynamics (Special Issue: Language and Development in Africa), 25*(1), 81–100.

Ochieng'-Odhiambo, F. (2010). *Trends and issues in African philosophy*. New York, NY: Peter Lang.

Ong, W. J. (1986). Writing is a technology that restructures thought. In G. Baumann (Ed.), *The written word; Literacy in transition, Wolfson College lectures* (pp. 23–49). Oxford, England: Clarendon Press.

Orey, D. C. (2000). The ethnomathematics of the Sioux tipi and cone. In H. Selin (Ed.), *Mathematics across culture: The history of non-western mathematics* (pp. 239–252). Dordrecht, The Netherlands: Kluwer Academic Publishers.

Pearson, P. D. (2015). Foreword. In Z. Babaci-Wilhite (Ed.), *Local languages as a human right in education*. Rotterdam, The Netherlands: Sense Publishers.

Pennycook, A. (1994). *The cultural politics of English as an international language*. Harlow, England: Longman Group Limited.

Pullum, G. K. (1991). *The Great Eskimo vocabulary hoax and other irreverent essays on the study of language*. Chicago, IL: The University of Chicago Press.

Rosa, M., & Orey, D. C. (2011). Ethnomathematics: The cultural aspects of mathematics. *Revista Latinoamericana de Etnomatematica, 4*(2), 32–54.

Rugemalira, J. M., Rubagumya, C. M., Kapinga, M. K., Lwaitama, A. F., & Tetlow, J. G. (1990). Reflections on recent developments in language policy in Tanzania. In C. M. Rubagumya (Ed.), *Language in education in Africa: A Tanzanian perspective* (pp. 25–35). Clevedon, England: Multilingual Matters Ltd.

Shizha, E. (2012). Reclaiming and re-visioning indigenous voices: The case of the language of instruction in science education in Zimbabwean primary schools. *Literacy Information and Computer Education Journal (LICEJ), Special Issue, 1*(1), 785–793.

Smith, N. (2005). *Language, frogs and savants: More linguistic problems, puzzles and polemics*. Oxford, England: Blackwell Publishing

Spring, J. (2000). *The intersection of cultures: Multicultural education in the United States and the global economy*. New York, NY: McGraw Hill.

UNESCO. (1953). *The use of vernacular languages in education, Monographs on fundamental education*. Paris, France: UNESCO.

Vygotsky, L. (1978). *Mind in society: The development of higher psychological processes*. Cambridge, MA: Harvard University Press von Eckhardt Klein, B. (1978). What is the biology of language? In E. Walker (Ed.), *Explorations in the biology of language* (pp. 1–14). Montgomery, Vermont: Bradford Books, Publishers.

Walibora, K. (2015). While we wallow in self-hate, Tanzania has decreed that Swahili is the language of the gods. *The EastAfrican*. Retrieved from http://www.theeastafrican.co.ke/OpEd/comment/Tanzania-has-decreed-that-Swahili-is-the-language-of-the-gods/-/434750/2645374/-/10g0333z/-/index.html

Warren, M. D., Sikkerveer, I., & Brokensha, D. (Eds.). (1995). *Indigenous knowledge systems: The cultural dimension of development. Indigenous knowledge systems*. London, England: Intermediate Technology Publications Ltd.

Wolff, H. E. (2006). Background and history—Language politics and planning in Africa. In H. Alidou, A. Boly, B. Brock-Utne, Y. S. Diallo, K. Heugh, & H. E. Wolff (Eds.), *Optimizing learning and education in Africa—The language factor. A stock-taking research on mother tongue and bilingual education in Sub-Saharan Africa* (pp. 26–55). Paris, France: Association for the Development of Education in Africa (ADEA).

Zeleza, P. T. (1997). *Manufacturing African studies and crises*. Dakar, Senegal: CODESRIA.

Ziman, J. (1978). *Reliable knowledge: An exploration of the grounds and belief in science*. Cambridge, England: Cambridge University Press.

Sam Mchombo
Department of African American Studies
University of California, Berkeley
USA

BRAD D. WASHINGTON

9. GLOBAL INTERSECTIONS OF ENGLISH LANGUAGE HEGEMONY AND TECHNOLOGICAL INNOVATION IN THE REPUBLIC OF THE PHILIPPINES

INTRODUCTION

Despite a rich heritage of an estimated 200 languages representing historical pride and community knowledge, multilingualism has been identified as a possible impediment to the continued educational, social, and economic growth of the Republic of the Philippines (Cenoz & Genesee, 1998; Smolicz, Nical, & Secombe, 2007; Tupas, 2014; Wa-Mbaleka, 2014). Progression in the expansion of educational opportunities has been framed as inseparable from the narrowing of the medium of instruction (MoI), identifying English as the sole language not only capable of offering a promising future to Filipino communities, but also the lone language in which the fields of Science, Technology, Engineering, and Mathematics (STEM) are taught at the nation's universities (Tupas & Lorente, 2014).

As a global curriculum shift has magnified the development of STEM programs and funding in higher education (Drew, 2011; Talaue, 2014), preparation for post-secondary studies at the secondary level continues towards aligning classroom curriculum and linguistic expectations with the perceived standards of STEM in students' next stage of academic development. The 2013 announcement by the Philippines' Commission on Higher Education (CHED) to have all instruction in the General Education Curriculum (GEC) in English by 2017 was justified, in part, by the 2-year expansion of high school/senior high school education teaching through the medium of the Filipino language, most commonly affiliated with Tagalog (Angeles, 2014). Under the guise that the Department of Education and CHED are inherently separate at all levels, there remained the idea that at the very least there would be two additional years of access to one's regional vernacular. The decision resulted in ongoing challenges in the judicial system of the Philippines, along with the stated disapproval and resistance of educators throughout the country (Fernandez, 2014). In 2015, efforts to preserve and expand language instruction in Philippine languages grow. Yet, the continued emphasis towards STEM education, specifically for the purpose of competitiveness in the global market, magnifies a false pretense that innovation and investment is inherently based upon shifting language, and ultimately culture.

Z. Babaci-Wilhite (Ed.), Human Rights in Language and STEM Education, 151–167.

The purpose of this study is to complicate, through a survey of Philippine education history and the narratives of Filipino educators and students in the Province of Bulacan, the interplay between access to STEM education with the necessity of English language instruction in addressing the educational goals of the Republic of the Philippines. I will actively explore how government and non-governmental organizations attain STEM-themed educational opportunities for marginalized and privileged Filipino communities, while at the same time navigating the juxtaposition of seeing an active and intentional reduction of instruction in first or heritage languages. The uncertainty of the lasting implications of language shift via institutional education will also drive this study.

In order to fully capture the goals of this study, it is critical to clarify the stance of the author, as well as how I situate my identity in the research. It is vital to detail that the purpose of this study is not to present, what in my measurement would be a flawed bilateral consideration, a framework where those promoting the multilingual preservation of languages in Philippine education are somehow angelic in posture and those vying for predominantly English-only policy are ultimately oppressive in intent. In mitigating the rationale of adopting the English language almost exclusively in higher education, I hope to highlight the implications for lost of intellectual prowess through not offering Filipino and vernacular languages widely in higher education.

Regarding my stance and identity, it is also important to identify myself not only as a professor at a private institution in the United States, but also the degree of subjectivity I have as an advocate for learning languages. Decidedly, I believe through scholarship and experience that language acquisition rarely if ever comes in a vacuum void of social, political, and racial paradigms (Block, 2003; Kubota, 2001; Ligett, 2014; Vandrick, 1997). At the same time, speaking with and learning from the voices of Filipino communities directly impacted by these educational changes does not remove me from the convenient distance in terms of real-life consequences that come with the daily effects of a policy that impacts the present and future of the Philippines. If I were to fail in articulating these confounding variables as an American professor writing a chapter exclusively in English about multilingual advocacy and the dangers of English language hegemony, I would strain the lengths of hypocrisy ultimately met with the cynicism of those I hope to engage. Thus, this chapter will be a critical review of the proposed changes in Philippine education that is informed by research and narratives of Filipinos, while naming the privileged position I occupy.

Finally, this study does not solely focus upon the preservation of linguistic history and the role of STEM education in the Republic of the Philippines. This chapter will include perspectives of how heritage Filipino languages also carry knowledge that is unique in expression and development, in addition to being vital in regional and global contexts.

Research Questions

The study will address the following research questions:

1. What are the historic, current, and potential future contexts in which the English language is used in STEM education for Filipinos?
2. How is STEM education relevant to the needs and interests of the Republic of the Philippines and its citizens?
3. How do Filipino and vernacular languages continue to evolve and remain active in practice at regional, national, and global spaces?

These central research questions will be addressed through the organization of chapter themes that will focus on:

- Surveying the current status of education in the Republic of the Philippines.
- Identifying the relationships among administrative education offices at the regional and national levels of the Philippines.
- Reviewing the expansion of educational opportunities to unreached and underserved children in the Philippines, and how language and STEM education are directly related to the ultimate success of this expansion.
- Documenting the narratives of Filipino educators in the City of Meycauayan.

Limitations

In relation to addressing the challenge of navigating the historical role of colonization, conflict, and suppression in the current status of education through the mediums of language and technology in the Republic of the Philippines, it is critical to the validity of the study to acknowledge what is not given sufficient review in this research. Namely, the historical narrative of how the vernacular languages that are still in use in what is now known as the Republic of the Philippines will not be covered. In addition, the origin of science, technology, engineering, and math knowledge as it pertains to Filipino tradition is scarcely reviewed. With a focus on the status of education in the Philippines in 2015, my contribution to the research looks through the prism of technological innovation as understood in a late 20th/early 21st century Western context, albeit if that context takes from civilizations throughout the world's history (Andresen, 1999; Selin, 2008).

Therefore, my study is based upon the possible implications of delivering STEM education exclusively in the English language from a Western context in Philippine education systems. This also includes a pretext of the purpose of curricula and financial investment driven by industrial and capitalistic models. This is not to necessarily advocate that the same approach to education and investment is transformed to being equitable and just by delivery through the medium of Philippine vernacular languages. However, a goal of the study is to further investigate the socio-political reach of formal education to the nation's ethnically and linguistically diverse population.

A SURVEY OF LANGUAGE AND EDUCATION IN THE REPUBLIC OF THE PHILIPPINES

As this section attempts to form a foundation upon which further synthesis can take place regarding the status of education in the Republic of the Philippines, it would be important to begin by reviewing components of Article XIV in the 1987 Philippine Constitution that covers Education, Science and Technology, Arts, Culture and Sports:

> Section 1. The State shall protect and promote the right of all citizens to quality education at all levels, and shall take appropriate steps to make such education accessible to all.
>
> Section 6. The national language of the Philippines is Filipino. As it evolves, it shall be further developed and enriched on the basis of existing Philippine and other languages.
>
> Subject to provisions of law and as the Congress may deem appropriate, the Government shall take steps to initiate and sustain the use of Filipino as a medium of official communication and as language of instruction in the educational system.
>
> Section 7. For purposes of communication and instruction, the official languages of the Philippines are Filipino and, until otherwise provided by law, English.
>
> The regional languages are the auxiliary official languages in the regions and shall serve as auxiliary media of instruction therein.
>
> Spanish and Arabic shall be promoted on a voluntary and optional basis.
>
> Section 10. Science and technology are essential for national development and progress. The State shall give priority to research and development, invention, innovation, and their utilization; and to science and technology education, training, and services. It shall support indigenous, appropriate, and self-reliant scientific and technological capabilities, and their application to the country's productive systems and national life.

These sections of Article XIV serve in outlining that the current evolution of educational policies and curriculum are neither isolated, recent events in time for the nation, nor driven by one individual or entity in determining linguistic and curriculum implementation for schools in the Philippines. Based off of a 1968 seminal study by the Language Study Center of the Philippine Normal College, Sibayan (1975) provides a narrative on the more than 6,000 surveys sent across the

Philippines in 7 languages to assess language attitudes throughout the nation, as well as how linguistic interaction impacted acquisition:

> English is the language of distance – of semi-formal and formal usage for the Filipino. The vernacular or ethnic language thus dominates the Filipino's speaking life. It is, to him, the language of intimacy or nearness…The data appear to support the common impression that Pilipino is becoming a lingua franca, gradually taking the place of English. (p. 118)

Sibayan goes on to report that though English at the time was used in very specific instances (e.g., writing letters), it was still the preferred MoI throughout the Philippine educational system. This was mainly due to the "Philippine experience with English…[being]…the language for acquiring an education…the language of government…[and] of science and technology" (p. 120).

At a time where research was intentionally embedded with barrios throughout the Philippines, there was also another critical factor in exploring the 21st century context of language policy and education that Sibayan's study accounts for in the 20th century:

> A seemingly disturbing finding is that while 23% [547] of the householders and 20% [468] of the teachers reported that Tagalog was their first language, the great majority of these native speakers do not seem to prefer Pilipino as the language of instruction. This information lends itself to many interpretations… It can be interpreted to mean that Filipinos should continue to use English as the main, if not only, medium of instruction; this is the position of Dean Leopoldo Y. Yabes of…the University of the Philippines. It can also be misinterpreted and used to condemn the respondents for being colonial minded or, worse yet, stupid, and incapable of knowing what is good for the nation… (p. 120)

In reviewing the question of language navigation and policy as related to education in the Republic of the Philippines, the constitution, as well as Sibayan's (1975) study complicates the narrative that a majority of educators are (have been) adamant about the retention of native languages in the national school system. However, the legacy of violent global and national politics from Japanese colonial rule, to U.S. government assimilation practices, to the Philippine government's oppression through Marshal Law (1971–1986), also played dominant roles on the suppression of information for most of the second half of the 20th century, possibly leaving an air of inconclusiveness as to the broader preference and interpretation of communication at that time. Though Dayag (2004) concludes a 2003 study by stating the positive expansion of the English language through media in facilitating the reemergence of human rights for the Philippines during the time of President Ferdinand Marcos, the author poses a now recurring question regarding the segregation of society and opportunity as a result of language: "[A]re English-language advertisements preferred by Filipinos who belong to the middle or upper class; conversely, are Filipino-language advertisements preferred by the poor? If so, how would this relate to the language issue in the country?" Reminding all of the core foundation from which English was introduced to the Philippines as a colonizing and dehumanizing force, Martin (2012) challenges how the national government has forsaken its role in being a steward of the future of education policy: "The Philippine government's formula for economic success has become painfully simplistic: English equals money" (p. 194). How do these questions of language access and utilization codify the status of education for children in the Republic of the Philippines in 2015?

Status of Filipino Education in the 21st Century

According to the United Nations Educational, Scientific, and Cultural Organization (UNESCO) (2012), there are an estimated 1 million youth who are designated as out-of-school in the Philippines. This includes children and young adults ranging from 5–17 years of age. Identifying dangerous weather and income inequalities as leading factors in the number of out-of-school youth, Diola (2014) details the low level of investment from the national budget (less than 15 percent in 2011) that has contributed to stagnation in educational opportunities. According to the Philippines' Department of Education, there are an estimated 21 million youth who are currently enrolled in school, accompanied by a 71% increase in fund allocation for education. As the reported numbers of youth discontinuing their education continues to rise, there is a need to further investigate not only the factors that lead to this division in educational access that threatens the future of the nation, but also identify areas of active resistance that may exist among youth who are historically absent from the Philippine educational system.

Two primary sources of information were reviewed to gain a general sense of the current status of education: The Department of Education for the Republic of the Philippines and the 2009 document, UNESCO Philippine Education for all 2015: Implementation and Challenges. The foundation, purpose, and goals of each are central to the purpose of this study in grasping the relevance of the English language, as well as the standing of vernacular languages in the Philippines.

According to the Department of Education (DepEd) (2001), its goal is "to protect and promote the right of every Filipino to quality, equitable, culture-based, and complete basic education…" Within this vision, specific aims in facilitating this goal is made relevant in relationship to students, teachers, administrators, staff, family, community, and other stakeholders. The mandate of DepEd is also expansive, as it defines accountability to both formal and non-formal educational spaces for private and public education. For example, from Kindergarten to Grade 3, the child's heritage language is the primary MoI. This is part of the 2013–2014 Mother Tongue-Based Multilingual Education (MTB – MLE) model, from which 19 languages are officially incorporated by the government system. It is important to note that even at the K–3 level, Filipino and English are introduced at an unspecified level (in terms of dedicated time of instruction) each year.

The implementation of the new senior high school (grades 11–12) identifies STEM as one of four specialty strands defined as Empowerment Technologies to build competencies through social/mobile technologies, as well as assistive media and online systems. In reviewing its contents, the curriculum is designed (at least in part) to consider the impact of technology within the historical and contemporary narratives of the Filipino diaspora.

As the goals and rationale for the vision of the Department of Education are clear, there remains the question of its effectiveness. Curriculum, instruction, and outreach in the planning of education for Filipinos is constructed in such a way to potentially support learners in different social, academic and economic spaces. However, there is substantial data that suggests despite these efforts, the most marginalized populations in the Republic of the Philippines still lack access to basic education.

According to United Nations Education, Scientific and Cultural Organization's report, Philippine Education for All 2015: Implementation and Challenges (2009), less than half of the estimated 2.3 million children who were 6 years of age in 2007 were enrolled in a school (p. 36). Within the same report, there was a stark comparison made in the retention of children beginning school based upon public versus private education: "Repetition rate(s) in public elementary schools was nearly 10 times higher than private ones (in 2005)" (pp. 42–43). As social and cultural factors (e.g., the age of a child attending school for many communities is older than age 6) were cursory reviewed, the focus of the report shifted to geographic and regional disparities that continue to exasperate educational and economic divides throughout the nation. These disparities included poverty, armed conflict, and extreme weather (typhoons and earthquakes). Even in a seemingly more ideal area where these factors are not prominent, many educational offerings at the elementary level are incomplete (e.g., K–5 schools in a region only having teachers/resources that run up to grade 3), presenting another interruption in the education of school-aged youth.

In terms of this study, what is the significance of detailing statistics for grade 1? Although there is a long, established policy and plan to extend preschool programs including an expansion of Kindergarten, grade 1 has been a barometer for a Filipino child's first encounter with institutional education. This reality becomes even more heightened when considering the number of children who were born in the Republic of the Philippines. Over the last decade, the Philippines has consistently ranked among the top nations in Asia in terms of birth rate, with 2012 data reflecting 24 births per 1,000 people (At a glance, 2014; Litke, 2014). The life-threatening consequences of sustaining quality education becomes even more pressing as many of these children are born into poverty, with young mothers having limited access to reproductive healthcare (Narang, 2015).

Globalization and Stability of Philippine Education

As the United Nations and the Government of the Philippines both cite efforts to recruit thousands of teachers each year to strengthen the infrastructure of the nation's educational system, the forces of globalization continue to impact those efforts, especially as it pertains to the most highly skilled educators. Over the past decade, thousands of Filipino educators ranging from principals, classroom teachers, and administrators have been recruited to teach in the most underserved and impoverished areas of the United States (Bartlett, 2013; Chua, 2014; Diaz, 2011; Dillon, 2009; Joseph, 2014). Often citing the opportunity of earning a living wage in making the decision to teach abroad, Filipino teachers often have to negotiate paying a substantial amount of their potential salary in the United States to attain a contract. There are numerous accounts of fraudulent practices in terms of the agreements the teachers make with recruiters based in the United States, further complicating the absence of needed instructors in the Republic of the Philippines. However, these reports have neither stemmed the volume of recruitment, nor the number of educators that engage in this exchange. This is an important paradigm to highlight, as educators are the conduit for leading classroom instruction in language and STEM. If the most seasoned and qualified educators in the Republic of the Philippines are in a position where they cannot be compensated at an equitable rate to support their own families, and have an opportunity to do so by teaching abroad, investment in education is only one component of expanding schools. Moreover, it is because the Filipino teachers utilize English primarily as a MoI, and that there is also a major investment in supporting underserved American communities in STEM education that they can be viable candidates to teach in the United States. Therefore, there is an immediate need for the retention of educators, as well as the retention of children in Philippine schools.

Universal Education

Over the past 6 years, the response by the Philippines to retain teachers and strengthen K–12 education has been to further articulate a process for STEM education, address the question of language instruction, and develop a national pre-school education plan. In reviewing broad guidelines from the United Nations to adopt and begin to incorporate measures to strengthen the school system in the Philippines in the year 2000 to be reviewed by 2015, the national government was reported to not begin that implementation until 2006. As cited in the draft 2015 UNESCO report for the World Education Forum, the agenda of President Benigno Aquino, III has collectivized a vision to address all areas of educational development at once. Among the points of President Aquino's educational reform agenda were plans to (1) offer access to preschool education for all, (2) reiterate the need for more students to pursue education in the STEM fields, and (3) broaden the scope of utilizing languages as it relates to MoI.

The goals for each of the three reforms highlighted are predicated on financial investment. According to the UNESCO report, offering access to preschool education will set a foundation to supporting children to become acclimated to attending school, and therefore lead to greater retention at the first grade level. Part of the retention efforts include providing nutrition and healthcare to children at schools to make it as much an extension of home as it is a place for education. As the proclamation from President Aquino was announced in 2012 to have universal kindergarten, the Philippine-based Alliance of Concerned Teachers (ACT) have challenged the viability of the plan since it was originally introduced, due to lack of teachers, buildings, and pay resulting in current compensation for teachers being at or falling under the minimum wage (Cruz, 2011). In terms of making STEM a priority in the direction and investment of education, its reiteration in the constitution, Department of Education, and global economy for the past century has made it the preeminent option for (versus a component of) education.

In terms of MoI, the current president has made a mandate to accelerate the development of resources and schools that allow for an expansion of materials beyond the recognized languages supported by the MTB – MLE; this reform is in line with recommendations from UNESCO. What makes this new call unique is that it explicitly states the need to extend vernacular languages beyond the home, though wanting in how to make relevant throughout a child's educational career. President Aquino states, "We should become trilingual as a country. Learn English well and connect to the world. Learn Filipino well and connect to our country. Retain your dialect and connect to your heritage" (Lee-Chua, 2010). As Lorente (2013) states in reference to Aquino's speech, "…its emphasis on the mother tongue as a means of connecting to one's heritage…is conceptually different from [multilingual education] which emphasizes the cognitive advantages of the mother tongue as well as to a lesser extent, the grassroots development that can stem from it" (p. 200). This illustrates the challenges of implementing a multilingual framework when the first language is in policy to assume a role of preservation versus in utilizing a role in instruction for acquiring the targeted language. It is under the former versus the latter role that national planning for supporting ethnic minorities by continuing the diversification of language instruction again became a focus for the department of education.

Alternative Learning Systems

Because of the unpredictable nature of weather, conflict, and poverty in the Republic of the Philippines, there is a separate division of education that accounts for multiple pathways to attain education for children and youth whose lived experiences do not allow for pursuing a linear model of education. The Department of Education's Alternative Learning Systems (p. 1987) is defined as being different from the formal education system through its vision of learning outside of the classroom.

ALS Non-formal Education happens outside the classroom, community-based, usually conducted at community learning centers, barangay multi-purpose hall, libraries or at home, managed by ALS learning facilitators, such as mobile teachers, district ALS Coordinators, instructional managers at an agreed schedule and venue between the learners and facilitators.

As of 2015, the 10 alternative programs ALS supports are (1) Program for illiterates: Basic Literacy Program (BLP), (2) Program for dropouts of formal Elementary and Secondary levels, (3) Program for Indigenous Peoples, (4) Program for Muslim Migrants, (5) Program for Hearing Impairment, (6) Program for Adolescents and Reproductive Health, (7) Parent Education, (8) Family Basic Literacy Program, (9) Radio-Based Instruction (RBI), and the (10) Program for Disadvantaged Children (ALS Programs, 2014). As there is potential overlap in the scope and demographic of the targeted learners among some of the programs, the uniqueness in the approach of supporting the estimated 41 million Filipinos (by 2008 estimates 45% of the nation's total population) further defines how education can become accessible (ALS FAQs, 2014).

Of the programs that make up the Alternative Learning Systems, only one identifies the communities it is developed to serve by religion first, versus economic or geographic marginalization. There are multiple reasons as to why identifying Filipinos who are of the Muslim faith in an ALS is critical to the discussion of language and STEM education in the Philippines. First, regardless of intent, there is an intentional political overtness to the rationale of outreach to Muslim communities in the Republic of the Philippines. In one of the most cohesive and active Catholic Christian societies in the world (Abanes et al., 2014; Andaya, 2012; Braülein, 2012; Cornelio, 2014), the manner in which the basis of education is interconnected with political representation, social mobility, and economic stability embodies the total lived experience of the Philippine Muslim community.

With a continued focus on the hegemony of the English language, the Philippine Constitution serves as a more recent reference of a centuries-old conflict between Christianity and Islam. Article XIV, Section 7 serves as an artifact of the linguistic colonial struggle within the Philippines, citing the English as a tentative national language, while Spanish and Arabic are cited as languages to be taught on a voluntary and optional basis. Admittedly, an initial takeaway would be to consider the necessity of including any component of societal identification in the legal foundation of a nation, including languages, if it is described as voluntary and optional. Yet, the pre- and post- colonial conflicts of ethnicity, culture, trade, religion and war are represented by the inclusion of Arabic and Spanish in the constitution.

Milligan (2003) frames his study with social identity theory to review access to education for self-identified Muslim regions in the southern islands of the Philippines. In regards to educational access and attainment, Milligan's study illuminates an unfortunate continuity with the rest of the nation, as resources are scarce and a centralized deployment approach of setting educational policy and distributing economic investment has failed in either stabilizing or shifting the trend towards enrollment in schools. However, there is an additional component of marginalization that Milligan identifies that further complicates the education of Muslim Filipinos as structured in a dominant Christian Filipino context:

> Of [texts meant for the classroom] that did include some mention of Muslim Filipinos, most contained no more than a few paragraphs. Furthermore, much of the information included was either erroneous or insulting to Muslims. While a few authors and publishers have attempted to include more of Filipino Muslim history and culture in recent textbooks, representation remains extremely limited and continues to portray Filipino Muslims as either irrelevant artifacts of early Philippine history or rebels threatening the peace and stability of the country. (p. 480)

Again, contemporary globalization comes into play. Though many Filipino educators are seeking employment abroad in the United States, Filipinos trained in STEM fields are working with companies with bases in the Middle East, and where Islam is the largest represented religion, incorporated into every component of life (Santos, 2014). As with Filipino teachers in the United States, the utilization of the English language plays a critical role in the employment of Filipino architects and engineers in the Kingdom of Saudi Arabia, Jordan, and the United Arab Emirates. According to McKenzie et al. (2012), an estimated 2 million Filipinos were Overseas Foreign Workers (OFWs) in over 180 countries. While the focus of this study is on English language and STEM education, it is critical to not only acknowledge that the vast majority of OFWs are categorized as domestic workers supporting the lives and families of communities in other countries, but that also this much larger representation of unskilled labor is comprised almost exclusively of women (Semyonov & Gorodzeisky, 2005). Of the Filipinos who are OFWs throughout the world but especially in traditionally Islamic nations, many are of Christian faith. Yet, in historical and in current affairs, the Republic of the Philippines has a global relationship with Muslim communities, and one predicated on social, cultural, and economic ties.

How do these connections with the Islamic world translate into the education of Muslim communities in the Philippines? As a part of the Alternative Learning Systems of DepEd, there is an educational policy in place described as the Arabic Language and Islamic Values Education (ALIVE) model. Though targeted at out of school youth, it holds at its core educational policies that "positively contribute to the peace efforts of [the Philippine] government in order to improve the quality of life of Muslim OSY and Adults" (ALS Programs, 2014). Yet, with over 90 percent of Muslim youth attending public schools, a UNESCO report identifies the still existing challenges in incorporating an effective and equitable curriculum for this community of learners.

NARRATIVES OF FILIPINO STUDENTS AND EDUCATORS IN THE PROVINCE OF BULACAN

The following section of this study aims to capture the narratives of Filipinos invested in education in the City of Meycauayan, Bulacan. In October of 2014, I had the privilege to travel to the Province of Bulacan, the largest of 80 provinces in terms of population in the Republic of the Philippines. Situated minutes north of Metro Manila, the nation's capital, the Tagalog language is by far the most prominent vernacular language in a northern region where Christianity is the dominant religion. The conversations documented (both through the MoI) are from leaders in two schools in Bulacan, focusing on the continued development of the Philippine educational system, the role of language in schools, and the continued impact of STEM learning in education reform.

Narrative from a Public School Perspective

The site chosen to document a current perspective of a public school is based in the City of Meycauayan. The school serves an estimated 1,300 students from first grade (1st–10th) to *fourth year* university students. Through discussion with the head of what will be referred to as *School X*, the conditions under which the institution was founded and its goals were discussed. First, *School X* was founded in 2003 to promote education for the development of infrastructure and employment in Bulacan. Because *School X* is public, the head of school is appointed by the provincial government, and his appointment is based as much upon the election cycle as it is on gauging the continued support of his leadership by the teachers.

As it is only a decade old, *School X* is still developing as an institution and is in the process of being defined by the youth who enroll there. As a part of its charge, the school is focused on the specific role of technology in constructing a new future for the Philippines. Its collegiate majors are limited, but focused on the fields of finance/economics as well as STEM, with all but "a few history courses" being taught in English. The students served are all from Meycauayan, as aligning with companies and organizations for the employment of its graduates in Bulacan is a major charge for the head of school. As of 2014, most of the student tuition is funded through the government as well as charitable donations from local business and Filipino expatriates abroad. Financial stability and support is critical, as the majority of students *School X* serves are from poor backgrounds, though many represent the second and third generation of their families to pursue and complete basic education with a university degree. Like the students, the teachers (an estimated total of 90) are also from Bulacan, and see their relationship with *School X* through its mission to serve their local community.

The following is an excerpt from an interview I had with the head of school. The following narrative is in response to a question about what his vision of a *School X* graduate should be:

> I believe this should be someone who understands the importance of paying forward his [or her] opportunity to gain a better education and life. Although there [is] no formal [policy] on what graduates need to do in returning an investment to [the school], they should be focused on having an orientation to

> be better members of society, to be responsible citizens to help their families, and a moral obligation to work towards a stronger nation. (October, 2014)

In considering a vision for *School X*, the head of school shared varying factors linked with governmental interpretations of accountability that would determine the ongoing functions and support of the institution:

> My vision for this college is to be able to expand and grow instrumental systems and procedures that would allow for accepting more students. We need to [consider] a commission and feasibility study towards the number of [additional] students we could admit now to maintain low or free tuition. In addition, our accreditation procedures are needed to improve our standards for evaluation like all other local colleges and universities. For 2015, I want to move more towards being evaluated to support transparency and gain government funds to grow the school. (October, 2014)

In my time with the head of school, I learned that the expansion of the *School X* was already in progress versus process, as the local government had moved to purchase additional land adjacent to the school grounds to begin construction on a new wing of the institution.

Narrative from a Private School Perspective

The conversation I had with the head of school at *School Y*, a private institution in Meycauayan, echoed what I learned at *School X* in terms of mission and goals. However, there were also distinct differences in socio-economic backgrounds of students, as well as heightened awareness for new educational reforms. *School Y* serves over 2,800 students, and has been consistently operational for nearly 90 years, a historical cornerstone that is a point of pride for the institution's leadership and alumni in reflecting upon its uncertainty during Japanese occupation and Marshal Law. Although *School Y* also reflects a curriculum and commitment to STEM education, it is unique in offering graduate level degrees in education and management. It is similar to the public *School X* in traditionally offering a grade 1 through grade 10 curriculum. However, *School Y* in 2014 was recruiting children for a new childcare center and kindergarten program, as well as appointing an administrator to specifically prepare for the incorporation of senior high school (grades 11 and 12) by 2016.

Educators at *School Y* often have an intimate level of familiarity with the province and the students. Although many of the teachers at *School Y* are not from principalities immediately surrounding the location of the school, they are in some cases alumni of *School Y* and return to teach. In addition, as the school has a record of serving generations of the same families, there is an additional connection with the local community unique to longtime educators in *School Y*. The two languages of instruction were identified as Tagalog and English, with special emphasis citing that families were demanding an *English-only* curriculum from as early as kindergarten. In speaking with the head of school, themes around student demographics and tuition were highlighted.

In response to being asked about the families *School Y* serves, the head of school expressed how the occupation of the parents demonstrated a clear picture of the school's diversity:

> Many of our students come from houses where they are the children of tricycle and jeepney [bus] drivers. These students are [often] eligible for scholarship[s]. [However,] we also have many students from families that are well off. This combination shows the different economic classes that make up the school, including day laborers, OFWs, and parents working in banks. (November, 2014)

Continuing the theme of affording an education at *School Y*, the head of school shared a narrative around the expense of private education that illuminated the complexity of access:

> With the growing cost of education, we rely on private funders and the families to bring students. We are much more expensive than public schools, so it is difficult to compete. We do not receive much financial support from the government, so we have to work harder in recruiting and keeping our students here. With the addition of the senior high school, that is also an additional expense of two more years, but necessary for our graduates to be globally competitive with those graduating in other countries. (November, 2014)

Although the challenges of funding a private education was a theme during my conversation with the head of *School Y,* there was also a level of curriculum freedom that was not prevalent at *School X.* Upon completing grade school, less than half of the students at *School Y* pursued collegiate studies at the institution, if at all. The head of school detailed how he and his team were able to surmount that an important reason for this trend was a lack of response to market/student need for employment. As a result, any new major and course content is informed by the career interest of students. The most recent example of this market response to education is a degree in criminology, which has led to a marked growth in enrollment and retention of new students within the province and beyond.

CONCLUSION

The purpose of this study was to survey the continued link of educational reform in the Republic of the Philippines being defined through the use of the English language and STEM education. Though often cited as contemporary neoliberal and colonial/post-colonial constructs, the underlying tensions and rationales for this direction of education in the Philippines has existed for centuries. Through reviewing the history of educational policy in the Philippines, it became apparent that schooling or education does not (and rarely ever) occur in silos, and is susceptible to economic and political pressures at local, national, and global levels.

This chapter concludes that though specific government policies exist to acknowledge and explore ways in which the multilingual and multicultural society of the Philippines can have equitable access to education, the implementation and investment are weighted heavily on *meaningful* education that can equate to competitiveness in the global marketplace. Holding this lens of economic viability for the rationale of education in the Republic of the Philippines has been rejected with direct and indirect resistance. In reference to the two schools in the Bulacan Province, DepEd and CHED standards for curriculum in terms of English as the MoI are being met, but with a vision that its graduates will foster development in their own communities upon graduating, and in fields of study beyond Science, Technology, Engineering, and Mathematics. Due to social tensions around language and culture, universal education meant for all children in the Philippines is not designed for all the identities they represent. Despite multiple national and global reports that have urged additional investment in the schools ranging from teacher salary to relevant literature in textbooks, there remains a delay in the implementation of these reforms at a time where required years for education is expanding, STEM-educated Filipinos are working outside of the Philippines, and the most qualified Filipino educators are opting to teach abroad. No matter the framework for accounting for the linguistic diversity of the nation, as well as the promotion of STEM in its future, reform in Philippine education will result in at best incremental growth in national education goals until a more inclusive, equal, and decentralized vision of education via language, heritage, finance, and curriculum takes hold.

REFERENCES

Abanes, M. S., Scheepers, P. L., & Sterkens, C. (2014). Ethno-religious groups, identification, trust and social distance in the ethno-religiously stratified Philippines. *Research in Social Stratification and Mobility*, *37*, 61–75.

ALS FAQs – Republic of the Philippines, Department of Education. (2014). Retrieved from http://www.deped.gov.ph/als/faq

ALS Programs – Republic of the Philippines, Department of Education. (2014). Retrieved from http://www.deped.gov.ph/als/programs

Andaya, B. W. (2012). Christianity in Southeast Asia: Similarity and difference in a culturally diverse region. In C. E. Farhadian (Ed.), *Introducing world Christianity* (pp. 108–121). Oxford, England: Blackwell Publishing, Ltd.

Andresen, J. (1999). Science and technology in non-Western cultures. *Journal of Religion & Science*, *34*(2), 345–352. doi:10.1111/0591-2385.00218

Angeles, M. (2014). Professors of Filipino breaking bad over CHED memo. *GMA Network.* Retrieved from http://www.gmanetwork.com/news/story/365618/lifestyle/artandculture/professors-of-filipino-breaking-bad-over-ched-memo

Bartlett, L. (2013). *Migrant teachers: How American schools import labor.* Cambridge, MA: Harvard University Press.

Block, D. (2003). *The social turn in second language acquisition.* Edinburgh, Scotland: Edinburgh University Press.

Braülein, P. (2012). "We are 100% Catholic": Philippine passion rituals and some obstacles in the study of non-European Christianity. *Journal of Religion in Europe*, *5*(3), 384–413. doi:10.1163/18748929-00503003

Cenoz, J., & Genesee, F. (Eds.). (1998). *Beyond bilingualism: Multilingualism and multilingual education.* Philadelphia, PA: Multilingual Matters.

Chua, K. (2014, October 2). Arizona looks toward the Philippines to fill teacher positions [Web log comment]. *Education Week*. Retrieved from http://blogs.edweek.org/teachers/teaching_now/2014/10/arizona-looks-toward-philippines-fill-teacher positions.html?intc=mrs

Cornelio, J. S. (2014). Popular religion and the turn to everyday authenticity: Reflections on the contemporary study of Philippine Catholicism. *Philippine Studies: Historical and Ethnographic Viewpoints, 62*(3–4), 471–500.

Cruz, T. (2011, January 30). Is the Philippines ready for universal kindergarten in 2011? *Asian Correspondent*. Retrieved from http://asiancorrespondent.com/47329/is-the-philippines-ready-for-universal-kindergarten-this-2011/

Dayag, D. T. (2004, February). The English-language media in the Philippines. *World Englishes*, *23*(1), 33–45. doi:10.1111/j.1467-971X.2004.00333.x

Diaz, R. (Producer & Director). (2011). *The learning* [Motion picture]. San Francisco, CA: CineDiaz and the Independent Television Service (ITVS).

Dillon, S. (2009, September 14). Schools look abroad to hire teachers. *The New York Times*. Retrieved from http://www.nytimes.com/2009/09/15/education/15teach.html?_r=0

Diola, C. (2014, January 23). Philippines' education crisis far from over – UNESCO. *The Philippine Star*. Retrieved from http://www.philstar.com/headlines/2014/01/23/1282186/philippines-education-crisis-far-over-unesco

Drew, D. E. (2011). *STEM the tide: Reforming science, technology, engineering, and math education in America*. Baltimore, MD: The Johns Hopkins University Press.

Fernandez, A. (2014, December 2). Filipino language advocates to seek SC help vs. new CHED curriculum. *GMA Network*. Retrieved from http://www.gmanetwork.com/news/story/390733/news/nation/filipino-language-advocates-to-seek-sc-help-vs-new-ched-curriculum

Joseph, G. (2014, May 26). Trafficked teachers: Neoliberalism's latest labor source. *In These Times*. Retrieved from http://inthesetimes.com/working/entry/16738/trafficked_teachers_neoliberalisms_latest_globalized_labor_source

Kubota, R. (2001). Teaching world Englishes to native speakers of English in the USA. *World Englishes*, *20*(1), 47–64. Retrieved from http://doi.org/10.1111/1467-971X.00195

Lee-Chua, Q. (2010, May 23). Noy Aquino's blueprint for education–Part II. *Philippine Daily Inquirer*. Retrieved from http://newsinfo.inquirer.net/inquirerheadlines/learning/view/20100523-271633/Noy-Aquinos-blueprint-for-educationPart-II

Liggett, T. (2014). The mapping of a framework: Critical race theory and TESOL. *The Urban Review*, *46*(1), 112–124. Retrieved from http://doi.org/10.1007/s11256-013-0254-5

Litke, M. (2014, August, 24). Amid population explosion, birth control access roils the Philippines. *PBS Newshour*. Retrieved from http://www.pbs.org/newshour/bb/reproductive-healthphilippines/

Lorente, B. P. (2013). The grip of English and Philippine language policy. In L. Wee, R. B. H. Goh, & L. Lim (Eds.), *The politics of English: South Asia, Southeast Asia and the Asia Pacific* (pp. 187–204). Philadelphia, PA: John Benjamins Publishing Company.

Martin, I. P. (2012). Diffusion and directions. English language policy in the Philippines. In E. L. Low & A. Hashim (Eds.), *English in Southeast Asia: Features, policy and language in use* (pp. 189–205). Amsterdam, The Netherlands: John Benjamins Publishing Company.

McKenzie, D., Theoharides, C., & Yang, D. (2012, April). Working paper: Distortions in the international migrant labor market: Evidence from Filipino migration and wage responses to destination country economic shocks (Forschungsinstitut zur Zukunft der Arbeit, No. 6498). *The Open Access Publication Server of the ZBW – Lebiniz Information Centre for Economics*. Retrieved from http://www.econstor.eu/bitstream/10419/58852/1/715979841.pdf

Milligan, J. A. (2003, November). Teaching between the cross and the crescent moon: Islamic identity, postcoloniality, and public education in the Southern Philippines. *Comparative Education Review, 47*(4), 468–492. doi:10.1086/379494

Narang, S. (2015, March 9). In this Manila 'baby factory,' why women put up with crowding four to a bed. *National Public Radio*. Retrieved from http://kbia.org/post/manila-baby-factory-why-women-put-crowding-four-bed

Republic of the Philippines Department of Education. (2001). Retrieved from http://www.deped.gov.ph/about/mandate

Santos, M. (2014, July 1). Filipino engineers stand out in Asean, says head of PH construction firm. *Inquirer.net*. Retrieved from http://newsinfo.inquirer.net/615957/filipino-engineers-stand-out-in-asean-says-head-of-ph-construction-firm

Semyonov, M., & Gorodzeisky, A. (2005, March). Labor migration, remittances and household income: A comparison between Filipino and Filipina overseas workers. *International Migration Review, 39*(1), 45–68. doi:10.1111/j.1747-7379.2005.tb00255.x

Selin, H. (Ed.). (2008). *Encyclopaedia of the history of science, technology, and medicine in non-Western cultures* (2nd ed.). Dordrecht, The Netherlands: Springer Publishing.

Sibayan, B. P. (1975). Survey of language use and attitudes towards language in the Philippines. In S. Ohannessian (Ed.), *Language surveys in developing nations. Papers and reports on sociolinguistic surveys* (pp. 115–143). Arlington, VA: Center for Applied Linguistics. Retrieved from ERIC database. (ED 104170)

Smolicz, J., Nical, I., & Secombe, M. J. (2007). Assimilation or pluralism? Changing policies for minority languages education in Australia and the Philippines. *Political Crossroads, 14*(2), 71–99. Retrieved from http://dx.doi.org/10.7459/pc/14.2.05

Talaue, F. T. (2014). Social equity and access to a Philippine STEM school. *Theory into Practice, 53*(1), 33–40. doi:10.1080/00405841.2014.862120

Tupas, R. (2015). Inequalities of multilingualism: Challenges to mother tongue based multilingual education. *Language and Education, 29*(2), 112–124. doi:10.1080/09500782.2014.977295

Tupas, R., & Lorente, B. P. (2014). A 'new' politics of language in the Philippines: Bilingual education and the new challenge of the mother tongues. In P. Sercombe & R. Tupas (Eds.), *Language, education and nation-building: Assimilation and shift in Southeast Asia* (pp. 165–180). London, England: Palgrave Macmillan.

United Nations Education, Scientific and Cultural Organization. (2009). *Philippine education for all 2015: Implementation and challenges*. Retrieved from http://planipolis.iiep.unesco.org/upload/Philippines/Philippines_EFA_MDA.pdf

United Nations Education, Scientific and Cultural Organization. (2012). *Philippines UNESCO: Country programming document: 2013–2016*. Retrieved from http://unesdoc.unesco.org/images/0022/002296/229694E.pdf

United Nations Education, Scientific and Cultural Organization. (2015). *Philippines: Education for all 2015 national review (draft report)*. Retrieved from http://unesdoc.unesco.org/images/0023/002303/230331e.pdf

Vandrick, S. (1997). The role of hidden identities in the postsecondary ESL classroom. *TESOL Quarterly, 31*(1), 153–157.

Wa-Mbaleka, S. (2014). Teaching English to speakers of other languages: The case of the Philippines. *International Journal of Academic Research in Progressive Education and Development, 3*(3), 64–78. doi:10.6007/IJARPED/v3-i3/952

Brad D. Washington
International and Multicultural Education Department
School of Education
University of San Francisco
USA

BASIL A. OKONKWO

10. THE IMPORTANCE OF LOCAL LANGUAGE TO THE DEVELOPMENT OF TECHNOLOGY

INTRODUCTION

Technology grows with demand and necessity. The different social activities of the world shape the necessity of what is constructed or developed. It is a known fact that we come from different cultures and therefore emphasize certain things differently. For instance, due to extreme heat in many regions of the world, people work to advance and develop forms of air conditioning. This coolant would never be necessary for cold regions, where individuals would rather invent a heater to counteract cold weather.

Social problems cannot be solved without use of available materials within the locality of a people. Since technology is borne out of the fact of solving problems, it then negates the possibility that society could be a society in isolation. Following this line of thought, Magda (1965, p. 73) says: "The world is not a thing, but man himself is worldish: he is, at the bottom of his being, world-disclosing, world-forming. Man alone is so, that he fore-goingly, a priori, understands his own being in a relational-whole, in which and from which he can meet other beings and understand them in their being."

The keyword here is 'communication'. Communication is the activity of conveying information from one person to a greater number of peoples. Information cannot be conveyed without language in a broad sense, whether in signs or in words. Here language becomes a key factor for development. Apart from the fact that language is the primary identity of a people or culture, it is very much the primary factor for societal development. In the light of this, in Babaci-Wilhite and Geo-JaJa (2014), quoting the Director of the National Kiswahili Council, Babaci-Wilhite notes that, "If African culture and language is not equated with the development, it will be impossible to encourage development on African terms." As previously observed, the need for technology is borne out of socio-cultural necessity, and the satisfaction of these necessities cannot be attained without the aid of language and communication. In communication people share ideas, offer solutions and are able to arrive at reasonable, efficient conclusions. It is then through language that the communality in communication can be utilized in the development of the society to serve our daily needs.

Z. Babaci-Wilhite (Ed.), Human Rights in Language and STEM Education, 169–177.

LOCAL LANGUAGE, ACTIONS AND TECHNOLOGY

In order enhance efficient productivity; we must translate our words into actions. When one communicates, it is expected that this act of communication is made through actions. For instance, when a group of people says, “Let’s sleep,” these words, communicated in clear language, is actualized through the act of sleeping. Thus, we translate what we know into action. According to Fromkin and Rodman (1974), we live in a world of words and, since these words are not mere signs and graphics symbols, the universe is given form and shape by the power of the words. One can summarize the relationship between language and action in noting that language is the soul of every action. Even uneducated individuals or those with limited literacy can communicate and the language communicated is actualized. Language is the centre of all communication and without language the world would have remained in a chaotic primitive state, b. But the presence of language has aided communication and has helped to actualize desired development.

According to Thomas (2009, p. 9) the term ‘technology’ in English, as well as its cognates in the other Germanic Languages, has referred most directly to treatises and published accounts describing various technical crafts. Language and technology are two different areas of knowledge but are very dependent on one another. In language, technology finds terminologies, ideologies and constructive communication, while, in technology, language skills are advanced or enhanced to enable the effectiveness of communication. The word technology, which is derived from ‘techno’ from the ancient Greek ‘techne,’ meaning ‘craftsmanship,’ and ‘logos,’ which connotes ‘word, speech, account, reason,’ simply means the word, speech, account or reasoning of craftsmanship.

By etymologically defining the word ‘technology’ we have arrived at two words: craftsmanship and word. Craftsmanship denotes artistry, construction and engineering while the logos or word denotes language and communication. Language is a primary object in reasoning. One cannot think outside the context of what his/her language can grasp. Every language has its societal adaptation. Language is a social character and without language, society fails to exist. From an Igbo cultural point of view, Okonkwo (2012, p. 71) identifies, ‘*nkana ’uzu*’ as the igbotechne. He argues that the concept of *nkanauzu* is one of the most common expressions to mark the Igbo essence of things that can be called into being through Igbo-human-existence, i.e that which precedes all that is a fact. For instance, in Nigeria, there are three major languages: Igbo, Hausa and Yoruba and each of these three languages is associated with a different major ethnic group. In Igbo culture, there might be fabrics, which are not used in Yoruba culture. This means that if *Akwaete* (a kind of fabric) is used in Igbo culture, it might not be available for the person in Yoruba culture because the conditions for production might not be the same. *Akwaete* is a thick fabric used to cover the body in cold weather, and it is made from raw materials, which come from that culture. With the loss of raw materials, we also risk losing specific language or terminology, which is closely linked to cultural identity.

THE LOGOS OF TECHNE AND THE NEED FOR DEVELOPMENT

Language can then be said to be the 'logos of techne'. An Igbo man cannot think like an American because he lives in a different sociocultural milieu. Each man's mode of rationality is culturally and geographically shaped in his different language. For instance the Igbo word '*akwukwo*' has many usage. This word is used to refer to vegetables (*akwukwo-nri*), textbooks (*akwukwoagumagu*), notebooks (*akwukwoedemede),* school (*ulo-akwukwo*), or to refer to educated individuals (*ndi guru-akwukwo*). The reason for the multiple applications of the term '*akwukwo'* is explained by the original use of the word to refer to leaves prior to the colonial era. The writing was done on walls. Consequently; there was no need for books, schools or other related terms. Education was done by apprenticeship and practice. The choice of '*akwukwo*''leaves' to refer to these terms pertaining to education, textbooks, and notebooks is due to the fact that 'paper,' which is central to the educational context, comes from plants (Papyrus).

With the advent of colonialism, culture seems to develop a new language to adapt to colonial terminologies and thought patterns. In the case of Igbo culture, local thought patterns are restricted and the thinking is narrowed by a colonial mentality. Innovation in the development of technology is also halted, and foreign materials are introduced in developing local problems. Language is the heart of technology because it is the embodiment of cultural thinking. No culture thinks outside its language unless its cultural thinking pattern has been altered by a foreign presence. Advancement in technology is born out of a society's quest for relief in work. In Igbo language, the words '*ahuhu*' (suffering), *Ibuaro* (heavy load), and *ije* (walking) could necessitate emancipation and it is words like these that necessitate emancipation. It is in words or expression such as these that language is linked to technological development.

The need for development and emancipation is the daily quest of man. This need can only be communicated with language. The origin of any technology originates in the question, 'Keduiheanyigaeme?'(What shall we do?). The quest to find a solution is the assemblage of different ideologies also born out of language and communication. This *'Keduiheanyigaeme*' is the initial language that sets the development of technology in motion. Mankind's quest to solve problems calls for a communal agreement to facilitate life in the society.Apart from the iheanyigaeme question, we also see the continuous zeal to get something done, and ask the question, "*keduetuanyiaga-esi me ya"* (How shall we get this done). This question highlights the curiosity to seek solutions and work towards development. Curiosity for development leads to what we may term 'cooperative communication'. This cooperative communication is centered on language and this language is a language of technology. We cannot achieve success in invention without the aid of communication.

IMPORTANCE OF LANGUAGE TO THE DEVELOPMENT OF TECHNOLOGY

Having established the link between technology and language through the use of 'cooperative communication,' we now focus on the importance of language to technology. We could list the following benefits: the development of local technology, rapidity in technology research, originality in invention and advancement in communication technology.

Language in the Development of Local Technology

Language could be said to be at the root of all technological inventions. As we previously noted, technological development is the product of man's curiosity to enhance and emancipate his condition. This process of emancipation cannot then come to life without language. Names and terms of materials and raw materials could be identified from the cultural background and usage of those items. For example, people who live in cities that are located nearer the North Pole experience cold weather and, due to limited exposure to the sun, they develop chimneys to warm up their homes. But due to the advancement of technology, an alternative to the chimney, a heater, can be found in homes using electricity. The development of a chimney or heater is inspired by the need for humans to fulfill this desire to answer questions that solve his problem. Advancement and development from the chimney to the electric heater would not been achieved without cooperative communication.

Rapidity in Technology Research

Language could be said to enhance studies in technology. A United Nations report (1953, p. 11) and several scholars assert that learning is more effective in one's mother tongue (NgugiwaThiong'o, 1994; Skunabb-Kangas & Philipson, 1995; Prah, 2003; Benson & Kosonen, 2013; Okonkwo, 2014; Babaci-Wilhite, 2015). In local and international contexts, language is an object of study. Close observation in countries like China, Japan shows that the use of local languages in school improves learning and the development of technological capabilities (Odinye & Odinye, 2012).

Based on individuals' local needs and their quest to solve local problems or to make their lives easier, institutes of technology have been the source of great support. Through the use of language, this skill of enhancing technology has transferred from one generation to another. And through cooperative communication research fellows and students of technology have been able to develop these needs of man as well as lighten the burden of labour that slows down the efficiency of the society.

Originality in Invention

Language is at the core of one's cultural background. Due to the influence of colonialism and perhaps globalization, most cultures have lost originality in the development of technology. Most people in responding to the question '*keduiheanyiga- eme*' (*What shall we do?)* have sought the aid of their colonizers to solve local challenges. Language, especially one's mother tongue, could enhance thinking and inventions to solve life's problems from an original perspective.

On this point/subject, Iwara (2011, p. 11) stresses that:

> Using foreign language such as English or French at this early stage of the child´s learning… severely restrict the involvement of the child in the learning process and forces him into a sort of passivity, as he is silenced by the limits of his vocabulary.

We can observe that in developing countries thought patterns are narrowed to those of the colonizers and consequently the thoughts of the colonized can never be original. This classification is exemplified by today's global terminologies, such as 'producing' and 'consuming' nations. This classification makes it look more like 'thinking' and 'tutored' nations. Prior to the advent of the colonizers, the soon-to-be colonized had a form of technology that was to be harnessed and developed. These technologies ceased to take shape due to the fact that most of these were either tagged 'inferior' or 'obsolete.'

It is therefore through language as a primary feature of cultural identity that humans identify problem-solving techniques that are locally acceptable and develop them to increase efficiency. The workability and productivity of this original technology could be of great help to the world at large.

Advancement of Communication Technology

Advancement of communication technology can be seen as the most concrete or physical importance of language to the development of technology. The quest to reach out as quickly and efficiently as possible has led mankind to develop several means of communication. In the past, birds, smoke or even metal gongs might be used to communicate, but this is not so in this present era. The quest to transfer, teach and understand all aspects of human language and communication has led to the question '*keduiheanyigaeme*?' (What shall we do?).

There are even innovations that could translate one cultural language to the other as used in international conferences to communicate to a multilingual audience. The difficulties faced by society are so numerous that even the quest to solve a problem is even a problem in and of itself. In order to seek knowledge and solutions to problems, man seems to have extended his cultural boundaries. These boundaries are not exceeded as a result of an inferiority complex of one culture vis-à-vis another but cooperative communication must exist between two cultures in order to find efficient solutions to problems. Language therefore, aids technological exchange among societies rather than dependency on another culture to solve our original cultural challenges.

LIMITATIONS IN THE DEVELOPMENT OF TECHNOLOGY THROUGH LANGUAGE

I have highlighted the importance of language in the development of technology in our society. There are then factors, which may hinder the efficiency of language in the advancement of technology with a focus discussion on three factors, namely, cultural monopoly, the dearth of invention, and the paradigm shift in the focus of language.

Cultural Monopoly

This can be identified as one of the factors with the potential to limit the efficacy of language in technological development. Cultural monopoly refers to the appraisal of one culture as superior to all others. This most time hides under the cloak of globalization. Ifesie and Ejobee (2013, p. 193) assert that:

> Although globalization is credited to technological, innovation, it would have been pretty difficult to contemplate a global village without the natural language. If nations could not communicate through languages, it would be difficult for globalization to begin.

Most cultures, especially in developing countries, are being lost. Highest casualties are found for indigenous languages from most developing countries. Some of these languages are either entirely lost or replaced by the language of colonizers or the varieties may have been Anglicized in Anglophone countries.

It is through language that actions are performed and with the loss of this, indigenous languages, cultural groups may begin to think in their adopted language. The problem with this is not just that a cultural identity is influenced in their adopted language but rather this culture can never live up to the thinking and cultural originality of the their adopted identity. Thus Fishman (2001, p. 6) writes:

> In our day and age, it is definitely the globalization of pan-western culture (and pop-consumer culture in particular) that is the motor of language shift. And since America dominated globalization (she) has become the major economic, technological and cultural thrust of worldwide modernization and westernization, efforts to safeguard threatened languages (and therefore, contextually weaker languages) must oppose the very strongest process and powers.

At the receiving end of this are indigenous technologies. Due to a shift in thinking, individuals have lost local understandings of how things are to be done to solve immediate, local problem. Consequently, lost languages become eternally dependent on the inventions and technological originality of a new adopted language.

Dearth of Invention

Reasoning from the idea of cultural monopoly, inventions have drastically decreased. That is not to say that there have not been technological innovations in the past years but rather we are looking at a society that may be heading towards the loss of Faradays, Archimedes, Stevenson, Fleming Gregor Mendel and many other inventors. What we have today is just a common remodeling of what has already been done, which is good in itself.

Through language our original actions are oriented towards making the world a better and more livable place. In other words, language should be redirected to train our minds, especially those of youth in schools, that many *issues demand attention in the world and language can be mobilized to respond and develop a course of action.*

Paradigm Shift in the Focus of Language

This is what we may call 'language revolution,' which tends towards a more negative approach in relieving a society's problems. This paradigm shift in language focus has led to a dearth of inventions. The general language of today emanates from technological inventions. In today's society, it is no longer language that shapes the societal thinking; it is rather the techne or technology that shapes society's language. Instead of language enhancing actions that bring about concrete realities, concrete realities shape the action that brings about language.

For instance, many of us have computers and mobile devices readily available. These are works of technological advancement in communication. They have facilitated and reduced the cost of our communications from one location to another (distant) location. On the other hand, people have turned to these devices as the determinant of daily language. Language has then been reduced to linguistic pleasantry rather a motivator for action. People spend time chatting on social networks about relationships rather than seeking concrete ways to solve their problems. The primary focus of language has now drifted away from its essence of reaching out to get things done or answer the question '*Keduiheanyiga- eme*?' (What shall we do?).

THE WAY FORWARD AND THE FUTURE OF TECHNELINGUAL COLLABORATION

From the foregoing analogy, we have observed the importance and the limitations of language in the development of technology. Language is the soul of technology and that all technology results from cooperative communication, which is encapsulated by the fundamental question: *Keduiheanyiga- eme*? (What shall we do?). The choice of using the Igbo language to express or ask this fundamental question, which provokes thought, enables dialogue and communication, and the quest for development and simple livelihood is more strongly felt when framed in a local perspective. One can then ask this same question in his or her own language and observe how personal and motivating this question can be. This is why the promotion of indigenous languages should be encouraged: humans think more efficiently in their own language, which they inhabit, than in a language of which they are still a learner.

The achievement of this 'technelingual' collaboration could lessen the downside of Information Technology in our present world where language lives in an abbreviated manner. Technology should not limit language; rather, language should enhance cultural assimilation of technology. Technology should be represented in people's identity and should encourage cultural exchange rather than a cultural independence.

Due to a lack of serious efforts to promote local languages, we hardly expect to witness the invention of local technology or its advancement in the global sphere. One learns to be creative through his or her cultural appreciation, but when one doesn't appreciate his or her culture, and especially his or her language, one lives to consume the finished pleasures of other cultures. In light of this, governments should encourage and enforce legislation that promotes local languages as official languages to be used in school. Foreign or more international languages could still be taught, but the language of instruction should be the local language. Local terminologies for raw materials, which might have otherwise been forgotten, will consequently be resuscitated. Further, this will enable all cultures to be explored in other to find solutions, which might be of help in other parts of the world.

CONCLUSION

Language is communication, and communication is a promoter of technology. The indivisibility of technology and language is a clear fact. Technology enhances the development of language, and language is the soul through which technology can be enhanced. We can acknowledge that the world seeks to break down barriers in order to solve problems, and the first step to breaking a cultural barrier is learning the core of a culture's identity, language. Language explains everything about a people and their immediate environment even without explicitly being taught.

We can observe that all forms of life have their own language even technology has its own language and terminologies. Every society or community seeks a language to communicate and with this they preserve their consciousness for the society and spirit to explore within its cultural or social context. The utility of language is not just of import for technological efficiency but for all spheres of development in life.

REFERENCES

Babaci-Wilhite, Z. (2015). *Local language of instruction as a human right in education: Cases from Africa* (pp. 1–154). Rotterdam, The Netherlands: Sense Publishers.

Babaci-Wilhite, Z., Geo-JaJa, M. A., & Lou, S. (2014). Localization of introduction of instruction as a right in education: Tanzania and Nigeria. Language-in-education's policies. In Z. Babaci-Wilhite (Ed.), *Giving space to African voices. Rights in local languages and local curriculum* (pp. 1–217). Rotterdam, The Netherlands: Sense Publishers.

Benson, C., & Kosonen, K. (Eds.). (2013). *Language issues in comparative education: Inclusive teaching and learning in non-dominant languages and cultures* (pp. 1–314). Rotterdam, The Netherlands: Sense Publishers.

Fishman, J. (2001). Why is it so hard to save a threatened language? In J. Fishman (Ed.), *Can threatened languages be saved?* (pp. 1–22). Claverdon, England: Multilingual Matters.

Fromkin, V., & Rodman, R. (1974). *An introduction to language*. New York, NY: Holt, Rinehart and Winston Inc.

Ifesieh, E. C., & Ejobee, L. (2013). Globalization and vanishing voices: A case study of Ughorbo. *Journal of Nigerian Languages and Culture, 14*(1), 191–212.

Iwara, A. U. (1981). Mother tongue education: Problems and prospects in a post-colonial African State: Nigeria. *Presence Africaine, 119*(3), 90–108.

Iwara, A. U. (2011). Language, culture and education in the Nigerian environment. *Journal of Nigerian Languages and Culture*, *13*(2), 1–20.

Magda, K. (1964). *Heidegger's philosophy: A guide to his basic thought.* New York, NY: The Macmillan.

NgugiwaThiong'o, J. (1994). *Decolonising the mind: The politics of language in African literature.* London, England: James Curry Ltd.

Odinye, I., & Odinye, I. (2012). Western influence of Chinese and Nigerian cultures. *A New Journal of African Studies*, *9*, 108–115.

Okonkwo, J. I. (2012). *OkwuDanahuOnu: The basic principle of Igbo philosophy of language* (Series No. 6). Owerri, Nigeria: Imo State University Inaugural Lecture.

Okonkwo, J. I. (2014). Appropriate language in education: The strategy for national development in Nigeria. In Z. Babaci-Wilhite (Ed.), *Giving space to African voices: Right in local languages and local curriculum* (pp. 131–146). Rotterdam, The Netherlands: Sense Publishers.

Prah, K. K. (2003). Going native: Language of instruction for education, development and African emancipation. In Z. Brock-Utne, Z. Desai, & M. A. S. Qorro (Eds.), *Language of instruction in Tanzania and South Africa (LOITASA)* (pp. 14–34). Dar es Salaam, Tanzania: E & D Publishing.

Skutnabb-Kangas, T., & Phillipson, R. (1995). *Linguistic human rights, overcoming linguistic discrimination.* New York, NY: Walter de Gruyther.

Tardinaco, S. (2015). Is social media sabotaging real communication? *Forbes.* Retrieved December 13, 2015, from www.forbes.com/sites/.../is-social-media-sabotaging-real- communication/Last

Thomas, M. (2009). History of technology. In J. K. B. Olsen, S. A. Pedersen, & V. F. Hendricks (Eds.), *A companion to the philosophy of technology*. West Sussex, England: Blackwell.

UNESCO. (1953). *The use of vernacular languages in education.* Paris, France: UNESCO.

Basil Amarachi Okonkwo
Nnamdi Azikiwe University
Awka, Anambra State
Nigeria

ABBAS RASHID, IRFAN MUZAFFAR,
FATIMA DAR AND SALAEYA BUTT

11. THE ISSUE OF ENGLISH AS A MEDIUM OF INSTRUCTION IN PRIMARY SCHOOLS IN PAKISTAN

Learning English, Mathematics or Science?

INTRODUCTION

Pakistan emerged in 1947 in a linguistically diverse cultural context. Among other things, the dynamics of politics and power ensured that English and Urdu assumed the role of the dominant languages (see Rahman, 1996; Rashid & Shaheed, 1993). English, as part of the colonial legacy retained its status as the language of the elite. Urdu, too, had been provided greater space by the British, particularly in Punjab, which, though not the majority province, assumed the dominant role in terms of the power configuration in the new country. Also, in the course of the struggle for Pakistan, Urdu had been given a special status as a marker of identity. Currently, the Constitution recognizes Urdu as the national language and English as one to 'be used for official purposes until arrangements are made for its replacement by Urdu' (Constitution of Pakistan, 1973). As for other languages spoken in the country, the Constitution states that 'without prejudice to the status of the National language,' a province could take steps to ensure 'the teaching, promotion and use of a Provincial language (Constitution of Pakistan, 1973).'

The discourse around language in education was anchored in this broader context. By way of a measure for national unity, Urdu was advocated as the Medium of Instruction (MoI) in mainstream public schools even as elite schools continued to provide English medium education to the select few. And, there was always the customary nod to the regional languages, particularly in the context of primary schooling. So, the significance of the mother tongue for initial schooling was recognized but without according it any real status. In many areas where operationally the mother tongue is the MoI, the materials are in Urdu. In a differentiated set of policy recommendation, the Hamood-ur-Rahman Commission in 1964 proposed Urdu as the MoI from Grade1–10 for Punjab but Sindhi in Sindh schools (Mustafa, 2011-for a new edition see Mustafa, 2015).

During the All Pakistan Education Conference in 1947, Urdu was proposed as Pakistan's lingua franca. The first advisory board set up in 1948 to deliberate on and make recommendations with regard to the MoI appropriate to different stages of the education system made a strong argument in favour of teaching in the mother tongue

Z. Babaci-Wilhite (Ed.), Human Rights in Language and STEM Education, 179–204.

in the initial stages, pointing out that it was "universally agreed that until a child has learnt one language well enough, he should not be made to learn any other. That this language should be the mother tongue needs no argument" (Mustafa, 2011, p. 39). In 1951 the second education conference endorsed the idea of mother tongue as the MoI at the primary stage while recommending the adoption of Urdu as the national and official language.

In 1959, the Sharif Commission on National Education again recommended that the MoI up to Grade-5 should be the mother tongue, suggesting that Urdu should be made compulsory after Grade-3 so that it could serve as the MoI from Grade-3 to Grade-12 (Mustafa, 2011, p. 40). The first democratically elected parliament reiterated the importance of Urdu and English through the 1973 constitution, Article 251 by declaring Urdu to be the national language and English as the official language. With the advent of General Zia-ul-Haq's military regime in 1977, its quest for legitimacy also resulted in an increased emphasis on Urdu and it was declared as the MoI in all the state schools of Pakistan. The teaching of English was ordained to start from Grade 4.

In 1989, with a democratically elected government, another change in language policy was effected with a greater emphasis on English also for regular, non-elite, government schools and the teaching of English was made compulsory from Grade 1. There was further re-calibration at the policy level in 1992 when the decision over MoI—national or provincial language or English—was left to the provinces.

There was a pronounced shift towards English under the Musharraf government underlined in the National Education Policy (2009), which while leaving the choice of MoI for the primary level up to the provinces, nevertheless stipulated that English would be the medium of instruction for science and math classes from Grade-4 onwards, providing a grace period of 5 years for the implementation of the policy. The Punjab government took the decision to introduce English as the MoI from Grade 1, in 2009, in government schools. It was a strategy to be implemented in phases. By 2011, all government schools, as a matter of policy (partially withdrawn in 2013) were supposed to be teaching in English.

The MoI has thus remained a standing issue since the inception of Pakistan. The constant alterations in language policies over the years illustrate the indecisive stance of various governments and the apathy and indifference with which such a major issue has been dealt with. Ayres (2003) is of the view that despite the diversity of languages and ethnicity in Pakistan, the government has paid little attention to language policies. In the official discourse, while Urdu has remained prominent, English has remained the language of the dominant class—a language which ultimately ensures empowerment and brighter futures. Rahman (1996) argues the same and reports, "Despite the efforts of the Urdu proto-elite to promote the use of Urdu and the ruling elite's apparent support of these efforts, it was English which continued to be dominant till the end of the fifties" (p. 232).

It should be noted that the emphasis on English as the preferred MoI is now no longer simply a matter of government policy; it has broad social demand behind it.

On the one hand, English remains the passport to better jobs, and mobility. On the other, there is the widespread perception that an early start in English, not just as a subject or language but as a MoI enhances the prospects of becoming proficient in English.

So, in order to better comprehend the implications for learning of this policy in Punjab, a study was conducted in 21 schools distributed across 6 districts of the province. A similar, though not identical, study was carried out in Sindh. The province of Sindh, in comparison to Punjab, has not opted to adopt English as the MoI in its schools at the primary level. However, going by its sector plan, in line with the recommendation of NEP, 2009, the Sindh government aims to ensure that English is taught as a subject in grade 1 and adopted as a medium of instruction in grade 4. Again, understandably, we have the Sindh government aiming for the objective of making students proficient in English. In order to study the readiness of the Sindh government to promulgate the envisaged policy a situational analysis was done in 9 schools of Sindh across 4 districts. Both urban and rural schools were selected in each district.

We have brought together the findings of both the studies to assess in one case the effectiveness of policy where English was adopted as the MoI from Grade-1 and in the other the readiness of the province to undertake the task of teaching English as a subject in public schools from Grade 1 and the MoI in Grade 4.

HISTORICAL BACKGROUND

Language acquisition is a process which according to Krashen (1981) develops slowly. According to him in language acquisition, "the best methods are therefore those that supply 'comprehensible input' in low anxiety situations, containing messages that students really want to hear. These methods do not force early production in the second language, but allow students to produce when they are 'ready', recognizing that improvement comes from supplying communicative and comprehensible input, and not from forcing and correcting production" (pp. 6–7). Wolgaardt (2001) elucidates the importance of exposure to target language and is certain that successful language acquisition takes place through exposure to the target language. The stated philosophy of Krashen is further authenticated by Ellis (2005) who also claims that if learners do not receive exposure to the target language they cannot acquire it. In general, the more exposure students receive, the more and the faster they will learn. It is, therefore, imperative to maximize the use of a second language inside the classroom. Ideally, this means that it needs to become the medium as well as the object of instruction.

In the domain of language acquisition, there are, however, differing opinions on the use of L1 (mother tongue) and L2 (the second or foreign language) in English as second or foreign language classrooms. Atkinson (1993) agrees with the need to maximize L2 usage, and he states that "every second spent using the L1 is a second not spent using English—and every second counts" (p. 12). However, he also describes how to achieve a proper balance of L1 in the learning process, which can be done without depriving learners of valuable L2 input. This balance is achieved by measuring out L1 use in carefully considered doses according to four factors: (1) the students' previous experience, (2) the students' level, (3) the stage of the course, and (4) the stage of the individual lesson. Atkinson (1993) realizes this vital link between this mixed interplay of native and second language and remarks that, "the L1 can be a vital resource, and there is certainly no reason why any teacher of monolingual classes should feel that it is somehow 'wrong' to make use of it" (p. 13). Nation (2003) also opines that whenever a teacher feels that a meaning based L2 task might be beyond the capabilities of the learners, a small amount of L1 discussion can help overcome some of the obstacles. Garrett et al. (as cited in Mattioli, 2004) further argue that in countries where English is seen as having an imperialistic role the use of the mother-tongue is a signal to the children that their language and culture have value, and this will have a beneficial effect on self-perceptions, attitudes, motivation and, consequently, on achievement.

It is, however, seen that the complex exchange between L1 and L2 is further open to debate when L2 is taught through translation in L1. Duff (1989) expounds on the merits of translation as a language learning activity and describes how translation can help develop flexibility, accuracy, and clarity. Nation (2001), on the other hand, comes up with a strategy to use an L1 translation in combination with the use of word cards for the initial learning of vocabulary. This makes the learners learn vocabulary speedily. On the contrary, Murrah (2001) suggests that there is evidence that if translation is used to facilitate comprehension, it becomes a signal of a communication breakdown. Murrah mentions what Polio and Duff (1994) also confirm—that these moments are "genuine opportunities for students to negotiate meaning in the (target) language and to develop strategies to correct and adjust their communication" (p. 7). Polio and Duff (1994) state that if teachers resort to translating difficult target language (TL) items, students will be less likely to attend to the TL forms.

In countries where English is taught as a second or foreign language, there is still a strong inclination to teach it through what Long (1991) refers to as a form focused approach. Ellis (2005) too argues that traditionally, language instruction has been directed at developing rule-based competence (i.e. knowledge of specific grammatical rules) through the systematic teaching of pre-selected structures. Myles (as cited in Ellis, 2005) reports that curriculum designers and teachers need to recognize that this type of instruction is as likely to result in students learning rote-memorized patterns as in internalizing abstract rules. Brown (2000) too argues that rote-learning may not have a meaningful situation to be related to, neither for association with the learners cognitive structures nor with anything else than performance of results in classrooms.

The need, therefore, is to formulate strong language policies to mitigate ambiguities in language teaching and learning. Blake and Kramsch (2007) prefer to "use a language policy where students are able to use another language in addition to their native language and also achieve an international, intellectual identity as well as a national one" (p. 249). Lo Bianco (as cited in Pufahl, Rhodes, & Christian, 2001) reports that in Australia, one of the most successful aspects of foreign language education relates to the National Policy on Languages (NPL) which provides a framework for language education. The NPL has initiated pluralism in the languages being offered, supported projects for indigenous and first language education, led to policy development in each Australian territory, and resulted in the near-universal introduction of languages at the primary level. Pufahl et al. (2001) further report that, "one of the most influential policies with respect to foreign language learning is the status of languages within the school curriculum. In all European countries and in Canada, Kazakhstan, Morocco, and Thailand, at least one foreign language is compulsory for all students" (p. 12). There is thus a strong advocacy of teaching more than one language from the primary grades except for the US where, according to Phillips (2007), 92% of US college-going students still do not study a world language.

Research, therefore, supports teaching more than one language from the primary grades in order to produce an educated citizenry aware of the role of language and culture in the world and in human cognition (Brecht, 2007). At the same time research strongly advocates indigenous or heritage language being employed as the medium of instruction in schools. According to Cummins (2000), the language in which education is conducted is the language in which basic skills and knowledge are imparted into the population and the language in which the production and reproduction of knowledge is done. In this regard, studies by Salami, Setati and Adler, Ncedo et al. (cited in Jegede, 2011) have advocated the use of LI to supplement English. The proponents believe that the learning sequence is best formed when constructed along the following trajectory: students acquire basic literacy skills in the first language and communicative skills in the second. They then learn to read and write in the L2, in effect, transferring to it the skills they have acquired in the first. The pedagogical principles behind this positive transfer of skills are Cummins' interdependence theory (Cummins, 1991) and the concept of common underlying proficiency, whereby the knowledge of language, literacy and concepts learned in the LI can be accessed and used in the second language once oral L2 skills are developed, and no re-learning is required. Adding to this, Mansoor (2005) argues in favor of using L1 as the MoI with a second language introduced only when sufficient competence has been acquired in the use of first language. Cantoni (2007) argues in favor of the mother tongue as MoI by saying that "The arguments given for mother tongue as MoI in schools were more linked to the child and its needs, the individual and the local context, factors such as concept formation, cultural identity, closer relation between school and home and practical use after primary school" (p. 5) Christian (2007) reiterates the same:

> Students who enter our schools with native-like proficiency in a language other than English suffer academically because they receive instruction only

> in English. Research shows that these students could benefit from continued development in their native language, both academically and cognitively because they could develop high levels of proficiency in their native language. (p. 271)

Cantoni (2007) thus claims that the use of English as a MoI hinders the full participation of the pupils because it does not seem to provide comprehensible input, it does not seem to work as a tool for constructing knowledge in the content subjects.

While the focus of MoI should remain the mother tongue, English can still be taught from the primary level but with a different pedagogical focus. Pufahl et al. (2001) advocate a communicative focus for the teaching of a second language. They add that in Denmark, Germany, the Netherlands, New Zealand, Peru, and Spain, a focus on communicative and intercultural learning has not only stimulated a productive discussion of teaching objectives, methods, and underlying rationales that are now reflected in curricula and textbooks, but has also resulted in increased oral and written proficiency for their students. Bot (2007) campaigns for another approach towards the teaching of a foreign language. According to him "a foreign language has to be acquired as a by-product of some other activity, not as the central focus of the subject matter" (p. 276). Christian (2007) further puts faith in mother tongue instruction and strongly recommends recognition through awards for those who study heritage languages. Olsen et al. (as cited in Christian, 2007) refers to the practice of offering diplomas to students to recognize their bilingualism and illiteracy. These they claim motivate students to maintain and develop their heritage skills. Christian (2007) goes on to claim that a language policy that values languages would generate incentives of many kinds, both material and attitudinal.

Coupled with the need to stick to mother tongue instruction the crucial factor of well trained teachers, giving second or foreign language instruction goes a long way towards the successful development of language skills among students. Blake and Kramsch (2007) rightly point out that student learning in extended sequences of instruction is directly related to the quality of teaching and to their teachers' level of professional preparedness. According to Pufahl et al. (2001) in Morocco, English teachers are among the best trained teachers in the country. After a 4-year degree in English from a university or teacher training college, including one year of specialization in literature or linguistics, students spend a year studying language teaching methodology and getting practical training. Ellis (2010) emphasizes the need to have solid teacher training programmes for English teachers and expresses the need to have topics in such courses which consist of 'ideas' rather than models. He also advocates the selection of comprehensible texts for SLA courses as teachers mostly lack technical knowledge about SLA courses. Moreover, he opines that the topics covered in SLA courses need to be demonstrably relevant to teaching. Morgan and Zhao (2004) thus sum up the principles of language instruction by stating that it should be concerned with the learners' level, and engage the learners in activities or situations that require adaptation, by using teaching methods that actively involve students and present challenges, taking into account each individual's own preferences over method and style.

RESEARCH CONTEXT

The research methodology for the Punjab study followed a mixed method approach while the Sindh study followed a primarily qualitative approach. Using such qualitative approaches researchers are able to access highly detailed and rich descriptions of human behaviors and opinion (Savenye & Robinson, 1996). In the Punjab study the quantitative approach was used in order to conduct a detailed analysis of the bilingual classroom discourse. The following research questions guided the Punjab study where how were teachers using English as MoI in English classrooms and what were head teachers, teachers and parent's perceptions about using English as a language of learning and teaching in the classrooms as well as what insights could be developed about the effects of the policy shift to use English as the MoI. The following research questions guided the Sindh study where we addressed what were the perceptions of the head teachers, teachers and parents about using English as the MoI, what methodologies did teachers use to teach English at the middle school level and where the teaching and learning conditions at the middle school level apt for the transition from Sindhi or Urdu to English medium. The following data collection procedures were used in the two studies:

Observation of Classes

For the Punjab study, 126 Grade 4 classrooms were observed of which 42 were English language classrooms, the rest being evenly split between the subjects of Mathematics and Science. The classroom interactions between teachers and students were observed and recorded for entire sessions. In the Sindh study, nine Grade 8 English language classrooms were observed. The purpose of the study was to assess the preparedness of the schools to teach English as a subject from Grade 1 and to subsequently adopt it as the MoI. Each classroom observation was audio-taped and transcribed. The observation was based on carefully designed instruments by subject specialists and university based researchers in the relevant field. The observation tools used in Punjab were piloted in four schools of Lahore, two rural and two urban.

Semi-Structured Interviews

Semi-structured interviews were conducted to aid in the investigation of unobservable phenomenon which impact the teaching of languages in schools (Wellington, 2006). In Sindh these interviews were conducted with head teachers, teachers and parents to gather their opinion on the MoI policy and the repercussions if English was made the MoI. In Punjab, two kinds of semi-structured questionnaires were designed. Head teachers were asked about their opinion on the recent language policy and its advantages and disadvantages, experiences of implementing the policy, its impact on learning outcomes, the language of assessment tools and their recommendations. Similar questions were used for the teacher interviews with some additional queries with reference to their teaching practices in the classroom. These tools were piloted and revised accordingly. Interviews were conducted with 21 head teachers and 38 teachers in Punjab and 9 head teachers and 23 teachers in Sindh.

Focus Group Discussions

In Punjab, Focus Group Discussions (FGD) were conducted in 21 schools with a total of 152 parents with an average of seven parents in each FGD. In Sindh, these group discussions were held with 8 sets of parents in nine schools. The instrument included prompts aimed at getting parents to talk about the pros and cons of using English as MoI. The parents were encouraged to provide rationale for their opinions on the policy shift.

Sample Selection Strategy

In Punjab, 21 schools were selected which were distributed across 6 districts of the province. Since the policy initiative introduced by the government in 2009 targeted children in Grade 1at the time, students from grade 4 were selected for this study. The main aim of the study was to gauge the efficacy of the MoI policy that was implemented. Out of 23 districts of Sindh, only 4 more urbanized districts were selected for the purpose of data collection i.e. Karachi, Jamshoro, Sukkur and Khairpur. The main aim of the study was to assess the readiness of the province to adopt English as the MoI in the final year of middle school. Since English is introduced as a subject in Grade 6, the study focuses on Grade 8 on the assumption that the teaching environment supports the learning of English by this stage.

ANALYSIS OF OBSERVATIONS AND DISCUSSION

English Language Use by Teachers in Classrooms of Punjab

The salience of Urdu or English in Teacher language use in English classrooms was determined by calculating the percentage of utterances in the different languages. The review of data assessed if patterns of language use supported learning of English language. On average, English-only utterances constituted just over one-third of the total utterances. Nearly a quarter of all utterances were Urdu-only, while 41% were mixed language use with inter and intra-sentential code-switching between Urdu and English. Moreover, the data for both the incidence of English and Urdu words was also reviewed. To this end, the number of English and Urdu words used by the teachers were counted. Teachers spoke a minimum of 6 and maximum of 957 English words. With respect to Urdu, the minimum words uttered by the teacher were 22 words and maximum were 1854 words. In over 75% lessons, teachers on average uttered 360 English words and 631 Urdu words. It was found that Urdu was used more than English in the classrooms. English words were embedded within Urdu sentences and intra-sentential code-switching was common.

Teacher talk was mostly restricted to textbook reading, 46% in English, and behavioral commands [25% in English]. Questions and explanations constituted less than a third of class conversation. The questions were mostly close-ended and the explanations were short and usually followed by a prompt. Urdu was predominantly used as a host language for all the communication taking place in the classrooms. In English classrooms especially, this compromised the learning of English as frequent use of Urdu detracted from a focus on the target language, i.e. English.

English Language Use by Teachers in Classrooms of Sindh

The findings from Sindh revealed that the main language of instruction by teachers in classrooms was the mother tongue, i.e. Sindhi. It was Sindhi in Sindhi medium schools and Urdu and Sindhi in Urdu medium schools. In the English language classroom, English language was hardly used by teachers except for the text. Out of the 9 schools observed for this study English was the majority language (69%) used in an English classroom in only one school.

The findings from the four districts of Sindh revealed that Sindhi was a predominant language of practice and instruction. It was the MoI in 87% schools across Sindh. With this huge dominance of Sindhi language, having English as the MoI would certainly be a tall order. The enforcement of policy by making English the MoI seems difficult is set to threaten the cultural, social and intellectual heritage of Sindh.

English Language Use by Students in English Classrooms of Punjab and Sindh

In the classrooms where there was evidence of considerable communication from the student, the student responses were reviewed in detail to understand the ways in which students put various languages to use in their expressions. Generally, the communication of students with respect to teachers was limited. In all 42 English classrooms of Punjab, student communication mostly constituted a very small percentage of classroom talk. To understand the purpose of student communication, the student utterances by language were reviewed. It was found that 59% of the total student communication was in the form of English sentences from textbooks. Other full sentences/utterances spoken in Urdu were about 20% with about 8% in English. There was 13% code-switching which meant a mix of Urdu and English sentences. There was no evidence of meaningful conversation, for instance, using "appropriate expressions to express regret," as stated in the SLOs for Grade 4. Students usually gave limited one word responses to questions. The findings of classroom observations in Sindh revealed the same trend. Students were found to be extremely weak in English language. The only instance of language use was reading aloud from the text and answering 'yes' or 'no' to close-ended questions.

English Language Use by Teachers in Mathematic Classrooms [1]

The language in use in by teachers in Math classrooms was mapped in 42 Math classrooms.

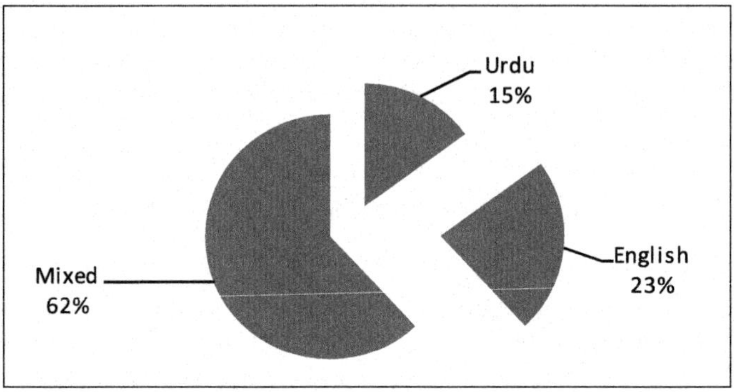

Figure 11.1. Teacher language use in Math classrooms

Figure 11.1 provides an overall distribution of languages used by teachers. A more detailed look at teacher utterances reveals that most teachers did not use English as frequently as Mixed language. In over 75% of lessons, teachers made no more than 14 English utterances. A mix of English and Urdu, however, was predominantly the de facto MoI. In Math lessons, this mix consisted of [mainly Mathematical] English terms inserted within Urdu sentences. Thus, the teachers were responding to the requirement of using English in teaching of Math mainly through the use of specialized English vocabulary.

Yet, as the data from interviews with teachers suggest, they remained somewhat reticent and uncomfortable in using English terms. An examination of teacher utterances in English and mixed languages suggests that they mostly consisted of reading from the textbooks, passing explicit instructions, or asking [mostly closed-ended] questions.

A snapshot of how teachers used English and Mixed language has been provided in Figures 11.2 and 11.3. Rather than explaining difficult concepts, teachers were prone to providing a simple set of instructions about how to perform particular Mathematical operations.

As can be seen from the figures, teacher utterances involving dictation of steps for the execution of particular Mathematical algorithms are most frequent; 55% of total English utterances and 67% of all mixed utterances consisted of such instructions. An illustrative example has been provided in Excerpt 1.

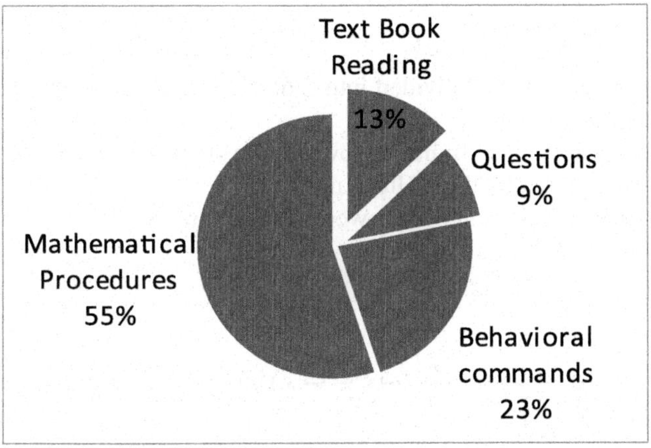

Figure 11.2. Types of teacher talk in English utterances (Math classrooms)

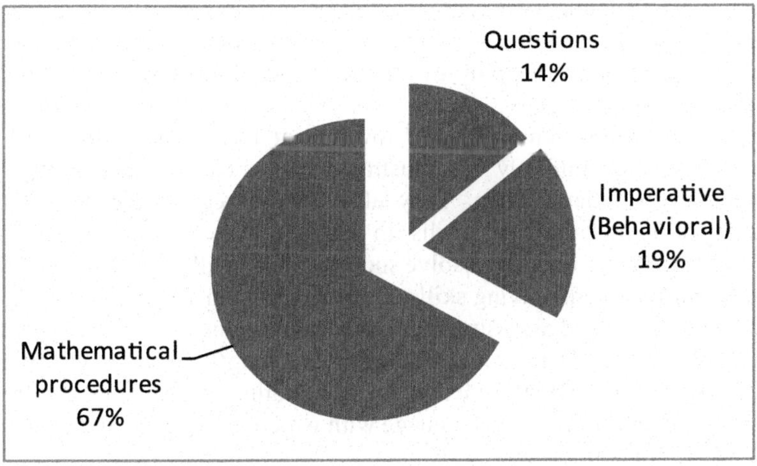

Figure 11.3. Types of teacher talk in mixed utterances (Math classrooms)

Excerpt 1

Teacher First of all we change the sign of division, firstly we replace the sign of division by multiply.
Interchange the numerator by denominator, interchange numerator & denominator

In other similar examples (refer to Excerpts 2 and 3), the teacher dictated exact steps to students to carry out division of one fraction by another fraction.

Excerpt 2

Teacher	Write down 7/10 divided into 5/6 Write down 3 are 2x3 are six 2x5 are ten 2x5 are 10 Write down here [pointing towards the board] three and you will multiply 7 multiplied by three Hurry up After cancellation our answer is twenty are by 25

Excerpt 3

Teacher	جب بھی آپ کریں گے تو آپ کی right side والی کی fraction ہوگی ۔ہو change ہوگی ۔

The choice of an excerpt from a lesson on division of fractions is illustrative of a trend whereby teachers use English to pass simple instructions to execute steps involved in Mathematical operations. Students are expected to remember these steps and follow them faithfully. There are no activities, no explanations, and no communications aimed at helping students understand the procedures that they are expected to implement.

The students, whose knowledge of division of fractions is restricted to steps, such as "invert and multiply" as illustrated in the excerpt, get no opportunity to develop deep understanding. They also do not acquire the capacity to use Mathematical knowledge and skills to solve real world problems involving division of fractions. Learning to solve such problems will require a whole range of knowledge and problem-solving skills. A good deal of instructional time should be spent on developing and assessing these important notions (refer to Figure 11.4 for further discussion of this issue). This requires meaningful communication between teachers and students. Yet such meaningful communication becomes difficult when attempting to communicate in a language with which the student, and in many cases the teacher, is not comfortable.

Figure 11.4 illustrates a problem in teaching at two levels: content knowledge and communication. The teachers do not appear to be encouraging students to think, develop conceptual understanding, and use problem-solving skills. In order to teach difficult concepts well, the teachers need to represent the concepts in terms of things that students are familiar with, make use of students' daily life experience to construct Mathematical problems, and provide multiple representations and explanations of difficult Mathematical concepts.

In addition to developing their subject knowledge, they also need to learn to communicate it in meaningful ways. Yet, this is where the problem also becomes intertwined with the choice of MoI. Meaningful interactions in the Math classroom require greater ability to fluently communicate in the language of learning and

Quality instruction must involve, at a minimum, following existing teacher guides. For every lesson, the guides encourage teachers to tailor instruction to respond to student learning outcomes. They are required to assess the existing knowledge of students through appropriate questions, design and implement purposive activities, conduct and scaffold classroom conversations in ways that meet the learning objectives for the lesson.

Reconsider the example of division of a common fraction by another common fraction. If the instruction largely involves dictating the steps in the division algorithm to students then this is what they are expected to remember and use. When presented with a problem in the following form:

$$\frac{18}{7} \div \frac{16}{21}$$

Students who remember these instructions will perform the inversion, convert the division form into a multiplication form, and find the answer as follows:

$$\frac{18}{7} \div \frac{16}{21} ==> \frac{18}{7} \times \frac{21}{16}$$

The ability to solve this problem depends entirely on student's ability to remember to invert the divisor and their knowledge of whole number multiplication. It does not require any conceptual understanding of division of fractions. Without such understanding, they are not prepared to solve problems of the form:

*How many halves are there in six fourth?*Or

The longest race on our tehsil sports day track is 3/4th of a mile long. The organizer wants to place a water cooler every 3/8th of a mile. How many coolers will be needed?

A proper development of concepts and relationships among fractions is essential for understanding operations on fractions. For this communication of concepts to take place, meaningful interaction in a language with which both teachers and students are comfortable with, is a pre-requisite.

Figure 11.4. Discussion of quality instruction in Math classrooms

teaching. Yet, if the current level of competence in English is any indication at all, it is highly unlikely that a switch to English as MoI will help remove hurdles in the way of quality of instruction in Mathematics.

Quality instruction must involve, at a minimum, following existing teacher guides. For every lesson, the guides encourage teachers to tailor instruction to respond to student learning outcomes. They are required to assess the existing knowledge of students through appropriate questions, design and implement purposive activities, conduct and scaffold classroom conversations in ways that meet the learning objectives for the lesson.

Reconsider the example of division of a common fraction by another common fraction. If the instruction largely involves dictating the steps in the division algorithm to students, then this is what they are expected to remember and use when presented with a problem in the following form:

$$\frac{18}{7} \div \frac{16}{21}$$

Students who remember these instructions will perform the inversion, convert the division form into a multiplication form, and find the answer as follows:

$$\frac{18}{7} \div \frac{16}{21} \rightarrow \frac{18}{7} \times \frac{16}{21}$$

The ability to solve this problem depends entirely on student's ability to remember to invert the divisor and their knowledge of whole number multiplication. It does not require any conceptual understanding of division of fractions. Without such understanding, they are not prepared to solve problems of the form:

How many halves are there in six fourth 7 Or The longest race on our tehsil sports day track is 3/4th of a mile long. The organizer wants to place a water cooler every 3/8th of a mile. How many coolers will be needed?

A proper development of concepts and relationships among fractions is essential for understanding operations on fractions. For this communication of concepts to take place, meaningful interaction in a language with which both teachers and students are comfortable with, is a pre-requisite. Data from SAHE's study (2013) was analyzed in greater depth with respect to math teaching and learning by Halai and Muzaffar (2015):

> Provided below is an extract from a lesson in class four. The teacher (T) introduced the topic of Highest Common Factor (HCF) by Prime Factorization and worked on the chalkboard to demonstrate to the learners (L) the procedure for deriving the HCF of 50 and 75 by Prime Factorization.

1. T: Bachonkal hum nay kyaparha? [Children what did we study yesterday?]
2. L: HCF (Chorus)
3. T: HCF kamatlabkyahai? [What is the meaning of HCF?]
4. L: Highest Common Factor (Chorus)
5. T: Aaj hum nay parhnahai HCF by Prime Factorization—kichotichotitajziyanbantihain. [Today, We have to study HCF by Prime Factorization – small small factors are made]

In line 2 above the teacher asked the learners to provide the "meaning" of HCF. But line 3 shows that learners simply gave the full name of the mathematical term HCF. Teacher's acceptance of the full name in English was symptomatic of an emphasis on learning mathematical names in English without necessarily probing the meaning that learners made of those terms. In line 5 the teacher made a pedagogic move by introducing the topic of "HCF by prime factorization." In the same line she stated that, "small small factors are made" (chotichotitajziyanbantihain). Presumably, "small small factors" referred to prime factors as 'small' because they cannot be further factorized. Of course, small (choti) can be interpreted in a number of ways and not all of them would lead to this conclusion. Additionally the word 'tajzian' has its root in tajzia that means to analyse or split apart. A use of tajzian could potentially provide the learners with a conceptual link to the notion of factors. It is noteworthy that an attempt to explain prime factorization, however limited and inaccurate, was made in Urdu.

To continue with the lesson above, interactions from line 6–25 (full transcript in Appendix A) showed that the teacher worked on the chalk board through the procedure of finding the HCF of 50 & 75 by taking their prime factors. She found the prime factors of 50 (2, 5, 5) and of 75 (3, 5, 5) and then the common factors (5, 5) and the highest common factor (25). Once completed she set the class to do similar work in their notebooks.

This is an instance of interactions that were dominated by the teacher and did not involve meaningful participation by the learners where they don't just learn mathematical procedures and their names in English, but could also have had an opportunity to learn concepts and mathematical relationships. For example, the teacher accepted the learners' response in line 5 and moved towards introducing the topic of the day 'HCF by Prime factorization.' However, it showed no evidence of learner's engagement with mathematics concepts, ideas and relationships around highest common factors and prime factors. While, some of the issues illustrated in the data are about pedagogy that emphasized procedures above concepts and relationships, they were compounded due to an additional effort required by the teacher and the learners to become familiar with mathematics terms in English.

English Language Use by Teachers in Science Classrooms

The use of mixed language at 53% was higher in comparison to English and Urdu utterances at 28% and 19% respectively. In over 75% of the classrooms, there were 25 or less English utterances and about 44 or less mixed language utterances.

Our coding of teacher talk data in Science lessons reveals similarities as well as differences with the use of language in Math lessons. As mentioned above, teachers used more English in Science lessons than in Math. However, on closer inspection this increase could be attributed to more frequent textbook reading in Science subjects. A difference was that teachers did actually try to explain the passages read out from the textbooks in Mixed language or Urdu. Hence more instances of inter-sentential code-switching were documented in the case of Science lessons.

Most English utterances consist of reading from the textbooks, while all explanations are given in Mixed language. These mixed utterances typically follow the reading of textbook passages in English and, therefore, representan instance of purposive or strategic code-switching (Moodley, 2007). Such explanations were about 57% of all mixed utterances, however, while teachers attempt to switch code to explain the textbook contents, the choice of words is not always appropriate. Teachers are not necessarily using the most appropriate English words, when using them within an Urdu syntax. There are also numerous instances of inappropriate insertion of English terms in instances of intra-sentential code-switching.

One manifestation of quality teaching is its ability to meet the Student Learning Outcomes (SLOs) stated in the National Curriculum Framework (2006). The SLOs typically require a great deal of communication from students. For instance in the case of a unit on movements of earth, the SLOs at Grade 4 level have been provided in Table 11.1.

Table 11.1. Movement of the earth

Earth	Describe the shape of Earth.
Earth's Spin Day and Night	Relate the Earth's spin with the occurrence of day and night.
Revolution	Define the term revolution. Identify that the distance between the Earth and the sun effects the time Earth takes to revolve around the sun.
Seasons	Explain that the Earth is titled on its axis and this tilt causes seasons.

The verbs relating to student learning in these SLOs are describe, relate, define, identify, and explain, whereas, the excerpts above suggest a teacher-centered instructional practice with limited opportunities for students to make their learning visible. This is the case regardless of the language in use. However the requirement to use English is apparently perceived by teachers as a serious issue to be addressed, pushing into the background the problems central to their instructional practices.

Teacher Talk Summary

In summary, the use of English is relatively greater in English classrooms but still only 36%, as compared to 23% in Math and 28% in Science classrooms (refer to Figure 11.5). Mixed utterances are about 41% in English, whereas there are about 62% in Math and 53% in Science. Urdu-only utterances are 15% in Math and 19% in Science in comparison to 23% in English.

Urdu was pre-dominantly being used as a host language for all the communication taking place in the classrooms in all the three subjects. In English classrooms especially, this is likely to compromise the learning of English as frequent use of Urdu detracts from a focus on the target language.

Teacher talk was mostly restricted to textbook reading [13 % in Math, 46% in English, and 61% in Science] and behavioral commands [23% in Math, 25% in English and 14% in Science] (refer to Figure 11.6 for details). Questions and explanations constituted less than a third of the talk. The questions were mostly closed-ended and the explanations were short and usually followed by a prompt such as "Have you understood?" Such prompts were usually met by "Yes miss" type of responses from the students. Such communication did not furnish any evidence about the actual

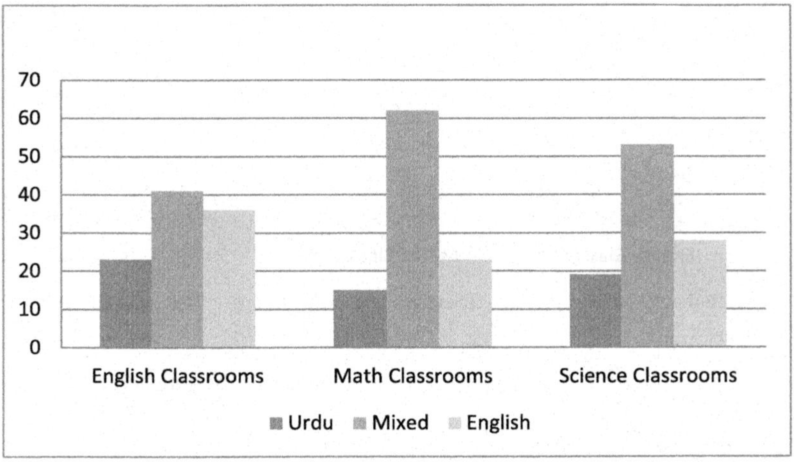

Figure 11.5. Teacher language use by subject

understanding of the students, nor did it appear to encourage them to express their understanding.

In English lessons, textbook reading was seldom accompanied by any discussion that would elicit meaningful responses from the students and enhance their ability in English. In Math, teachers used simple instructions by way of inserting Mathematical terminology in Urdu sentences to explain Mathematical operations without attempting to explain difficult concepts. In Science, teachers did try to explain the textbook passages after reading them but this was more in a form of translation without any meaningful discussion.

Teachers frequently switched back and forth between Urdu and English in all the three subjects. There is more inter-sentential code-switching in the English lessons and greater intra-sentential code-switching in the Math and Science classrooms. Given that in this study inter-sentential code- switching appeared mostly in instances of textbook reading, the greater proportion of this activity in English classrooms may explain this trend.

They also attempted to use specialized vocabulary i.e. certain subject specific terms inserted in Urdu sentences. However, the use of English words in Urdu syntax does not appear to be strategic or purposive. In fact when teachers code-switch, they are not necessarily using the most appropriate English words, within Urdu syntaxes then.

The competence of teachers in English came across as a major issue. Even when some teachers were able to explain grammatical structures in the English lessons, they remained reluctant to communicate in English. In the case of Science and Math, teachers' subject knowledge deficit combined with their lack of facility with the MoI to compromise the quality of teaching.

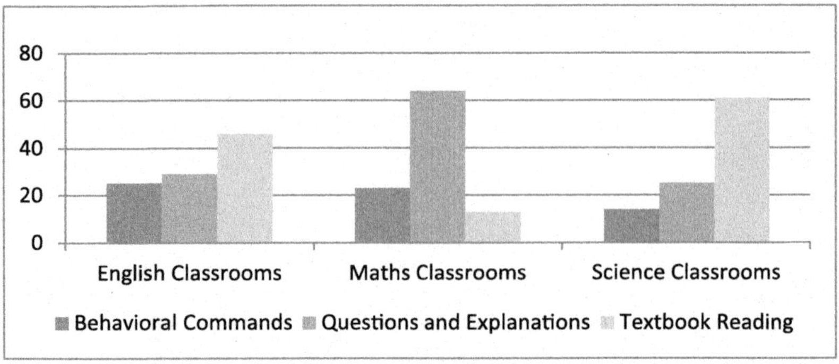

Figure 11.6. Teacher talk types in English across subjects

Perceptions of Head Teachers Regarding Medium of Instruction

In Punjab, head teachers had mixed perceptions of the benefits of the language policy for students. For a majority of head teachers the beneficiaries were mostly students of 'middle class' and 'educated parents'. Over 50% of head teachers interviewed suggested that only children coming from better-off families were at an advantage with this policy. Some head teachers also felt that 'intelligent' children were also benefitting from this policy. Understandably, children, whether of middle class or educated parents or naturally gifted, were few and far between. So a majority of children, according to the head teachers, were not benefitting from the policy. In Sindh, 8 head teachers were interviewed. While 7 head teachers forcefully recommended that the mother tongue should be the MoI, one of them put faith in English together with instruction in one's mother tongue. The head teachers considered Sindhi to be a language of identity and thought that education could best be given in it. Moreover, they considered the existing circumstances to be non-conducive to the shift from one's mother tongue to English due to:

- lack of teacher preparation
- lack of structural and human resources
- political interference in the hiring process of teachers
- weak primary education structure
- lack of innovative teaching strategies to teach English
- student absenteeism parental disinterest in child education
- lack of conducive and encouraging environment to learn English

Perception of Teachers Regarding Medium of Instruction

More than half of the teachers in Punjab (53%) interviewed were not supportive of the policy. One of the reasons provided for favoring Urdu or one's mother tongue as MoI included student's rural background, which according to teachers inhibited their ability to learn English without support from home. As one teacher said, "No, I will not favour this policy of enforcement to all, things here in urban areas are different and in rural areas are quite different." In Sindh, 23 teachers were interviewed. Out of them only three considered that English should be the MoI. These teachers were from districts Sukkur and Khairpur. The three teachers considered English to be a language of progress and human development. They also felt that English as the MoI in schools would remove the age old discrepancy between schools and colleges. In colleges, where the MoI was English, the students from Sindhi and Urdu medium schools faced an academic setback because of lack of English language awareness. However, teachers who favoured one's mother tongue reiterated the same problems as head teachers towards the transition from learning in one's mother tongue to instruction in English language. They strongly blamed the weak primary education structure as the main cause behind the lack of language competence among students together with lack of opportunities for teacher training. Most of the teachers revealed that students in Grade 6 did not know the basic concepts of key subject areas.

Perception of Parents about Medium of Instruction

In Punjab, 152 parents were asked to voice their opinion about the policy to use English as the MoI. Over 71% agreed with this policy (57% strongly agreed and 14% agreed), whereas, 22% of the parents did not approve of the policy. Those who approved of the policy appeared cognizant of the role of English in higher education and in the job market. In Sindh, 8 sets of parents were interviewed for the purpose of the study. It was interesting to note that out of them 6 set of parents strongly recommended English to become the MoI. Two sets of parents who advocated mother tongue to be the MoI were from urban Karachi district. The reasons that parents came up with in favour of English as the MoI were:

- better job opportunities for children
- better job prospects abroad
- the value of English as the language of power
- the importance of English for progress in life
- better alignment with college studies where the MoI was English

It was, therefore, interpreted that parents considered English to be a vital means for social and financial growth. For them it was a superior language the knowledge of which was sure to bring better prospects for their children.

Analysis of Classroom Observations

The observations of all classes in Punjab and Sindh revealed that there was "overuse" of L1 where the teacher and the learners having once switched to L1 continued to speak in L1 and therefore could not have a sustained interaction in the target language. In Sindh specifically, teachers mainly did not follow a lesson plan. Mostly, they taught through a teacher centered approach and used the following strategies to conduct classes:

- Teaching English through translation
- Teaching of English grammar through mother tongue
- Reading aloud of text both by the teacher and student
- Explaining of difficult words in the mother tongue
- Writing spellings of new words in note books
- Writing meanings of English words in mother tongue

The above mentioned strategies clearly pointed to the subject-based approach to teach English rather than a skill-based approach. Teaching English grammar through translation, explaining difficult words in one's mother tongue and writing meanings of new words in mother tongue spoke volumes of the language inadequacy of both the teachers and students. Moreover, they reflected the inability of the teacher to teach language through a skill-based approach where various skills of listening, speaking, reading and writing were used in a cohesive manner to augment language skills. Here the emphasis seemed only on reading and writing which indicated the use of grammar translation method towards the teaching of English. Yule (1996) described grammar translation method as one that involved learning of grammar rules. He further remarked that, "the emphasis on learning about L2 leaves students quite ignorant of how language is used" (p. 193). Krashen (1981) further argued that grammar-translation method provided little opportunity for acquisition and relied too heavily on learning.

It was also observed that reading aloud was considered the only method to ensure comprehension of the text. This misconception needed a comprehensive clarification, as reading aloud was not a strong determinant of comprehension. Scott and Ytreberg (2002) considered reading aloud to be an inefficient way to use class time. Moreover, they found it, "as a way of training and checking rhythm and pronunciation" (p. 58). Reading comprehension was ensured through silent reading. Scott and Ytreberg (2002) elaborated the point by saying that, "use the textbook to concentrate on conscious language development, but let your pupils read books for understanding and for pleasure" (p. 60). The strategies used in the classrooms clearly indicated a lack of opportunities for teacher training, a dearth of resources in schools, and a lack of language competence of teachers and students.

According to Krashen (1981) language acquisition did not require extensive use of conscious grammatical rules, and did not require tedious drill. Real language acquisition developed slowly. In the presence of traditional and ineffective methods employed by teachers it seemed unlikely that the proposed policy could be successfully implemented.

Analysis of Perception of Head Teachers, Teachers and Parents

A very high percentage of parents, i.e. 70% in Punjab and 75% in Sindh, approved of the new government policy, equating the greater emphasis on English with increased prospects for their children to learn English, a skill they understandably valued highly. However, a majority of teachers and head teachers found the policy of teaching in English from Grade 1 problematic. Those favoring the policy of English added a qualifier for the success of the policy: the government must put in place an enabling setting to ensure the success of this policy This put a question mark against their approval and which might be more indicative of their concern at the implications of dissenting from official policy as opposed to genuine endorsement of the policy within a pedagogical context.

CONCLUSION

One of the chief impeding factors that might arise towards the successful implementation of the stated policy would be teacher competence. The findings and analysis of the studies revealed the same and considered lack of teacher preparedness and training as the main hurdle behind the transition to English mode of teaching and learning. The teachers, specifically of English, neither had the desired skills nor the pedagogy to teach English. Language skills, both English and Urdu, were taught through grammar translation method. In the absence of the use of innovative and skill based methods to teach language, students displayed weak knowledge of language patterns.

Students in Sindh Would Not Be Able to Cope If Medium of Instruction Is Changed to English

In all the classes observed, the students were found extremely deficient in English language. The students had very poor English language skills and unable to follow English instructions without translation in one's mother tongue. The head teachers, teachers and parents revealed the same in their interviews and candidly pointed at the inability of students to cope with English language. One of the defining factors that led to weak language skills were poor primary education standards, dearth of innovative methods to teach English and lack of language competence of teachers.

The study also led to the conclusion that in Sindh especially students were not only weak in English but also in Urdu and Sindhi, primarily their L1s. This conclusion came as a result of observations and also authenticated further by head teachers and teachers in their interviews. Sindhi medium students started with Urdu instruction from Grade 3. Although they had a separate easy Urdu textbook to follow, they were still seen unable to cope with the language. The teachers pointed further that the same students had very weak Sindhi reading and writing skills. This struggle with languages gave clear indications of poor teaching and learning conditions, use of obsolete teaching methodologies and lack of teacher competence as possible reasons for this major linguistic breakdown.

Lack of Resources to Support Policy Implementation

The head teachers and teachers pointed out in interviews that schools lacked resources to initiate such policies. They disclosed that schools had ill equipped libraries, teaching materials and trained teachers. Most of the head teachers pointed at the fact that schools were not supplied with adequate number of teachers. The schools, therefore, could not fulfill the teaching demands. Teachers with specializations in other subject areas were seen taking language classes. In the absence of proper and adequate resources, the implementation of such policies seemed a far cry.

RECOMMENDATIONS

The studies elaborated in this chapter clearly pointed at the fact that the MoI must remain the mother tongue of the students. The teaching and learning conditions in those schools were not ready for an abrupt transition from the use of one's mother tongues to English language instruction. Mustafa (2014) supported the use of one's mother tongue as the MoI by stating that, "It is better still that children begin schooling in a language they understand, i.e. their home language. In this way, their cognitive development is facilitated and they also learn how to think" (p. 9).

A drastic change would further imperil the already faltering and emaciated education structure in the province. It is, therefore, recommended that the students' mother tongue be kept the MoI and meanwhile:

- Arrangements should be made to strengthen the primary education system
- Employ better trained teachers to teach all subject areas
- Employ teachers who are fluent in and have the necessary English language skills
- Schools be equipped with better structural and linguistic resources
- Libraries should be made well-resourced with books and interactive language learning materials
- Self-Access Centres be provided in schools for better access to learning materials
- Efforts be made to start or strengthen adult literacy programmes in the province to raise the literacy standards of parents
- School absenteeism should be controlled

It is strongly recommended that schools be equipped with better resources in terms of provision of well-resourced libraries, adequate teaching and learning materials, and better furnished classrooms. Moreover, the schools need to be equipped further with better qualified and trained teachers, and trained particularly in the teaching of English as a second or foreign language, if the policy is to be implemented in letter and spirit.

Appropriate curriculum and high quality textbooks need to be in place to help students learn the language. For this the curriculum needs to be reviewed again together with textbooks and examinations with the objective of teaching English as a second language. Easy to follow English textbooks with proper skill representation should be presented to students. Otherwise, no progress can be envisaged for such students.

The findings from parents' interviews reveal a strong desire for the learning of English. Some teachers also highlight the power factor attached to English language acquisition. The NEP, 2009 also makes a strong point in favour of English instruction as a way to remove social inequalities in the country. It is, therefore, recommended that English be taught as a communicative function from Grade 1 and not as a compulsive necessity. Hiep (as cited in Hunter, 2009) regards Communicative Language Teaching approach to be the best. The major problem lies with the age old 'custom' of teaching languages with a form focused approach with an abundant stress on developing language accuracy alone. Ellis (2005) too argues that language instruction has been directed at developing rule-based competence (i.e. knowledge of specific grammatical rules) through the systematic teaching of pre-selected structures – what Long (1991) has referred to as a focus-on-forms approach. Through this approach, learners stand a low chance of using it proficiently. In order to ensure prolific language use, it is important that it is taught with a communicative focus in initial years. The form can be stressed in later years. Skehan (1998) opines that proficiency in an L2 requires that learners acquire both a rich repertoire of formulaic expressions, which cater to fluency, and a rule-based competence consisting of knowledge of specific grammatical rules, which cater to complexity and accuracy.

An approach, therefore, which advocates language fluency in the initial years followed by accuracy later will thus ensure a more meaningful and expressive use of language. Richards and Rogers (2001) for this purpose advocate the communicative way of teaching language as the most successful approach as it aids continuous improvement in language. For this, the government of Pakistan may have to fulfill some vital preconditions of the provisions of adequate language learning resources, trained teachers with adequate knowledge of the required skills and a highly language-stimulating environment.

NOTE

1 CQE/SAHE, Policy and Practice: Teaching and Learning in English in Punjab Schools, pp. 21–27.

REFERENCES

Aryes, A. (2003). The politics of language policy in Pakistan. In M. Brown & S. Ganguly (Eds.), *Fighting words: Language policy and ethnic relations in Asia* (pp. 51–80). Cambridge, MA: MIT Press.

Bell, T. (1998). A description of the skill-based EAP training for pre-departure students at the British council in Jakarta. *The Internet TESL Journal, IV*(9). Retrieved May 24, 2014, from http://iteslj.org/Articles/Bell-EAP.html

Blake, R., & Kramsch, C. (2007). The issue: National language educational policy. *The Modern Language Journal, 91*(2), 247.

Bogdan, R., & Biklen, S. (1982). *Qualitative research for education.* Boston, MA: Allyn & Bacon.

Bot, K. D. (2007). Language teaching in a changing world. *The Modern Language Journal, 91*, 274–276.

Brecht, D. (2007). National language education policy in the nation's interest: Why? How? Who is responsible for what? *The Modern Language Journal, 91*(2), 264.

Brown, H. D. (1994). *Teaching by principles: An interactive approach to language pedagogy*. Englewood Cliffs, NJ: Prentice Hall Regents.

Brown, D. (2000). *Principles of language learning and teaching.* White Plains, NY: Longman.

Cabera, M. P., & Bazo, P. (2002). Teaching the four language skills in the primary EFL classroom. *The Internet TESL Journal, VIII*(12). Retrieved May 21, 2014, from http://iteslj.org/Techniques/Bazo-FourSkills.html

Cantoni, M. (2007). What role does the language play for a successful education? A case study of the impact of language choice in a Namibian school. *The Modern Language Journal.*

Christian, D. (2007). Charting a path to bilingualism: Policy and practice support. *Modern Language Journal, 91*(2), 271–273.

Cummins, J. (1991). *Language development and academic learning* (G. D. L Malave, Ed.). Clevedon, England: Multilingual Matters.

Cummins, J. (2000). Educational implications of mother tongue maintenance in minority language group. *Canadian Modern Language Review, 38*, 503–516.

Dat, B. (2003). Localizing ELT materials in Vietnam: A case study. In W. A. Renandya (Ed.), *Methodology and materials design in language teaching: Current perceptions and practices and their implications* (pp. 170–191). Singapore: SEAMEO Regional Language Centre.

Donato, R., & Tucker, G. R. (2007). K-12 language learning and foreign language education policy: A school based perspective. *The Modern Language Journal, 91*, 256–258.

Dudley-Evans, T., & St. John, M. J. (1998). *Developments in English for specific purposes: A multidisciplinary approach.* Cambridge, England: Cambridge University Press.

Duff, A. (1989). *Translation.* Oxford, England: Oxford University Press.

Ellis, R. (2005). Principles of instructed language learning. *Asian EFL Journal, 7*(3), Article 1. Retrieved from http://www.asian-efl-journal.com/September_05_re.php

Ellis, R. (2010). Second language acquisition, teacher education and language pedagogy. *Language Teaching, 43*(2), 182–201.

Gebhard, J. G. (1996). *Teaching English as a foreign or second language*. Michigan, MI: The University of Michigan Press.

Halai, A., & Muzaffar, I. (2015). In A. Halai & P. Clarkson (Eds.), *Teaching and learning mathematics in multilingual classrooms. Issues for policy, practice and teacher education* (pp. 63–64). Rotterdam, The Netherlands: Sense Publishers.

Hunkins, F., & Ornstein, A. (1998). *Curriculum: Foundations, principles, and issues* (3rd ed.). Boston, MA: Allyn & Bacon.

Hunter, D. (2009). *Communicative language teaching and ELT journal: A corpus based approach to the history of discourse* (Unpublished doctoral dissertation). Centre for Applied Linguistics, University of Warwick, England.

Jegede, O. (2011). Code switching and its implications for teaching Mathematics in primary schools in Ile-Ife, Nigeria. *Journal of Education and Practice*, *2*(10), 41–54.

Kitao, K., & Kitao, S. K. (1997). Selecting and developing teaching/Learning materials. *The Internet TESL Journal, IV*(4).

Krashen, S. D. (1981). *Principles and practice in second language acquisition* (English Language Teaching series). London, England: Prentice-Hall International (UK) Ltd.

Lenneberg, E. H. (1967). *Biological foundations of language*. New York, NY: Wiley.

Mansoor, S. (2005).*Language planning in higher education: A case study of Pakistan*. Karachi, Pakistan: Oxford University Press.

Mattioli, G. (2004). On native language intrusions and making do with words: Linguistically homogeneous classrooms and native language use. *Forum English Teaching Online*. Retrieved February 7, 2014, from http://exchanges.state.gov/forum/vols/vol42/no4/p20.htm

McMillan, J. H. (2008). *Educational research: Fundamentals for the consumer*. Boston, MA: Pearson Education Inc.

Murrah, S. (2001). *To use or not to use L1: Does native language have a role in the foreign language classroom?* Unpublished manuscript. Georgia State University, Atlanta, GA.

Mustafa, Z. (2011). *Tyranny of language in education: The problem and its solution*. Karachi, Pakistan: Ushba Publishing International.

Mustafa, Z. (2014, March 19). Rethinking the medium: Our ill-conceived language policies have divided society. *The DAWN*.

Nation, I. S. P. (2001). *Learning vocabulary in another language*. Cambridge, England: Cambridge University Press.

Nation, P. (2003). The role of the first language in foreign language learning. *The Asian EFL Journal*. *5*(2). Retrieved Februray 7, 2014, from http://www.asian-efl-journal.com/june_2003_PN.php

Penfield, W., & Roberts, L. (1959). *Speech and brain mechanisms*. Princeton, NJ: Princeton University Press.

Phillips, J. K. (2007). Foreign language education: Whose definition? *Modern Language Journal, 91*, 266–268.

Polio, C. G., & Duff, P. A. (1994). Teachers' language use in university foreign language classrooms: A qualitative analysis of English and target language alternation. *Modern Language Journal, 78*(3), 313–324.

Rahman, T. (1996). *Language and politics in Pakistan*. Karachi, Pakistan: Oxford University Press.

Rashid, A., & Shaheed, F. (1993). *Pakistan: Ethno-politics and contending elites* (Discussion Paper No. 45). Geneva, Switzerland: UNRISD.

Richards, J. C., & Rogers, T. S. (2001). *Approaches and methods in English language teaching*. Cambridge, England: Cambridge University Press.

SAHE. (2013). *Policy and practice: Teaching and learning in English in Punjab schools*. Lahore, Pakistan: SAHE.

SAHE. (2014). *Teaching and learning English in Sindh schools*. Lahore, Pakistan: SAHE

Savenye, W., & Robinson, R. S. (1996). Qualitative issues and methods: An introduction for educational technologists. In D. Jonassen (Ed.), *Handbook on research in educational communications and technology* (pp. 1171–1195). New York, NY: MacMillan.

Schweers, C. W. Jr. (1999). Using L1 in the L2 classroom. *English Teaching Forum, 37*(2), 6–13.

Scott, W. A., & Ytreberg, L. H. (2002). *Teaching English to children*. New York, NY: Longman.

Skehan, P. (1998). *A cognitive approach to language learning*. Oxford, England: Oxford University Press.

Wellington, J. (2006) *Educational research: Contemporary issues and practical approaches*. London, England: Viva-Continuum.

Wiersma, W., & Jurs, S. G. (2009). *Research methods in education: An introduction* (9th ed.). Boston, MA: Pearson.

Wolfaardt, D. (2001). *Facilitating learning: An investigation of the language policy of Namibian schools*. Cape Town, South Africa: University of Western Cape.

Yule, G. (1996). *The study of language* (2nd ed.). Cambridge, England: Cambridge University Press.

Zhao, A. H. Q., & Morgan, C. (2004). Consideration of age in L2 attainment – Children, adolescents and adults. *Asian EFL Journal, 6*(4), 1–8.

Abbas Rashid
Society for the Advancement of Education (SAHE)
Lahore, Pakistan

Irfan Muzaffar
Education and Social Research Collective (ESRC)
Lahore, Pakistan

Fatima Dar
Department of Education at the Shaheed Zulfiqar Ali Bhutto Institute of Science and Technology (SZABIST)
Karachi, Pakistan

Salaeya Butt
Society for the Advancement of Education (SAHE)
Lahore, Pakistan

PART IV

HUMAN RIGHTS IN MATHEMATICS AND STEM EDUCATION

UBIRATAN D'AMBROSIO

12. CHANGE IN SPACE, URBAN CULTURE AND ETHNOMATHEMATICS

ETHNOMATHEMATICS AND THE STATE OF THE WORLD

The advances in Ethnomathematics are shown in the amazing amount of publications and in the rich programs of five International Conferences on Ethnomathematics.[1] The *ISGEm/International Study Group of Ethnomathematics* was founded in 1985. Many books about Ethnomathematics are produced, journals and sites are created and events all over the World are organized.

Ethnomathematics is a rich research area. Its importance for Education in general is unquestionable. It is remarkably trans-disciplinarian and transcultural. It relies on research in various disciplines, particularly anthropology, ethnography, cultural studies, cognitive sciences, history and social dynamics. We are fascinated with the beauty and with the recognition and analyses of mathematical ideas present in professional and everyday practices, in folklore, in artisanship, in dressing tapestry and basketry, in games, music and dance, and even in cults of native, both in cultures situated in the periphery of the most developed centers, but also in the dominant cultures. This is fascinating. The books by Claudia Zaslavsky, Marcia and Robert Ascher, Paulus Gerdes and some others are classical references about the recognition and theoretical discussions of mathematical ideas present in artifacts from different cultures.

The increasing interest in the mathematical study of cultural artifacts results in various dissertations, publications, journals and sites. There are many results of careful research that do not get published, that are not even known to most researchers. But I will not comment on the beautiful advances of Ethnomathematics. Instead, I will reflect on the way I see the relation of Ethnomathematics with society as a whole and with the State of the World.

In an interview, conceded in 2010, the mathematician Gromov[2] reflects on the State of the World. He presents a vision that is really disturbing. Coming from such a renowned scientist, it deserves our attention. Gromov (2010) says:

> Earth will run out of the basic resources, and we cannot predict what will happen after that. We will run out of water, air, soil, rare metals, not to mention oil. Everything will essentially come to an end within fifty years. What will happen after that? I am scared. It may be okay if we find solutions, but, if we don't, then everything may come to an end very quickly! Mathematics may

Z. Babaci-Wilhite (Ed.), Human Rights in Language and STEM Education, 207–219.

> help to solve the problem, but if we are not successful, there will not be any mathematics left, I am afraid!

Even more frightening than the exhaustion of resources tomorrow is the destruction we are witnessing every day. Again, in the same interview, Gromov (2010) says:

> Being inside our *ivory tower*, what can we say? We are inside this ivory tower, and we are very comfortable there. But we cannot really say much because we don't see the world as well enough either. We have to go out, but that is not so easy.

Since we all are concerned with the survival of civilization on earth with dignity for all, we have to leave our Ivory Tower and get close to what is being done by the people in different cultural contexts, not only with the achievements and beauty of mathematics or Ethnomathematics. Mathematics has been the decisive intellectual instrument for weaponry, for an unrestrained form of capitalism and for strategies of domination. It has been used for greed, aggression, destruction and *killing*. We must look for a Mathematics that leads to an ethics of respect, solidarity and cooperation. This is the main concern of the *Center for Global Nonkilling*.[3] The volume on *Towards a Nonkilling Paradigm,*[4] has a chapter on "Nonkilling Mathematics?"

The Program Ethnomathematics (D´Ambrosio, 2006) is an analysis of the *tics* (techné, arts and techniques) of *mathem* (explaining, understanding, dealing with) in different *ethnos* (natural and socio-cultural environments).

The Program Ethnomathematics is a research program in the History and Philosophy of Mathematics and its pedagogical implications, focused on how and why the human species

- generates,
- organizes,
- transmits and diffuses

knowledge, particularly Mathematics.

The research methodology of the Program Ethnomathematics is based in three steps:

- how do *ad hoc* practices and solutions to deal with situations and problems develop into methods?
- how do methods develop into theories?
- how do theories result in innovation?

The three steps

$$\textit{ad hoc} \rightarrow \text{methods}$$
$$\text{methods} \rightarrow \text{theories}$$
$$\text{theories} \rightarrow \text{innovation}$$

synthesize the research program.

Research in Ethnomathematics draws from recent scientific advances:

- from Jean Piaget, Lev Vigotsky, Alison Gopnik (2009) and others, we learn to observe and listen to children;
- from neuroscience we learn to monitor synapses;
- from history we learn the dynamics of cultural encounters.

But also relies on mathematical ideas recognized in oral and written narratives, some forgotten, lost or repressed and others implicit, although unnoticed, in academic sources, in folk tales and artifacts, in mythologies and in fiction and, of major importance, in the concepts of time and of space occupation and in urban planning.

In this paper I reflect on space occupation and on urban culture.

THE ETHNOMATHEMATICS OF URBAN SPACE

From Antiquity through Middle Ages, in the development of the earliest cities in Mesopotamia and in Egypt, as well as in all the regions of the World, there was an equilibrium between the rural area, where the vast majority of the population was engaged in substance agriculture and cattle raising, and the urban areas. Emergent cities where small groups of populations engaged in economic activity consisting primarily of trade at markets and small scale manufactures and artisanship. The emergence of Modern Science and of the industrial revolution affected agriculture, demanding less rural population. This reflected in the migratory flux of population from the countryside to the urban areas. This migratory pattern intensified and continues until nowadays. This requires planning of urban growth.

Architects, planners, and sociologists analyze the way people live in urban areas from many perspectives, including a sociological perspective, and how technological developments in transportation and communication affect the urban mode of living. Planning urban development includes infrastructure and special attention to aesthetics, lifestyles and labor structure, modeled in geometric organization of space. Quantitative factors, such as time of displacement in the city, access to public and service areas, and socio-political issues, such as social and racial division, intervene in urban development.

The rapid growth of population is faster than planning urban development and has an effect of a sort of spontaneous urbanization, as discussed by von Osten (2009). According to him, the unplanned growth focuses primarily in residential structures for individuals and families, which are built not necessarily in legal property. This is a serious concern both in developed and developing countries. To build residential structures, material resources are dealt with a variety of ways of doing and knowing. It is to be also noticed that modern city planning is bound to colonialism and imperialism. Many technical developments were even tested and realized on colonial ground, subordinated to global political and economic structures, pressing for the adoption of capitalist forms of production. Modern cities reflect the administrative, governmental, productive and commercial sector.

The geometrically planned and surveyed town, with retangular blocks of houses and regular streets at right angles, was an elaboration of mathematical concept, such as the way to deal with space and forms, with measurements and quantities. These concepts tended to disregard social and environmental limitations, which are present in the natural evolution of the primitive village. This was also present in the Greco-Roman concept of *polis/urbs*, as it will be shown below. This leads us to observe the reality of unplanned growth and new forms of dealing with space and forms, with measurements and quantities. This leads to new mathematical concepts, which I call Ethnomathematics, which will be discussed below.

I had the personal opportunity of noticing the difference of styles of urbanization and the colonial influence back in 1970 when I was engaged in the educational project, called Project Mali-1, sponsored by UNESCO in the Republic of Mali. I visited many regions of the country and I was impressed by the urban concept of the Dogon villages, as well as the majestic urban structure of Djenne. This was in strong contrast with the colonial urbanization in cities like Bamako and other colonial administrative centers. Although notspecifically related to the the of this work, I believe it is of interest to relate my experience as faculty of the graduate program in the Project Mali-1, since it was a successful initiative in higher education in Africa, serving as a model for the entire world, in particular at the graduate level, with considerable reflexes at the undergraduate level. I will treat this as an Appendix to this work. My early ideas for the Program Ethnomathematics are a result of the experience. Occupation of land has been, and continues to be, a major characteristic of cultures. Towns and urban design are arranged focusing religious and political hierarchy and play a fundamental role in society. Since early dwelling until nowadays, the mutual influence of urbanization and mathematics is evident. Both *polis* and *urbs* reflect power structure and political organization.

The Greek *polis* and the Roman *urbs* are concepts of citizenship, as seen by contemporaneous chroniclers, not as a sort of fortresses, as seen by some later observers. The classic book *De Architectura*, by Vitruvius (1st century BCE), deals with urban planning and reveals concepts of time and space, as well as the power structure, of the Roman Empire. The way in which urban centers are understood, planned and described throughout history reveal the supporting science and technology, particularly Mathematics, of the period. Urbanization reflects the organization of space and time, particularly by giving sacred attributes to space and time. Descartes (1637, p. 83) comments on urbanization:

> The old cities, which were at their beginnings nothing but small towns, and have become in time large cities, are generally not well designed as by compass, at the cost of the regularity of their squares that an engineer traced following his fantasy on the plane.

Such remarks can be very well illustrated by design of cities from different periods of civilization, with characteristics that reflect the religious, the social and the political circumstances. Mathematics may even be inspired by urban design.

For example, Descartes studied in the *Collège de la Flèche,* spending there his youth. The design of the gardens of the *Collège* might have been a source for inspiring the coordinate methods of Descartes.

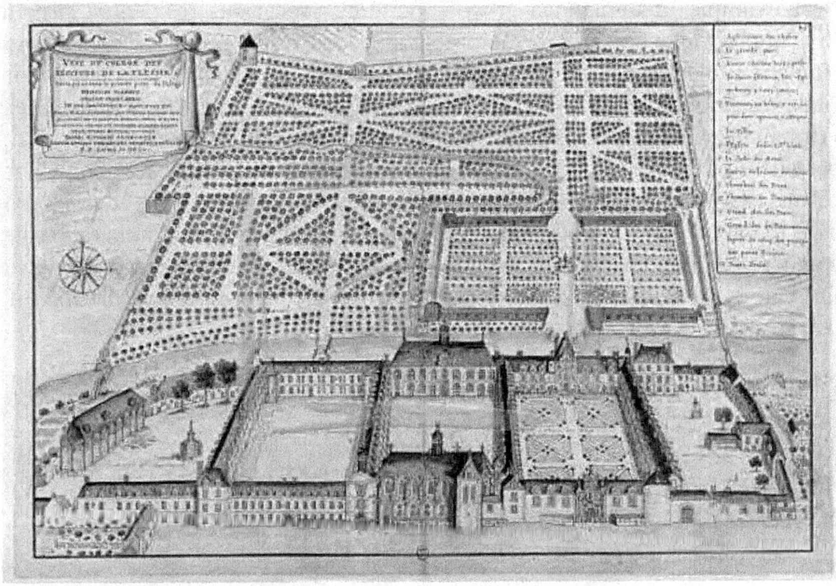

> The grounds and buildings of the *Collége de la Flèche* [where René Descartes studied], perfectly symmetrical and square in their design, seemed constructed as if by straightedge and compass. Descartes' writings later in his life … make it clear that he was impressed by symmetry and straight lines in the design of buildings and towns. (Aczel, 2005, p. 32)

As mentioned above, the geometrically planned and surveyed town, with rectangular blocks of houses and regular streets at right angles, was an elaboration of Mathematical concept, such as the way to deal with space and forms, with measurements and quantities. These concepts, differently than the natural evolution of the primitive village, disregards social and environmental conditions.

The concept of urban planning is not part of any Mathematics curriculum, although the perception and the consequences of the urban reality in daily-life affect every individual, children and adults. Reflections about urban reality stimulate concerns and imagination.

I will give some examples of research about the concerns and imagination resulting from living in specific urban realities. The imagination may lead to reflections of a Mathematical nature. The important research of Sonia Maria Clareto (2009) illustrates the subordination of common sense in urban planning to economic interests. She studied the emergence of the community of *Laranjal do Jari*, founded in 1967, in the State of Amapá, in the rain forest, in the margins of the River Jari, in Brazil, as a village to house workers of an enormous industrial complex. The workers built their houses over the River Jari. The community was legally established as a city in 1987 and today has a population estimated in 50,000 inhabitants. This is a clear example of unplanned growth, building precarious residential structures, with inappropriate material resources, a response to pressures of an ambitious capitalistic project. The industrial complex was part of the ambitious project of the billionaire Daniel Ludwig (1897–1992), established in 1967, with the agreement of the Brazilian Military Government, then in power, to build what would be the largest paper mill in the World. The area, in the Amazonian rain forest, was equivalent to the State of Connecticut. The Project became known as Project Jari. It was a project similar to the project of the fictitious RDA Corporation to explore mineral resources of Pandora (*Avatar, the movie*). As a result of unplanned urban and family growth, large sectors of the population are children that live in what plays the role of the "main street."

The exemplary research of Clareto was to understand children perception of space and of geometry in this urban environment. We can guess that something similar to Aczel's (2005) observation about the possible inspiration of Descartes in creating the coordinates system, may affect the future of these children. Their perception of space is unique. A basic question is: how do these children build their perception of space in such an environment? How these children, with no or only minimal schooling, generate the Mathematical strategies for survival in that environment?

We may pose the same question about children living in a modern urban environment, like São Paulo. Jesuit missionaries founded the city in 1554. It is now the largest metropolis of the Southern Hemisphere, and the most important financial, corporative and mercantile center of Latin America. The metropolitan region of São Paulo has a population of about 20,000,000 inhabitants.

As a result of unplanned urban and family growth, children are a large sector of the population. Many live in the streets, in miserable conditions, subject to criminality, drug dealing and drug addiction.

The question we pose is not different than the one posed by Sonia Maria Clareto in her thesis. How do children, with no or only minimal schooling, generate the Mathematical strategies for their survival in the streets? How do they socialize these strategies? How do they deal with time and space, with quantifying and measuring, with money? In other words, what is the Ethnomathematics of street children in a large city?

Mônica Maria Borges Mesquita (2011) did extensive research on the life of street children in São Paulo.[5] Besides Clareto and Mesquita, I mention also the dissertation of Joselita Macedo Filha, in which she establishes a methodological dialogue with the movie *Dances with Wolves* (Kevin Kostner, 1990). She applies Ethnomatematics as a research methodology to analyze the behavior and knowledge of street children and adults living "permanently" in a public plaza, in the city of Salvador, Bahia, founded 1549, which has now a population of about 3,000,000 inhabitants. This dissertation was printed as a book in Portuguese with a title meaning *Dances with Wolves – The street of street children.*[6]

Another strand for Ethnomathematics of the urban scenario results from immigration. Legal or illegal, immigrant population is a fact and immigrant children are a reality in the school systems all over the World. The immigrant population asks perspectives of Education for their children, particularly in Mathematics Education, since Mathematics is perceived as the door to social access, as a key instrument for citizenship. This leads, naturally, to reflections about the knowledge associated with the cultural roots of the immigrants, hence about their Ethnomathematics. There is much research on Ethnomathematics of immigrant populations (see endnotes).[7]

PERSPECTIVES FOR THE FUTURE

The world is incessantly transformed by the revolution in information and communication. World Wide Web establishes an entirely new appropriation of time and space and new human connections. The network society raises new issues, problems and possibilities for the economic, cultural and political world scenario. Indeed, networking characterizes a new civilization.

Individuals today acquire, virtually, new social, political and cultural habits of mind and ways of being. Traditions are challenged by new world-wide perspectives. People save, spend and optimize time in a different way. Design and implementation are no longer spatially and temporally fixed.

The concepts of time and space, which I mentioned as essential in the beginning of this chapter, are changing rapidly. Traditional ways of dealing with time, which evolved thanks to the development of the clock, are changing rapidly. Traditional ways of communicating location is changing. The @-addresses compete with postal addresses. Time, which refers to when, and space, which refers to where, acquire new meanings. We all know that time and space are essential components of mathematics. Hence, new concepts of space and time, which are the imprint of civilization, imply a new Mathematics.

Globalization acquires a new meaning, since it instantly overcomes boundaries and language barriers. The new Mathematics must respond to the new meaning of globalization and will, necessarily, be trans-disciplinarian and transcultural. Ethnomathematics may be the response.

APPENDIX: THE PROJECT MALI-1 OF UNESCO IN BAMAKO

The project known as Project Mali-1 of UNESCO, was conducted in the decade of 1970 to establish the *Centre Pédagogique Supérieur de Bamako*. This was a successful initiative in higher education in developing countries, in particular at the graduate level, with considerable reflexes at the undergraduate level. Since its formal inception in 1970, the program granted many *Docteur de Spécialité*, with a level higher than the French *3ème cycle* French or than a strong degree of *Master of Science*, with total identification with the educational issues of with their role in the process of development of Mali.

I had the pleasure of sponsoring seven students who finished their doctorates.[8] I always remember my surprise when I received, in 1970, an invitation from UNESCO to act as a consultant in a pilot project aimed at developing an *in-service* (*sur place*) graduate program in Mali. The invitation came from an officer of the Department of Education of UNESCO, the laureate Congolese poet Tchicaya-U-Tamsi (his official name in UNESCO was Gérard Félix-Tchicaya). Although not formally trained as an educator, Tchicaya had a most impressive educational vision. Allied to the creativity typical of a great poet, he was able to understand the needs and the desire of Malians and to promote a very original project that became known, in UNESCO headquarters, as Project Mali-1. I was fortunate to share friendship with Tchicaya and to learn much about the History and Philosophy of Africa and to appreciate the the dimension of a poet who was an humanitarian educator.

At that time I was Director of Graduate Studies in Mathematics at the State University of New York at Buffalo and was very concerned with teacher training for universities located in less developed regions of the United States, especially the so-called "black colleges," responsible for much of college education in Southern States. The formula usually adopted was to provide fellowships for the black students and send them to the best universities. They were treated in a "special" way and were granted a degree and returned home. Those who were very good, which was rare in view of the inadequate training offered by their universities of origin, received special "invitation" to stay in the best universities, amidst the known demand for blacks faculty in American universities. Others or gave up and returned, frustrated and poorly prepared, to their universities of origin, continuing the cycle. Concerned with this approach to the problem, I had submitted to some foundations a plan to offer them a program that would be taylored to the special needs of the students. Although highly praised, the proposal did not get the needed resources. So, I was extremely happy when realized strong similarities between my plan and the pilot project which later became known as "CPS-Bamako."

A quick analysis of the situation with regard to the African higher education was sufficient to identify the similarity with the above mentioned framework. There was still the case that the African intellectuals sought to preserve its cultural tradition amid Western values, pressured by the urgent need to provide their people with technical and scientific baggage necessary to overcome the barrier of underdevelopment without destroying those precious values of their culture. This is difficult to succeed when an individual stays away of the reality of his/her country for many years, normally in an age bracket in which the individuals begins to have personality and ability to analyze problems that affect their social and natural environment. After many years away from their countries, the students become "dépaysées," in the language of the UNESCO experts.

Designed within this structure, the *CPS* project was presented as a possible solution to the need to prepare teachers for the *École Normale Supérieure*, institution to prepare teachers for secondary school in Mali. At that time, the *École Normale*

Supérieure was largely dependent on European universities, mainly from France, who offered regular support for high education through *coopérants.* This was the continuity of colonial times. An alternativa was a program of scholarships for training of Malian students in Europe, which did not produce the expected results, even after ten years of independence. In most cases, Malian students educated abroad returned without adequate training to assume their responsibilities under completely different from those they had found at the University where they had studied. Consequently, they soon felt disenchanted with the profession. Thus, the number of Malian teachers at the *École Normale Supérieur* was only a small portion of those who had done their graduate studies in developed countries, since the vast majority remained abroad or assumed other functions in Mali. There is no doubt that this group plaied a key role in the country, but the problem of providing faculty for the *École Normale Supérieure* was not resolved. This was a problem not limited only to the *École Normale Supérieure*, since other higher education institutions were in the same situation.

Creative and courageous, Malians decided to tackle the problem forming, in their own country, the future faculty for its higher education institutions. To do this, they have adopted a formula of intensive training at the graduate level, largely dependent on the collaboration of visiting professors, which would remain in Bamako for three or four weeks three or four times per year. The commitment of the authorities of Mali awakened the enthusiasm of the UNESCO staff and the project get started in the late 1960s.

The *Centre Pédagogique Supérieure* was created as the Graduate School of the *École Normale Supérieure*, which aimed at the training of teachers for secondary and technical schools. In addition preparation of research personnel for some government offices was conducted. Admission to the *École Normale Supérieure* required passing a comprehensive high school exam (*baccalauréat*), and its graduates receive a *diplôme de l'ENS* at the end of four years of study. It was structured as an multidisciplinary faculty, with eleven areas of specialization: French and Literature in General, Philosophy, Psychopedagogy, History and Geography, English, German, Russian, Biology and Geology, Mathematics, Physics and Chemistry, and Natural Sciences. Further, the *École Normale Supérieure* represented an option taken by the Malian authorities in the profound educational reform of 1962, following the independence in 1960. Rather than adopt the classical formula for universities, they chose to create schools that prepared personnel to attend some of the most urgent needs of the country and to assume academic and professional roles. The *École Normale Supérieure*, initially designed as a three-year higher education program, was raised to four years in 1966. Starting from an initial number of twenty students, the *École Normale Supérieure* reached, in 1975, a record-breaking of 1,500 students enrolled. Installed in a modern building, the *Ecole Normal Supérieure* had all the features of an excellent level.

Upon this structure, the *CPS / Centre Pédagogique Supérieure* de Bamako was conceived with the objective to train teachers for higher education institutions in

Mali, as well as scientific researchers. The plan proposed a course of three years after obtaining *diplôme de l'ENS* or equivalent, at the end of which the student would receive the degree of *doctorat de spécialité.* This degree gave its holders a status in public service corresponding to the *doctorat 3ème cycle* in French educational system.

The curricular structure had that *CPS* from traditional university structure, since the Center was planned to support a faculty comprised of visiting professors. In the first year of the course, the main goal was to offer a basic, but specialized training, through *fast thematic courses*, seminars, courses, reading and independent studies, leading to research which should result in a thesis. Although it remained a basic training course, the first year of *CPS* was oriented as preparation for in depth research in a particular area. During the second and third year, the activity was mainly research with a view to producing a thesis, as well as less formal lectures and independent study designed to complement the training required in the field of thesis. The thesis was defended before a committee of invited specialists, from reputable universities around the world. In fact, in the first thesis defenses in Physics in 1973, the committee was chaired by Alfred Kastler, Nobel Laureate in Physics.

Although the lines of research were chosen solely on the basis of their relevance to the country, this did not prevent that some of the results obtained had sufficient scientific interest to justify its publication in international scientific journals. But this was just one consequence, and not a goal.

The disciplines that initially offered by the *CPS* were: Mathematics (Analysis, Geometry, Probability and Statistics, and Applied Mathematics), Physics (Energy and Optical), Biology (Botany, Zoology and Entomology), Ecology, Geomorphology, Organic Chemistry, Comparative Literature and Applied Linguistics. The choice of themes within each discipline varied from year to year to meet the demands of the institutions of higher education and the research institutes of the country.

Regular meetings were organized to discuss priorities of research in the country, to decide which options should be offered by the *CPS* and to analyze the overall structure and the curricula of the *CPS*, as well as of other higher education institutions and research institutes in the country. The visiting faculty served as an advisory board to the Minister of Education. The interdisciplinary atmosphere of the faculty prevailed in all activities of the *CPS*. The enormous difficulties arising from the lack of physical facilities, laboratories and libraries were shared and faced with creative solutions. In this model *CPS* unemployment, or the misuse of human resources, did not occur in the system. In this respect, the solution adopted by Mali represents a considerable advance over worn out formal model. The program rejected open recruitment and the preparation of personnel relied much on *in-service*. This way, these programs draws on the enormous potential of living experience of their students, not to mention the motivation, which has a very important mobilizing effect. It avoids, in a natural and spontaneous way, the stagnation which results of poorly adjusted socio-economic and productivity compromises. Repeating, the strong characteristic of the model *CPS* is the balance between the need of manpower and the adequate training of human resources, without the distortion of overproduction of graduates for areas which are not priority of the country.

Regrettably, most emerging countries are turning to the American/European model, when a schema as the *CPS*, with its natural and inherent flexibility, seems to be the right solution for the training at the graduate level. The model *CPS* is easier to implement where there is already a strong University structure. In this case, universities or research centers with some degree of development can act as agents to establish *CPS*-like structures in new universities, in particular in distant and less developed regions. Unfortunately, experiences like the *Projet Centre Pédagogique Supérieur de Bamako* are faced with disbelief and even resistance of universities and traditional research and education centers. The fact that, in the independence, the leadership of Mali opted for not creating an university modeled on French universities created te opportunity to implement such an innovative proposal as the *Projet Centre Pédagogique Supérieur de Bamako.*

The experience of Mali has shown its effects. Most of the *docteurs de spécialité* remained in the country, teaching at the *École Normale Supérieure, École Nationale d'Ingénieurs* and at the *École Polytechnique Rurale de Katibougo*, and later in other institutions of higher education, and conducted research relevant to the country. As an example, I look at the subject area of Mathematics. The themes covered are of various natures, covering virtually all undergraduate curricula of the various institutions of higher education, reducing the need of the traditional *coopérants*. The new teachers trained by *CPS* were closer to their students and could better understand their problems and difficulties and, most importantly, they shared the needs and aspirations of these students. The effects for the educational scenario of the country are noticeable. And this is, in fact, one of the most important goals of graduate studies in developing countries.

NOTES

1 These took place in Granada, Spain (1998), in Ouro Preto, Brazil (2002), in Auckland, New Zealand (2006), in Towson, USA (2010) and in Maputo, Mozambique (2014).

2 The mathematician Mikhail L. Gromov (1943–), Professor of the *Institute des Hautes Études Scientifiques* de Bûres-sur-Yvette, France, was awarded, in 2009, the Abel Prize (equivalent to the Nobel Prize), for his revolutionary contributions to Geometry.

3 http://www.nonkilling.org

4 http://www.nonkilling.org/pdf/volume_toward.pdf

5 Her doctoral thesis, entitled *Children, Space, and the Urban Street: An Ethnomathematics Posture*, has been submitted, in 2008, to the Faculty of Science and Technology of the *Universidade Nova de Lisboa.*

6 Joselita Nena Macedo Filha: *Dança com Lobos. a rua dos meninos e meninas de rua*, Lauro de Freitas – Bahia – Brasil, 2010.

7 I mention only three, with which I am more familiar: Cláudio Cadeia research deals with the Ethnomathematics of a gipsy community in Portugal, focusing on their capacity for mental calculation, His dissertation was submitted to the Universidade do Minho, Portugal, in 2005. Stathopoulou Xara (Charoula) researched the Ethnomathematics of Romany gipsy children in Greece. Her Doctoral dissertation was published as a book, in Greek, in 2005. Milton Rosa's ten years work with classes

of immigrants at Encina High School, in Sacramento CA, USA, was the basis of his doctoral thesis on *Understanding the Perceptions of School Leaders about the ELL Students' Achievement Gap: The Case of Mathematics* at California State University Sacramento, in the USA, in 2010.

[8] 1. Bakary TRAORÉ , 1973, Thesis: *Théorie du Controle Optimal et courbes generalisées de Young*;
2. Niamanto DIARRA , 1975, Thesis: *A propos des ensembles flous;*
3. Oualy KONTÉ , 1975, Thesis: *Les fonctions convexes et la théorie de la dualité*;
4. Kalilou MAGUIRAGA , 1975, Thesis: *Sur les rélevements des chemins dans les feuilletages de co-dimension un;*
5. Massiré SANGARÉ , 1975, Thesis: *Une théorie complète des systèmes dynamique généraux*;
6. Ibrahim FOFANA , 1977, Thesis: *A propos de l'opérateur gradient sur une hypersurface de Rn+1*;
7. Gaoussu TRAORÉ , 1977, Thesis: *Sur la variation seconde de l'aire*.

REFERENCES

Aczel A. D. (2005). *Descartes's secret notebook: A true tale of mathematics, mysticism, and the quest to understand the universe*. New York, NY: Broadway Books.

D'Ambrosio, U. (2006). *Ethnomathematics. Link between traditions and modernity*. Rotterdam, The Netherlands: Sense Publishers.

Descartes, L. (1637). *Le Discours de la Méthode*. Leiden, The Netherlands: January Mayor.

Gopnik, A. (2009). *The philosophical baby: What children's mind tell us about truth, love, and the meaning of life*. New York, NY: Farrar, Straus and Giroux.

Gromov, M. L. (2010). Interviewed by M. Raussen & C Skau. *Notices of the AMS*, *57*(3), 391–409.

Mesquita, M., Restivo, S., & D'Ambrosio U. (2011). *Asphalt children and city streets: A life, a city, and a case study of history, culture, and ethnomathematics in Sao Paulo*. Rotterdam, The Netherlands: Sense Publishers.

Von Osten, M. (2009). *Architecture without architects – Another anarchist approach*. Retrieved from www.eflux.com/journals

Ubiratan D'Ambrosio
State University of Campinas/UNICAMP
São Paulo
Brazil

EMMANUELLE LE PICHON AND ELLEN-ROSE KAMBEL

13. CHALLENGES OF MATHEMATICS EDUCATION IN A MULTILINGUAL POST-COLONIAL CONTEXT

The Case of Suriname

INTRODUCTION

In spite of Suriname's national independence, the Dutch language has remained the primary medium of instruction used in the school system, and the only medium that may open the access to upward social mobility in Suriname. However, Dutch is one of the home languages for only a small minority of the pupils. A high percentage of children, especially in the interior of the country, may have never heard a word of Dutch until they start school (Arends & Carlin, 2002, p. 285). In these circumstances, the development of bilingual education would be warranted. This approach has been proven to be most beneficial to the ultimate achievement of pupils speaking different languages at home (see for instance Skutnabb Kangas et al., 2009). Suriname is confronted with at least two additional challenges: first, the diglossic situation of the country, i.e. the Dutch language is still associated with prestige and upper class (Carlin et al., 2014), second, the number of languages involved in the linguistic repertoire of the Surinamese youth. At present, in Suriname more than twenty languages are actively used (see Annex). Although exact numbers are not available, it is estimated that 30–60% of the Suriname youth leaves school prematurely, without a diploma, with the highest percentages of dropout in the rural areas (Government of Suriname, 2005). Using local languages of instruction play a key role since the language used at school excludes the majority of the pupils the moment they enter the school system. Children are required to radically adjust to a monolingual Dutch language environment. There is thus an urgent need to identify the most successful educational strategies, which can improve the education of all children.

In 2010, at the request of indigenous leaders, parents and teachers, a bilingual experiment was initiated in four indigenous communities (Galibi, Donderskamp, Powakka and Washabo) and two Maroon villages in Suriname (Jaw-Jaw and Kayapaati). The original goals were (1) to enhance the quality of education in these rural areas, in particular math and numeracy (2) to support the revitalization of indigenous and maroon languages and cultures, and (3) to enhance the awareness of the parents, teachers and policy makers with regard to the advantages correlated with bilingual education (Kambel, 2013). From a scientific perspective, we wanted to examine the role of dual language support at primary school level. According to

Z. Babaci-Wilhite (Ed.), Human Rights in Language and STEM Education, 221–240.

the literature, even a limited introduction of mother tongue instruction may bring advantages in terms of language development, academic success, literacy skills, self-confidence and cognitive skills (Babaci-Wilhite, 2015; Benson, 2009; Brock-Utne, 2012) and the absence of any support in learners' home languages may result in a deficit in all languages (Youssef, 2002, p. 183). Hence, as a first step, we examined the feasibility and acceptance of a dual language program. Additionally, we explored to what extent the language used when assessing a child influences the mathematics performance scores. In three of the six villages, an assessment of academic achievement (mathematics, language and reading skills) was carried out. We will first start with some general information on the linguistic and education situation in Suriname, focusing on the challenges in indigenous and maroon schools in the hinterland of the country. We will then present the results of the math assessment, and explore the implications for future educational policies in Suriname.

LANGUAGES AND EDUCATION IN SURINAME

Suriname became a Dutch plantation colony in 1667. Between 1668 and 1830, the Dutch forcibly transported an estimated 215,000 Africans to Suriname to work on the coffee, sugar and cacao plantations (Van Stipriaan, 1993, p. 314). When slavery was abolished in 1863, the colonial administration arranged for some 70,000 contract labourers from China, India, Indonesia (Java) to replace the enslaved persons (Arends, 2002). In more recent times, gold miners from Brazil, as well as immigrants from Haiti, neighbouring Guyana and Chinese immigrants have settled in Suriname, adding to the diverse mix of languages and cultural traditions that currently make up Surinamese society (see Table 13.1 for an ethnic breakdown).

Table 13.1. Suriname population by ethnicity (total population: 544,000). Central Bureau of Statistics, 2012

Ethnic group	*Percentage of total population*
Hindustani	27.4
Maroon	21.7
Creole/Afro-Surinamer	16.4
Javanese	13.7
Mixed	13.4
Indigenous	3.8
Caucasian	0.3
Other	1.3
Don't know	0.3
Nothing	0.3

An English based creole language (*Sranan*, also called *Sranan Tongo*) which was used by the enslaved Africans to communicate with each other (Arends, 2002; ibid), was later adopted as the *lingua franca* by other ethnic groups and is currently spoken throughout Suriname, although less so in the rain forested hinterland areas (also called 'the Interior'). The Interior is the home of the indigenous population (Amerindians) and the Maroons whose African ancestors fled the plantations in the 17th and 18th centuries and established communities deep in the Interior of the country.

The Status of the Dutch Language in Suriname

Despite being under Dutch ruling, Suriname – or rather the capital city Paramaribo and the plantations in the northern coastal area – was a multilingual society from its very beginning: European colonists probably spoke more British, French, Portuguese and German than Dutch throughout the 17th– 18th century (De Kleine, 2002, p. 210). This changed in the 19th century when an elite group of free non-whites (called Creoles) was formed who preferred to speak Dutch at home and who sent their children to the Netherlands for education. Until 1844, the slave population was not allowed to receive any education at all and when they were educated this was in Sranan. For the free population, education was provided in Dutch. In 1876, after the abolition of slavery in 1863, the colonial government introduced compulsory education for every child with Dutch as the only medium of instruction. 'This meant that after two hundred years of reserving the Dutch language for a small elite, the colonial government now went to the other extreme by forcing a Dutch-only policy onto the Surinamese, actively suppressing the use of Sranan in education' (De Kleine, 2002, p. 214).

De Kleine explains that by the 20th century, a clear distinction had emerged between those who acquired Dutch as a first language (middle and upper class Creoles and Dutch-born whites) and those who acquired Dutch as a second language. The latter group included the former enslaved who migrated to the city and formed a lower-class Creole group speaking mainly Sranan. There was an even sharper distinction between the city and the plantation areas (the rural districts). The new Asian immigrants who replaced the former enslaved in the plantation areas lived in relative geographical isolation and spoke hardly any Dutch or Sranan. Well into the first half of the 20th century, teachers in the districts were allowed to teach in the students' first languages (Hindi and Javanese). Coupled with low school attendance rates among these groups meant that they had little exposure to Dutch (De Kleine, 2002, pp. 213–216).

After the second world war, the Hindustani and Javanese population started to migrate to the city, increasing their exposure to Dutch. As De Kleine asserts 'the discrepancy between city and countryside continues to exist till the present, as does the gap between the lower classes, using predominantly Sranan and/or their other respective group languages, and the higher classes, among whom Dutch is more common' (2002, p. 216).

While this accurately describes the situation of Paramaribo and the urbanized areas in the coastal districts (referred to as 'countryside' or 'urban rural'), the situation in the hinterland or Interior of the country shows a different picture. However, there are no detailed studies of the languages spoken by indigenous and maroon school age-children in the Interior. Before we discuss the education in the Interior, we provide a brief background of the status of indigenous peoples and maroons in Suriname.

INDIGENOUS PEOPLES AND MAROONS

According to the most recent census from 2012, there are approximately 117.500 people (or 21,7 %) who self-identify as Maroon and 20.500 (3.8%) as Indigenous in Suriname.[1]

The indigenous peoples of Suriname may be separated into two groups: the Kari'na (or Caribs) and the Lokono (or Arowaks) who are based in the northern coastal area; and the Tiriyo and the Wayana in the south, located close to the border with Brazil (see map above). In the south there are also a few remaining speakers of languages related to Trio. Although the Kari'na and Lokono speak languages that are completely unrelated (Kari'na belongs to the Cariban language family, whereas Lokono is part of the Arowakan languages), their communities are situated close to each other and partly because of this, there are few cultural differences at present. Indigenous communities throughout Suriname elect or appoint their own village council and are either led by a village leader (*kapitein*, or chief) or a grand chief (*gran man*) who enjoy some form of recognition by the central government.

A process of modernisation is taking place in which indigenous people – especially in the northern coastal area – are adopting urban life styles, partly forced by pressures on their territories resulting from logging and mining, which leaves less opportunities to live from their traditional activities. Such activities include hunting, fishing and rotational cultivation on forest plots ('*kostgrondjes*'), the preparation of an alcoholic drink made of cassava (*'kasiri')* weaving of hammocks and other artisanal work. Cash necessary to buy clothing, household items, and other western manufactured goods, is earned with the sale of bush meat, fish, or produce. Villagers may also find limited and low paid employment in logging or mining industries, as government employees or in eco-tourism (Kambel, 2006).

A language shift is taking place in the northern coastal indigenous communities, in which the indigenous languages are replaced by Sranan and to some extent Dutch. Arguably, Sranan is currently the most spoken language among indigenous youth, with only one Kari'na village left in Suriname where Kari'na is spoken by all generations (Galibi):

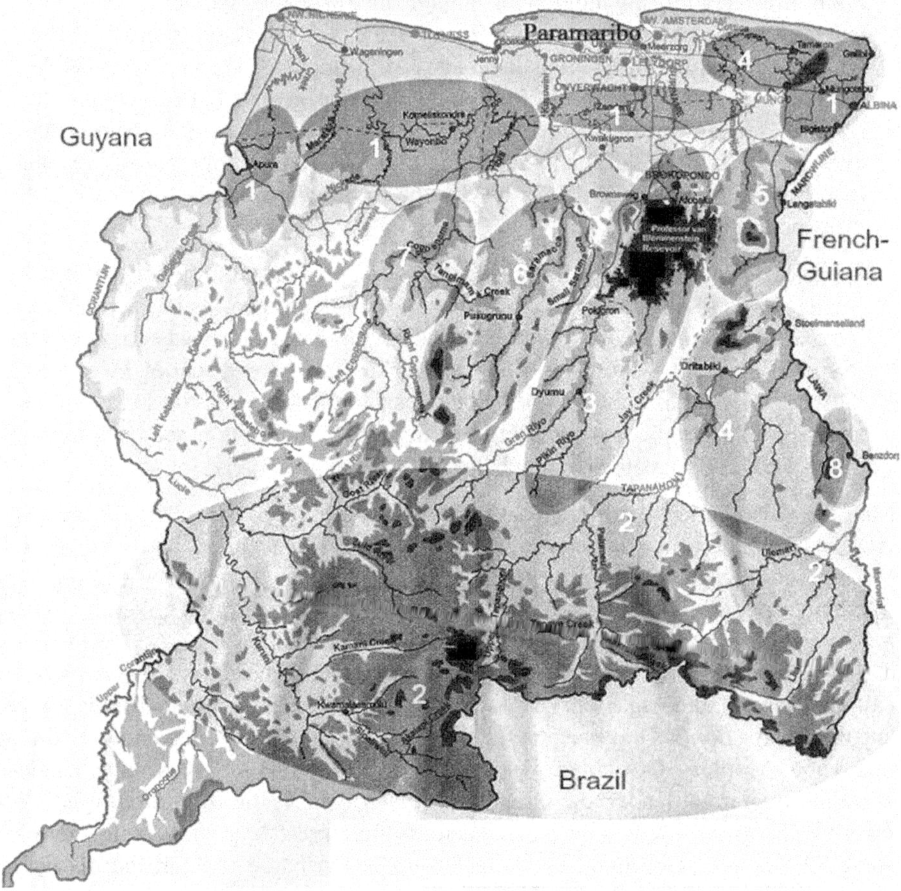

Indigenous:	Maroon:
1 Kari'na and Lokono	3 Saamaka
2 Trio, Wayana and related groups	4 Aukaner (Ndyuka)
	5 Paramaka
	6 Matawai
	7 Kwinti
	8 Aluku/Boni

Figure 13.1. Indigenous and maroon territories in Suriname.
Sources: Kambel, 2006; Carlin & Boven, 2002, p. 36; Smith, 2002, p. 144; Rotary Club Paramaribo

> I am from Galibi, an indigenous village in Suriname, where I have been teaching at the elementary school for 37 years. In Galibi, we speak Kari'na, which is one of the indigenous languages of Suriname. Everyone here speaks the language, from young to old. Besides Kari'na, we also speak Dutch, the official language of Suriname and we speak Sranan – the lingua franca that everyone speaks in Suriname. All education and all school books we use here [in Galibi] are in Dutch. It used to be forbidden to speak Kari'na in the classroom.
>
> I speak Kari'na now. But I did not in the past. I was raised in the city. When I was 6 years old, my father took me to stay with foster parents in Paramaribo. No one spoke Kari'na and I did not speak Dutch or Sranan. I was not allowed to speak in my own language. I felt inferior to the others at school. Because I did not understand. I always got bad grades and the teachers said I was stupid and lazy, that I could not learn and that my father should take me back to the village. (Interview with school teacher by Kambel in 2013)

Maroons (also called Bush Negroes or *bushi nenge*) are descendants of enslaved Africans who fled the plantations in the 18th and 19th century and formed their own communities in the rainforested interior of Suriname, eventually winning their freedom through peace accords established with the Dutch colonial government long before slavery was abolished in the coastal area (Kambel & MacKay, 1999). There are five distinct maroon groups (or tribes): the Saramakaners (or *Saamaka* as they call themselves), the Aukaners (*Ndyuka*), the Paramakaners (*Paamaka*), the Kwinti and the Aluku (*Boni*). They each have their own languages and – in common with the indigenous peoples – have their own form of governance with basyas, captains, head captains and a great leader, the Granman (*gaama*) who is the formal representative vis à vis the government. The Saamaka form the largest Maroon group. There are some 72 Saamaka communities along the upper-Suriname river in the district of Sipaliwini (Kambel, 2005). The Saamaka language is a mix of Portuguese (as many plantation owners were Portuguese Jews), African languages, Dutch and Sranan (Smith, 2002). Although many Saamaka have migrated to the capital, there seems to be far less of a language shift taking place in traditional Saamaka territory than in the northern indigenous communities. Saamaka remains the most spoken language on the upper-Suriname river.

While historically, ethnically and linguistically, Indigenous peoples and Maroons are radically different from each other, they share some important commonalities, distinguishing them from the other ethnic groups in Suriname: first of all their socio-economic, cultural and spiritual ties to the forest, second their self-governing system, and third their legal status: both indigenous peoples and Maroons fall under the protection of the international human rights legal regime pertaining to indigenous and tribal peoples.[2] This means among others that they have a recognized right to self-determination, which includes the right to educate their children in their own languages and in accordance with their own cultures and traditions (see, for example, article 13 and 14 of the United Nations Declaration on the Rights of Indigenous Peoples) (Kambel, 2014; Babaci-Wilhite, 2015).

Since 1992, indigenous village leaders of the approximately forty indigenous communities are member of the Association of Indigenous Village Leaders in Suriname (or VIDS, i.e. Vereniging van Inheemse Dorpshoofden in Suriname). The VIDS initiated the bilingual math project in 2010 and was joined in 2011 by the Association of Saamaka Maroon leaders (VSG, i.e. Vereniging van Saamaka Gezagsdragers), the Rutu Foundation for Intercultural Multilingual Education and the Foundation for Intercultural Bilingual Education in Suriname (ITOS, i.e. Intercultureel Tweetalig Onderwijs in Suriname). For both the VIDS and the VSG, the bilingual math project was seen as a way to improve the educational success of indigenous and maroon children in the Interior, but more importantly, it was considered as a way to reverse the ongoing loss of their languages, cultures and identities.

THE EDUCATION GAP BETWEEN THE CITY AND THE INTERIOR

There is a consistent educational achievement gap between children living in the Interior and their urban peers. In 2007, the United Nations Committee on the Rights of the Child (CRC) noted with concern 'that significant disparities exist in the quality and delivery of education between the coastal areas and the interior of the country and that a large number of primary schools in the Interior are managed by teachers with limited training' (United Nations Committee on the Rights of the Child, 2007). Passing rates for secondary school are 25–30% lower in the hinterland district of Sipaliwini compared to the national average (MINOV, 2008, p. 22; MINOV, 2011, p. 6). Indigenous and maroon children repeat grades almost twice as often than children in urban areas (MINOV, 2011, p. 8) and they take much longer to complete primary school: in 2008, only 1.2% of indigenous and maroon children completed primary school before the age of 12, compared to 24% of urban children (MINOV, 2008, p. 21). Exact numbers on dropout or school attendance are not available. Yet, in 2005, the Government reported to the CRC that almost sixty per cent of children between four and fourteen were out of school in the Upper Suriname river (Government of Suriname, 2005). This is the area occupied by Saamaka Maroons (see the map above) and the location of two of the pilot schools of the bilingual math project discussed in this chapter.

There are numerous factors that may explain the gap in education achievement between indigenous and maroon children, and children of other ethnic groups. One of the most significant ones are: geographical location (associated with fewer qualified teachers), the lack of early childhood programmes and the language of instruction.

Geographical Location

The lack of roads or other adequate infrastructure means that the communities furthest away from the capital city are difficult and expensive to reach. Few villages have 24/7 hours electricity, piped water or adequate housing for the teachers. Health care services are also at a much lower level than what is provided in the coastal area (Kambel, 2006). This partly explains why there are so few qualified teachers in the Interior. In fact, the majority of teachers in the Interior are not or not fully qualified to teach at a preschool or primary school.[3] For example, none of the schools participating in the bilingual math project had a fully qualified school team. Most schools had only one or two teachers with a diploma that would allow them to teach in the capital of Paramaribo and surroundings (Rutu Foundation, 2012). There are very few opportunities for further teacher training in the Interior.

Lack of Early Childhood Programmes

A UNICEF survey on early childhood development revealed that seventeen percent of children in the Interior attend early childhood programmes, compared to almost fifty percent of children in Paramaribo (UNICEF, 2013). The survey also measured the extent to which children between three and five years old in Suriname are developmentally on track:

> Children are identified as being developmentally on track based on whether they can identify/name at least ten letters of the alphabet, whether they can read at least four simple popular words, and whether they know the name and recognize the symbols of all numbers from 1 to 10. If at least two of these are true, then the child is considered developmentally on track. (UNICEF, 2013, p. 119)

In the Interior, 7.4% of the children were found to be developmentally on track in the literacy-numeracy domain, compared to 28.1% of children living in Paramaribo. Disaggregated by ethnicity, the survey showed that 11.1% and 11.5% of Indigenous and Maroon children, respectively, were developmentally on track, as compared to 31% of Creole children, 26.8% of Hindustani children and 26.8% of Javanese children (see Table 13.2).

Language of Instruction

The Ministry of Education stated that at least half of the children in the Interior enter school without any knowledge of Dutch (MINOV, 2008, p. 33) and in its latest report to the UN Committee on the Rights of the Child (CRC), the government conceded that: 'a major barrier in education [of indigenous and maroon children] is the language barrier, in particular the extreme discrepancy between the mother tongue of children in the interior, learned at home and the use of Dutch in schools' (Government of Suriname 2013, p. 53). The language of instruction is also mentioned in a study on early school leaving among indigenous youth:

Table 13.2. Percentage of children aged 3–5 years developmentally on track in literacy/numeracy (UNICEF, 2013, pp. 119–120)

By district/area	*Percentage of children developmentally on track in literacy/numeracy domain*
Paramaribo	28.1
Marowijne	10.2
Para	9.7
Sipaliwini	6.7
Urban	26.7
Rural Interior	7.4
By ethnicity:	
Indigenous	11.1
Maroon	11.5
Creole	31
Hindustani	26.8
Javanese	26.8
Mixed	29.4

> The home language and the street language is Sranan Tongo, while the language of instruction is Dutch. Especially in primary school many pupils have problems with this. […] They speak Dutch, but they don't always understand it. […] This is because they have limited access to information such as television, internet, the newspaper, library. (Michiels, 2011, p. 44)[4]

In 2009, another United Nations human rights body, the Committee on the Elimination of Racial Discrimination (CERD), commented on the lack of measures by the Surinamese government to support indigenous and maroon languages in Suriname and recommended that the State 'seek strategies with a view to introducing bilingual education' (CERD, 2009, p. 22). It was against this background that a pilot programme was initiated for indigenous and maroon children.

A BILINGUAL MATH PILOT PROGRAMME FOR INDIGENOUS AND MAROON PRIMARY SCHOOL CHILDREN

The bilingual math project started in October 2010. The project was a follow-up of an indigenous education festival organised by teachers, parents and grandparents in which over 350 indigenous Kari'na and Lokono children participated in workshops on traditional knowledge and language. This was a response to their concern about the increasing loss of their languages and cultural heritage. The UNESCO Atlas of Languages in Danger lists all indigenous languages of Suriname as severely or critically endangered.[5] Communities have realized that formal education plays an important role in the loss of their language and traditions, as children have to attend school now and cannot be taken along to the forest or the agricultural plots for long periods of time, as parents used to do in the past (De Jong & Kambel, 2006, p. 108).

The teachers who had participated in the indigenous education festival indicated that the children experienced many difficulties with numeracy and math. This was observed during a survey among indigenous early school leavers in one of the participating communities (Powakka). Each youth who was interviewed said that math was 'the most difficult' and the 'least enjoyable' subject. Teachers and policy makers also recognized that math was a national problem (Michiels, 2011, p. 29). The researcher who stayed in Powakka during her fieldwork, made some insightful remarks:

> […] I noticed the kind of difficulties the children experience with this subject. They get problems to solve and they have to do math exercises, while they do not master the basics. They cannot do tables, addition, subtraction, so they cannot solve even one of the problems. The pupils also have no idea what they are learning. One of the children in grade 4 did not know what a meter, decimetre and centimetre were until I showed her. But she was solving problems with them. (Michiels, 201, p. 29)

In 2012, a national math assessment organized by the Ministry of Education, showed that children, even those in the capital city, struggled with math. The assessment was performed among school children in Paramaribo in grade 5 who scored an average of 4.7 points out of 10. The report of the assessment was not made public, but the government employees who were interviewed about the results said for example, that only 19% of the children was able to correctly draw a circumference.[6]

The objectives of the project were, thus, threefold: first, to enhance the children's numeracy and math skills; second, to protect and encourage the indigenous and maroon languages and cultures and third, to support awareness about the benefits of bilingual mother tongue based instruction among policy makers, teachers and parents. The project included a number of project activities:

- Design of mother tongue based bilingual math materials (for preschool and grades 1–2) in three languages (in addition to Dutch): Kari'na, Lokono and Saamaka;
- Training of teachers in intercultural education and math didactics;
- Survey on the attitudes of teachers, parents and stakeholders regarding bilingual mother tongue-based intercultural education.
- Assessment of math and reading/listening comprehension skills in Dutch and the mother tongue (Kari'na and Saamaka)
- Publicity and organisation of meetings and seminars with stakeholders and policy makers.

Application of the Early Grade Math Assessment (EGMA)

The testing was done to examine the numeracy and math skills of the pupils at the moment of the test and included an exploration of their actual mathematical level. The achievement of adequate mathematical levels is a challenge for most of the developing countries even though much effort has been put into enhancing this issue (Valverde & Näslund-Hadley, 2010). Considering that in bilinguals, mathematical concepts are accessed more efficiently in the language in which the person learned simple arithmetic [...] than in the person's other language' (Salillas & Wicha, 2012, p. 746), we hypothesized that the pupils would score better in the language in which they learned to count at school, i.e. the Dutch language. However, we also hypothesized that the more the test would rely on verbal reasoning, the better the children would score in their own language, i.e. Saamaka. To test these hypotheses we collected data in one hundred pupils from two Saamaka Maroon villages: 51 pupils attending grade one and 49 pupils attending grade two, age 6–12 years (mean age 7.55, SD=1.2), 46 girls.

We chose to use the Early Grade Math Assessment, abbreviated EGMA. The EGMA is an internationally recognized assessment of early mathematics learning that can produce a snapshot of children's knowledge of fundamental skills in early grade mathematics. The skills assessed included number identification, reasoning about magnitude, recognition of number patterns, addition, subtraction, and word problems. The test was administered by the same assessors in both languages (Dutch and Saamaka), individually and orally. Importantly, the assessors who tested the pupils in the maroon villages were fluent in both languages. In order to test arithmetic skills that rely on verbal reasoning, four additional EGMA word problems were developed by the consultants. Each child was assessed twice, once in Dutch and once in Saamaka.

We examined the proportion of children who scored better or worse in a particular language. The goal was to understand the possible influence of the language of the assessment on performance scores. We distinguished two types of mathematical tests; first, simple arithmetic tests like additions which rely primarily on basic counting skills and do not require extensive verbal reasoning, and second, arithmetic tests which rely more heavily on verbal reasoning because the problem to be solved is presented in a narrative (word problems). For the addition subtest, the pupils of grade one were asked to state the sum of 20 addition problems with addends 1–10 in a limited time (level 1). In grade 2, they were asked to state the sum of 5 addition problems with up to two-digit addends. For the word problem test, the children's ability to apply problem-solving skills to real-world problems was assessed. To enhance comprehension, the assessors adapted the texts of the problems to the cultural knowledge of the children (for instance, busses were replaced by boats). An example of a word problem is: *'There are five children in the boat. Three children get in. How many children are in the boat altogether now?'*

In line with previous studies and with our expectations, we found that the overall performance in addition problems was relatively low. In grade 1, we found that around 30 percent of children obtained the lowest score possible (0), regardless of the language used. Intriguingly, in grade 2 an advantage of the school language emerged as about two thirds of the children scored 0 when assessed in their home language against one third when assessed in the school language. Thus, the children's capacity to solve simple calculations in their home language decreased substantially from grade 1 to grade 2.

When assessing the more verbally oriented test, word problems, a similar picture was observed in grade 1 with almost one third of children scoring 0 in the school- and in the home language. Here, a substantial improvement became manifest in grade 2, with, in accordance with our expectations, an advantage for the assessment in the *home language* (4% scored 0 when assessed in their home language against 10% when assessed in their *school language*). From these results, we conclude that a substantial improvement occurred between first and second grade.

In order to better understand the influence of the language of assessment we compared, for each child, the score obtained in the home language to that obtained in the school language. This way we were able to calculate the proportions of children (1) who scored better in their home language, (2) who scored better in the school language or (3) who obtained equal scores in both assessment conditions.

For the simple arithmetic tests (additions), 51% of the children obtained the same score in both languages in grade 1, while 16% of the children scored better in Saamaka and 33% scored better in Dutch. Remarkably, the advantage of the school language became even more outspoken when similar addition problems were assessed in grade 2 (at a grade 2 level) with 39% obtaining higher scores in the school language against only 2% with higher scores in their home language.

With regard to arithmetic tests which rely more heavily on verbal reasoning (word problems) an important shift from grade 1 to grade 2 emerged in the reverse direction. In grade 1 the same proportion of children (25%) scored better in either languages, while in grade 2, 37% of the children scored better in their own language against only 6% with better scores in the school language. In other words, a strong advantage for the use of the home language over the use of the school language emerged with regard to the more verbal word problem.

From these first observations, we could already draw several relevant conclusions. First, that the overall test performance for math problems was low in both schools, despite a significant improvement from grade 1 to grade 2. More relevant to the issues discussed in this chapter, is our finding that the language of assessment clearly influences the children's test performance. We find that, with regard to the simple arithmetic tests (less verbal tests), more pupils performed better when taking the test in the school language as compared to their performance when taking the test in their home language. On the contrary, on the verbal reasoning tests, they performed better in their home language (i.e. Saamaka) than in the school language (i.e. Dutch). Finally, this 'language assessment effect' appears to be more substantial in grade 2.

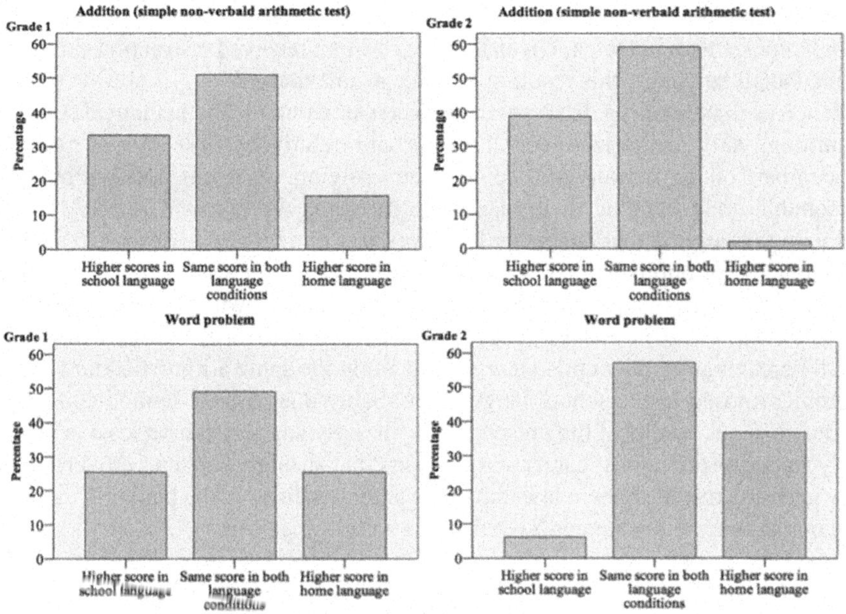

Figure 13.2. Influence of the language of assessment per test

CONCLUSION

The purpose of this study was to understand the effect of language choice in elementary education on the testing of the pupils. Results of the present study show that an important proportion of pupils did not master the basic numeracy skills. These findings confirm the notion that the level of mental arithmetic is overall very low in these regions. In fact, due to floor effect (a proportion of pupils may actually perform lower that the lowest performance score of the test), most of the subtests may not have been suitable measures. For instance, for the word problem, in Dutch and in Saamaka, half of the children failed the test, scoring 5 or lower in both languages.

However, results showed that pupils scored higher in Dutch as compared to their performance in Saamaka when the tests were less verbal (simple arithmetic) while they scored better in Saamaka when the arithmetic test relied heavily on verbal reasoning skills. This result confirms the hypothesis that young children favor the early schooling language for mental arithmetic (Dehaene et al., 1999; Spelke & Tsivkin, 2001). Moreover, the difference between the performances in the different languages was overall greater in Grade 2 than in Grade 1. In other words, even though most pupils scored higher when taking the addition test in Dutch than when taking it in Saamaka, the difference between the languages was even more pronounced in Grade 2. This shows that pupils improved from grade 1 to grade 2. The results on the word tests are particularly interesting because they indicate that pupils, particularly those in grade 2, are better able to solve word problems in their own language than in Dutch. Given that the children received instruction exclusively in the Dutch language, this result may strike as unexpected.

It seems that there is a discrepancy between automatized mathematical sequences (counting, and memorized calculations) and relatively little verbal reasoning in comparison to narrative-based problem solving. Problem solving implies a reasonable understanding of the language in which the problem is posed and the capacity to reason in that language. However, it is remarkable to observe that almost forty percent of pupils were able to transfer the arithmetic knowledge that they had acquired in Dutch to a problem posed in Saamaka, their own language, without any previous education in that language. But the other side of the coin is that only a small percentage of the pupils were able to apply the same arithmetic knowledge to a problem posed in the school language probably due to their limited competence in Dutch. Thus, results of the present study strongly suggest that, at least in the very early stages of the pupils' career, assessments that strongly rely on verbal reasoning may generate results that are lower than the actual abilities of the pupils. This implies that pupils may be evaluated lower than their real capacities.

DISCUSSION

Challenges in Introducing Mother Tongue Based Bilingual Instruction

Given the current high academic dropout rate in Suriname (see Government of Suriname, 2005), in particular in the rural areas, these results are decidedly relevant. Against this background it is essential to ensure that the children are able to access the level of education that corresponds to their cognitive possibilities in order to reduce early school leaving and to support excellent schooling attainment. Which recommendations may be given in early schooling of dual language learners such the Saamaka pupils in the rural areas in Suriname? As stated in the introduction, to reduce subtractive effects (loss of the home language, underperformances and, ultimately, school dropout) bilingual education is warranted. However, the diglossic situation of Suriname as well as the extreme diversity in the languages that are used may complicate its realization.

Multilingualism among School Children Is the Norm

In recent years, two studies were published on the languages spoken by school children in Suriname (Kroon & Yagmur, 2012; Léglise & Migge, 2014). Both studies claim that Dutch is the language spoken by the majority of school children in Suriname and that Dutch is the first language for most children. A closer look reveals that these findings do not apply to indigenous and maroon children in the Interior. Léglise and Migge (ibid) acknowledge that their survey may be limited because the interviews were carried out in schools and in the official school language (Dutch) which may have resulted in children echoing the school's views about language. Furthermore and more importantly, the Léglise and Migge survey excluded most children in the southern part of the country because of the high costs involved (in particular, the Saamaka children from the Upper Suriname river were excluded). The study by Kroon and Yagmur is more problematic, given that their survey was not only carried out in the Dutch language and at school, but also by teachers who distributed the forms in the classrooms (Kroon & Yagmur, 2012, p. 26). Considering that in some schools it is still practice to punish children for speaking another language than Dutch in the classroom, we have doubts to what extent their results reflect linguistic realities, rather than socially desirable answers. Finally, both the Kroon and Yagmur (ibid) and the Léglise and Migge surveys were limited to children who were in school, excluding out-of-school-children who may have no or very little exposure to Dutch.

Nevertheless, both studies show convincingly that multilingualism is the norm among Surinamese school children, at least in the northern part of the country (Kroon & Yagmur, 2012, p, 50; Léglise & Migge, 2014, p. 9). In the Léglise and Migge survey, 65 percent of the participating children said that they speak at least three languages, whereas one percent indicated that they only spoke one language. This is confirmed by our own observations, and by a small survey we carried out in a school in Paramaribo (in a suburb predominantly occupied by Maroons) where the majority of children said that they spoke at least three languages: one or two maroon languages, Sranan and Dutch.

Translanguaging as a Powerful Multilingual Strategy

Results of the present study suggest several strategies that merit further study. The most optimal but perhaps most challenging approach would require the introduction of mother tongue based bilingual education in the schools, aimed at achieving balanced bilingualism (children become fully bilingual and biliterate). This already poses a challenge for wealthy countries, e.g. Canada, the United States of America and Scandinavian nations, who have better educated teachers and better resources in schools. For a developing country with a small population struggling with its colonial heritage, the challenges are even greater.

However, what our results show is, that the least one could do, is to include assessments in the languages of the children. The systematic introduction of tests in both languages (the school and the home languages) would allow the teacher to distinguish between poor performances due to lack of school language proficiency and real cognitive challenges. This bilingual practice would start with raising awareness among Surinamese teachers that performance results may vary as a function of the language used during the assessment and would allow the introduction of a multilingual educational strategy, also called translanguaging.

The core of the educational strategy referred to as 'translanguaging' is to welcome the mother tongue of the pupils in the classroom where pupils are exposed to a new language, in order to maximize communicative as well as cognitive potential development. Translanguaging encourages the development of multilingual and multiliteracy developments. This approach is not primarily based on languages but on the practices of bilinguals (Garcia, 2009, p. 140) and it targets the transfer of knowledge between the languages. Pupils are thus encouraged to exchange bilingually on school subjects. It may be realized in pupil-pupil interactions if the teacher does not speak the languages of the pupils. As appeared from the results presented here, children may be able to transfer skills from one language to another without much help (in the present study, only the tests were presented in the pupils' mother tongue, whereas all prior school instruction was in Dutch). They only need to be supported and encouraged to do so.

In mathematics education, recent reports about Content Language Integrated Learning (CLIL) show the need to include the mother tongue of the pupils in the learning process to support basic skills development in mathematics. According to Tavares, in Hong Kong, most schools choose mathematics as one of the subject to be taught in the second language. However, these studies have shown mixed results: most students improved in language proficiency but at the expense of academic content learning (Tavares, 2015, p. 2). According to Tavares, the strategic involvement of translanguaging practices in the classroom would facilitate metalinguistic awareness, development of the school language and the content learning of the subject. This practice would allow pupils to become competent learners in the language in which they are taught, enhancing the development of cognitive skills from the very first year of school. At the same time, it would help to avoid a shift from the mother tongue to school language dominance as has been reported in number of studies (i.e. Sheng et al., 2011).

ANNEX: LANGUAGES SPOKEN IN SURINAME

(Kroon & Yagmur, 2012, pp. 11–12. The asterisks indicate the languages that were added by Léglise & Migge, 2015).

Indigenous languages	
Arowakan languages	*Arowak, Lokono, Loko dyan*
	Mawayana (nearly extinct)
Cariban/Carib languages	*Carib, Kari'na, Kalinya*
	Trio, Tirió
	Wayana
	Akuriyo
	Sikiana
	Tunayana-Katwena
	Waiwai
Warao languages	*Warao (believed to be extinct)*

Creole languages	
Plantation/City Creole	*Sranan, Sranan Tongo*
Maroon languages	*Saramaccan, Saamaka*
	Matawai
	Aukan, Ndyuka
	Aluku, Boni
	Paramaccan, Paamaka
	Kwinti
	Ndyuka-Trio
Caribbean Creole	*Guyanese Creole**
	*Haitian Creole**
European languages	
	Dutch, Surinamese Dutch
	English, Guyanese English
	Portuguese, Brazilian Portuguese
	Spanish*
	*French**
Asian languages	
Asian Surinamese	*Hakka Chinese*
	Sarnámi, Hindustani
	Surinamese Javanese
Other Asian languages	*Lebanese, Lebanese Arabic*
	Mandarin Chinese
	Northern Chinese, Wu, Min, Yue, Kejia (Chinese Dialect Groups)*

NOTES

1 This is a spectacular rise for the Maroons who numbered 72,000 in the 2004 census (Kambel, 2005, p. 10).

2 Although the Maroons are not indigenous to Suriname, they are recognized as 'tribal peoples' by the Inter-American Court for Human Rights with the same rights as indigenous peoples under international human rights law (MacKay, 2006, 2010).

3 To teach in the interior, a special 'Bushland Diploma' is sufficient which does not qualify to teach in the capital or coastal districts (MINOV, 2008, p. 20).

4 This was a study among 84 youth from the district of Para, including indigenous youth in Casipora, Redi Doti, Pierrekondre and Powakka (Powakka is one of the pilot schools of the bilingual math project).

[5] UNESCO Interactive Atlas of Languages in Danger, (last visited: 16 May 2015).
[6] De Ware Tijd, 15 mei 2012, 'Resultaten diagnostische toets hemelschreiend' [Results Diagnostic Test Disastrous'].

REFERENCES

Arends, J. (2002). The history of the Surinamese creoles I. A sociohistorical survey. In E. B. Carlin & J. Arends (Eds.), *Atlas of the languages of Suriname* (pp. 115–130). Leiden, The Netherlands: KITLV Press.

Arends, J., & Carlin, E. (2002). Epilogue. The languages of Suriname today and tomorrow. In E. B. Carlin & J. Arends (Eds.), *Atlas of the languages of Suriname* (pp. 283–286). Leiden, The Netherlands: KITLV Press.

Babaci-Wilhite, Z. (2015). *Local languages as a human right in education: Cases from Africa*. Rotterdam, The Netherlands: Sense Publishers.

Benson, C. (2009). Designing effective schooling in multilingual contexts: Going beyond bilingual models. In T. Skutnabb-Kangas, R. Phillipson, A. Mohanty, & M. Panda (Eds.), *Social justice through multilingual education* (pp. 63–84). Bristol, England: Multilingual Matters.

Brock-Utne, B. (2012). Language policy and science: Could some African countries learn from Asian countries? *International Review of Education, 58*(4), 481–503. doi:10.1007/s11159-012-9308-2

Carlin, E. B., & Boven, K. (2002). The native population. Migrations and identities. In E. B. Carlin & J. Arends (Eds.), *Atlas of the languages of Suriname* (pp. 11–46). Leiden, The Netherlands: KITLV Press.

Central Bureau of Statistics. (2012). *Census 2012*. Paramaribo, Suriname: Author.

De Jong, C., & Kambel, E.-R. (Eds.). (2006). *Marauny Na'Na Emandobo Lokono Shikwabana. 'Marowijne: Our Territory'*. Moreton-in-Marsh, England: Forest Peoples Programme.

De Kleine, C. (2002). Surinamese Dutch. In E. B. Carlin & J. Arends (Eds.), *Atlas of the languages of Suriname* (pp. 209–232). Leiden, The Netherlands: KITLV Press.

Dehaene, S., Spelke, E., Pinel, P., Stanescu, R., & Tsivkin, S. (1999). Sources of mathematical thinking: Behavioral and brain-imaging evidence. *Science, 284*, 970–974.

Garcia, O. (2009). *Bilingual education in the 21st century: A global perspective*. New York, NY: Blackwell/Wiley.

Government of Suriname. (2005). *Second periodic report to the United Nations committee on the rights of the child* (CRC/C/SUR/2). Retrieved May 15, 2015, from http://ow.ly/OjyNI

Government of Suriname. (2013). *Combined third and fourth report on the convention on the rights of the child of the republic Of Suriname 2007–2012* (CRC/C/SUR/3-4). Retrieved May 15, 2015, from http://ow.ly/OjytK

Kambel, E.-R. (2006). *Indigenous peoples and maroons in Suriname, Inter-American development bank, economic and sector study series* (RE3-06-005). Washington, DC: Inter-American Development Bank.

Kambel, E.-R. (2013). Bilingual numeracy education and indigenous rights: A community based initiative from Suriname. *Intercambio, Bulletin of the IDEA Education Research Network, 5*(5).

Kambel, E.-R. (2014, April). *Resisting the one-size-fits-all education model for indigenous peoples in Suriname: A human rights analysis*. Paper presented at SOGIP/Forum Conference: Education, learning and indigenous peoples' rights: what knowledge, skills, and languages for sustainable livelihoods? Tromsø, Norway.

Kambel, E.-R., & MacKay, F. (1999). *The rights of indigenous peoples and maroons in Suriname* (IWGIA Document No. 96). Copenhagen, Denmark: IWGIA.

Kroon, S., & Yagmur, K. (2012). *Meertaligheid in het onderwijs in Suriname. Een onderzoek naar praktijken, ervaringen en opvattingen van leerlingen en leerkrachten als basis voor de ontwikkeling van een taalbeleid voor het onderwijs in Suriname*. Rapport in opdracht van het Ministerie van Onderwijs en Volksontwikkeling en de Nederlandse Taalunie. [Multilingualism in Surinamese education. A study on the practices, experiences and opinions of students and teachers as the basis for the developing of a language policy for the education in Suriname. Report commissioned by the Ministry of Education and the Dutch Language Association]. Den Haag, The Netherlands: Nederlandse Taalunie.

Léglise, I., & Migge, B. (2014). Language practices and linguistic ideologies in Suriname: Results from a school survey. In E. B. Carlin, I. Léglise, B. Migge, & P. B. Tjon Sie Fat (Eds.), *In and out of Suriname. Language, mobility, identity*. Leiden, The Netherlands: Brill. Retrieved from http://dx.doi.org/10.1163/9789004280120

MacKay, F. (2006). *Moiwana zoekt gerechtigheid. De strijd van een marrondorp tegen de Staat Suriname.* [Moiwana looking for justice. The struggle of a maroon village against the State of Suriname]. Amsterdam, The Netherlands: KIT Publishers.

MacKay, F. (2010). *Saramaka. De strijd om het bos. Inclusief de uitspraak van het Inter-Amerikaans Hof voor de rechten van de mens.* [Saramaka. The Fight for the Forest. Including the judgment of the Inter-American Court for Human Rights] Amsterdam, The Netherlands: KIT Publishers.

Michiels, L. (2011). *Skoro na lasteng, Onderzoek naar voortijdig schoolverlaters in Para* (Unpublished master thesis). Katholieke Hogeschool Leuven, Leuven, Belgium.

Ministry of Education and Development, MINOV. (2008). *Implementation plan education for the interior 2008–2015. Plan of action to realize the right to quality education for every child in Suriname.* Paramaribo, Suriname: Author.

Ministry of Education and Development, MINOV. (2011). *Data map research planning and monitoring.* Paramaribo, Suriname: Author.

Rutu Foundation. (2012). *Survey on attitude of teachers, parents and stakeholders on intercultural bilingual education*. Amsterdam, The Netherlands: Author.

Salillas, E., & Wicha, N. (2012). Early learning shapes the memory networks for arithmetic evidence from brain potentials in bilinguals. *Psychological Science*, *23*, 745–755.

Serra, C. (2007). Assessing CLIL at primary school: A longitudinal study. *The International Journal of Bilingual Education and Bilingualism*, *10*(5), 582–602.

Sheng, L., Lu, Y., & Kan, P. F. (2011). Lexical development in Mandarin-English bilingual children. *Bilingualism: Language and Cognition*, *14*, 579–587.

Skutnabb-Kangas, T., Phillipson, R., Mohanty, A. K., & Panda, M. (Eds.). (2009). *Social justice through multilingual education. Linguistic diversity and language rights*. Bristol, England: Multilingual Matters.

Smith, N. (2002). The history of the Surinamese creoles II: Origin and differentiation. In E. B. Carlin & J. Arends (Eds.), *Atlas of the languages of Suriname* (pp. 131–151). Leiden, The Netherlands: KITLV Press.

Spelke, E. S., & Tsivkin, S. (2001). Language and number: A bilingual training study. *Cognition*, *78*, 45–88.

Tavares, N. (2014). How strategic use of L1 in an L2-medium mathematics classroom facilitates L2 interaction and comprehension. *International Journal of Bilingual Education and Bilingualism*, *18*(3), 270–288.

UNESCO. (2013). *Interactive atlas of the world's languages in danger*. Retrieved May 16, 2015, from http://www.unesco.org/languages-atlas/index.php

UNICEF. (2013). *Suriname: Multiple indicator cluster survey 2010*. Paramaribo, Suriname: Author. Retrieved May 16, 2015, from http://www.childinfo.org/files/MICS4_Suriname_FinalReport_Eng.pdf

United Nations Committee on the Rights of the Child. (2007). *Concluding observations to Suriname's second periodic report* (UN Doc. CRC/C/SUR/CO/2).

Valverde, G., & Näslund-Hadley, E. (2010). *The state of numeracy education in Latin-America and the Caribbean*. Washington, DC: Inter-American Development Bank.

Van Stipriaan, A. (1993). *Surinaams contrast: Roofbouw en overleven in een Caraïbische plantagekolonie, 1750–1863* (KITLV, Caribbean Series 13). Leiden, The Netherlands: KITLV Press.

Youssef , V. (2002). Issues of bilingual education in the Caribbean: The cases of Haiti, and Trinidad and Tobago. *International Journal of Bilingual Education and Bilingualism*, *5*(3), 182–193. doi:10.1080/13670050208667754

United Nations Committee on the Elimination of Racial Discrimination (CERD). (2009). *Concluding observations Suriname* (CERD/C/SUR/CO/12). Retrieved May 15, 2015, from http://ow.ly/OjAyU

Emmanuelle Le Pichon
Department of Modern Languages
Utrecht University
The Netherlands

Ellen-Rose Kambel
Rutu Foundation
Intercultural Multilingual Education
Utrecht
The Netherlands

LADISLAUS M. SEMALI

14. WHY DO INCONSISTENCIES PERSIST IN CHILDREN´S RIGHTS TO "GOOD" EDUCATION, HERITAGE EDUCATION AND STEM EDUCATION?

INTRODUCTION

The question of *children's rights* in general and of children's education, in particular, has been for over 20 years the focus of comparative education literature as well as in the development literature (Babaci-Wilhite, 2014; Jonsson, 2001). Champions of children's rights, emphasize that children everywhere consist of a vulnerable age group and therefore, need protections. The United Nations (UN) is known globally to protect children and for its work toward making human rights the fulcrum of its development goals and strategies because, as articulated by the Secretary General, "the promotion and defense of human rights is at the heart of every aspect of our work and every article of our Charter" (Annan, 1999).

In particular, UNICEF, a UN development agency charged with protecting the *rights of children,* views child development as the core of sustainable development, and thus, has been in the forefront of efforts to bring development theories and human rights principles together in a strategy capable of realizing the rights of children. Human rights of children are further specified in the *Convention on the Rights of the Child* (CRC). Collectively, these documents (the UN Charter and CRC) lay the foundation of the arguments contained in this chapter that examines: Why do inconsistencies persist towards children's rights in "good" education, heritage education and Science, Technology, Engineering and Mathematics (STEM) education? The goal is to interpret state obligations in relation to economic, social, education, and cultural rights (UN, 1987). For example, Paragraph 4 of the CRC points these obligations.

> States parties shall undertake all appropriate legislative, administrative and other measures for the implementation of the rights recognized in the present Convention. With regard to economic, social and cultural rights, States Parties shall undertake such measures to the maximum extent of their available resources and, where needed, within the framework of international co-operation. (CRC, Article 4)

Equally, the UN Declaration of Human Rights lays out the rationale and need to respect these universal rights. The Rights state:

Z. Babaci-Wilhite (Ed.), Human Rights in Language and STEM Education, 241–254.

> Everyone has the right to education. Education shall be free … Education shall be directed to … the development of human personality and to the strengthening of human right and fundamental freedom. (UN, 1987)

Respect of human rights is based on another characteristic of human rights, that is, these rights are inalienable, which means that they cannot be taken away. Consequently, it is imperative to respect and safeguard human rights against violations, abuse, or neglect (Jonsson, 2001, p. 15).

The arguments enlisted in the aforementioned quotes from CRC and UN illustrate a related principle that human rights must be enjoyed by *everyone* without discrimination based on either: attributes over which a person has no choice (such as gender, age, or ethnic origin), or attributes that, if denied, would result in the infringement of other human rights (such as religion and political ideology).

The assumptions underlying this inquiry on children's rights, and this chapter as a whole, point to the human rights of children specified in the CRC and uses a *human rights* lens to examine the inconsistencies of children's rights to education in Southern Africa. In this particular case, inconsistencies refer to national and global policy statements that advocate for *universal education* for all children but instead they fall short of the political will, budgetary (fiscal) support, and largely lack overall system-wide educational planning and evaluation (assessment) of school infrastructure, curriculum in content areas, teaching staff and basic statistics to monitor progress. For example, the year 2015 was set as the target to meet the *UN Millennium Development Goals* and the *Education for All* targets but according to the UN Global Monitor reports few countries in Africa will meet these goals (UNESCO, 2005; UNESCO, 2010, 2012).

In this chapter, I take the position to speak for children's rights to "good" education, heritage education and STEM education, and in this discussion, I argue that these rights are intertwined with children's health and overall wellbeing. For example, allowing rampant low birth weight (LBW) is seen as a violation of children's right to be born healthy, and of women's right to give birth to healthy children. When a nation's "*duty-bearers*" (actors responsible for human rights realization) lapse in meeting obligations in relation to LBW, they deny children and women adequate government health services, particularly ante-natal care; maternity leave, adequate supervision and monitoring of Village Health Workers; adequate staff (numbers and skills) at all levels, to implement and monitor programs on ante-natal care, malaria, parasitic infections and anemia (Jonsson, 2001).

Because young learners are vulnerable and need protections, inconsistencies in *Children's Rights* to "*good*" education, heritage education, and STEM education threaten children's wellbeing and future adult life. In this chapter, I discuss the need to match rhetoric with action and consequently examine the implementation of both national and global policies such as the *Education for All* (EFA) goals and the UN *Millennium Development Goals* (MDGs) to reveal the arguments and opinions that support children's rights to *good* education, and in the process, pinpoint inconsistencies and contradictions that seem to push back educational reform in Southern Africa. For example three UN MDGs recognize and mandate universal children's education (UNESCO, 2010). These goals are:

- *Goal 1:* Expanding and improving comprehensive early childhood care and education, especially for the most vulnerable and disadvantaged children.
- *Goal 2:* Ensuring that by 2015 all children, particularly girls, children in difficult circumstances and those belonging to ethnic minorities, have access to, and complete, free and compulsory primary education of *good quality*.
- *Goal 3:* Ensuring that the learning needs of all young people and adults are met through equitable access to appropriate learning and life-skills programs.

Elsewhere (Semali, 2007, 2014), I have discussed examples of weaknesses in the implementation of these UN goals in national education policies and showed that inconsistencies in providing "good" education exist in most African countries (Semali, 2008) and these discrepancies put children at risk. Governments failed to provide universal primary education to *all children* in spite of the target dates of 2000 (as part of the Jomtien EFA targets) and target of 2010 (as part of Dakar Framework for Action targets), to provide health protections (e.g., immunizations) to all children (UNESCO, 2012), and in particular those children under five years, and to establish viable mechanisms that offer vocational training and employment to the youth—the largest growing group in the population—to engage them in productive work that can generate earnings for decent livelihoods (Shostack, 1984; Semali, 2007).

A *Human Rights Approach* to children's rights to "good" education, heritage education and STEM education places primary emphasis on the intricate web of national capacities in meeting duties and obligations, and focuses primarily on accountability and process (Jonsson, 2001). The study presented in this chapter, has interest to pursue this approach to examine the extent to which children's education goes beyond existing 'Human Development Approaches' by recognizing that to achieve human development outcomes, human rights must be realized by those whose development is at stake. Thus, *Human Rights Approaches* (the UN Charter and CRC) demand a high quality process, in the belief that the process by which education rights are realized is just as important as the outcomes (Jonsson, 2001). As pointed out in Johnson's report, *Human Rights Approaches* focus on "accountability and identify those actors responsible for human rights realization (e.g., duty-bearers), whose capacities to meet their responsibilities must be strengthened" (p. 7).

There is a distinction to be made, however between the focus of human development approaches to "good" education and human rights approaches. The human development approach was primarily based on social and economic development as an outcome of development efforts (Fägerlind & Saha, 1983; Rostow, 1971). These approaches were less concerned with the quality of the process by which outcomes were achieved. Instead, Human Rights Approaches place primary emphasis on the complex web of *duties* and *obligations*, and focus primarily on *accountability and process*.

The practice of Human Rights Approaches goes beyond Human Development Approaches by recognizing that to achieve human development *outcomes*, human rights must be realized by those whose development is intended (UNDP, 2000). Similarly, this approach goes beyond education from a humanist perspective, which focuses on developing rationality, autonomy, empowerment, creativity, affections and a concern for humanity (Veugelers, 2014). This concern for humanity expresses the relation to other people. This social component can range from empathy to solidarity, and from the own community to the global world (Veugelers, 2014). Appreciating diversity and democracy are humanist ways of living together as human beings and concern for humanity can enrich Human Rights Approaches.

In sum, humanity is the condition that gives people the possibility of developing human capabilities: of being a reflective and dialogical person, of getting the sources to live a good life, of living together ruled by moral values, of helping others to live a good life too. Hence, the cultural development of young people and the pedagogical ideas and educational arrangements to support this development is seen as child's right. It includes the responsibility to educate children to become global citizens, that is, the social, political and religious domains, as well as cognitive, emotional and action oriented content. The concept of citizenship has extended from being a pure political judgment, to include the social and interpersonal dynamics of young people entitled to moral education. Morality has become a multifaceted and highly diversified construct that now includes cultural, developmental, situational and professional aspects. Citizenship and moral development are connected with the identity constitution of the next generations. A caring and supporting learning environment can help them to grow as responsible citizens who can participate fully in society.

Clearly, the *rights in education* concept works to shift the paradigm from how learning takes place in school towards the moral duty imposed through the international consensus of human rights. As noted in Babaci-Wilhite (2014), rights in education includes the principle that every human being is entitled to a decent education and gives priority to the intrinsic importance of education, thus implying that governments and their duty-bearers need to mobilize the resources to offer quality and necessary education (Babaci-Wilhite, 2014; UNICEF, 2009).

Human Development Approaches can enrich Human Rights Approaches. First, the Human Development Approach is based on a scientific analysis of causality and includes a vigorous assessment and analysis of the impact of different policy choices (Sen, 2000; UNDP, 2005). Second, although there should not be a hierarchy of human rights, in the real world—characterized by scarcity—action to realize rights must be prioritized (Jonsson, 2001). Human Development analysis helps to show how different choices result in different impacts and different costs. The UNDP Human Development Approach 2000 describes the relationship between the two approaches as follows: "[they are] harmonious enough to be able to complement each other and diverse enough to enrich each other" (p. 7).

THE CONTEXT OF RIGHTS TO "GOOD EDUCATION"

The discussion in this chapter emphasizes the need to be sensitive to children's rights to education, and in the process, develop a sense of autonomy in children and a sense of justice that protects children's development, including gender equity, access to schools, and a healthy growing environment. Autonomy development should be embedded in social change processes (Freire, 1993). For education, this means that learning is not a technical-instrumental rationality but identity development in a reflective and dialogical way in a social context; it is morally social constructivism. In Semali (2014) I have discussed in detail children's right to quality education. However, in some of the documents I examined for the current study, there is a general assumption that "good" education is "quality" education, and consequently, pledging to provide good education for all children is a promise and a pledge to all children's right to education, as outlined in the *Education for All* global goals (UNDP, 2000; Nkonongwa, 2012).

This chapter references numerous situations in which children are denied the opportunities of education that would enable the learner to acquire knowledge, values, attitudes and skills needed to face the challenges of the 21st century, and in particular, the looming requirements of globalization, technology expansion and global competition. Strikingly, inconsistencies persist in five areas that pertain to: (1) learners, (2) content, (3) teaching and learning process, (4) teaching and learning resources (environment), and (5) learning outcomes.

Inconsistencies in the Area of 'Learners'

Inconsistencies in the area of '*learners*' persist when learners are denied opportunities to live in a healthy environment with essentials of life—food, clothing and shelter. These essentials are necessary for the learner to be well nourished to be prepared to face the task of learning that occurs during the school day, and equally to meet the requirements of completing classroom tasks as well as homework assignments. A sick child cannot learn; a hungry learner cannot think or complete homework, and a child whose family is displaced because of war, famine, or natural disasters cannot find the school or education s/he deserves or that which his/her peers take for granted.

In the Sudan for example, where there has been continuous conflict since 1983, three generations of primary school children would have received no education at all had there not been several local and international organizations running basic education programs in conflict zones in southern Sudan. Such programs are designed to be sustainable, and involve curriculum development, teacher training, and institutional capacity building with local authorities—programs usually thought of as development work. These programs are necessary because short of which they would have dire consequences, namely, "children who grow up without ever receiving education and only knowing conflict are more likely to perpetuate human rights abuses and conflict" (Jonsson, 2001, p. 116).

The result is that children and local communities everywhere that encounter these inconsistencies are denied their rights and abilities to discover knowledge and implement culturally appropriate strategies that they require to address relevant issues in the local situation, globalized economy, and coincidentally, in the EFA's global goals (See Global Monitoring, UNESCO, 2005, 2011).

Inconsistencies in the Areas of 'Content' and Teaching and Learning'

Inconsistencies in the areas of '*content*' and *'teaching and learning process*' persist when learners are denied access to subject and discipline content that is essential for relevant pedagogy and instruction in the classroom. Access to relevant curricular materials and qualified teachers to deliver instruction in classrooms is essential for young learners to acquire basic knowledge that becomes the foundation of subsequent years of education. When the educational system denies learners well educated and dedicated teachers, it amounts to a breach of children's rights to quality education. The first grades are foundational for children to acquire literacy skills, math skills, and basic knowledge. Completing primary school education without such skills is tantamount to condemning the learners to a "dead-end" adult life. This is true in part for rural children, remote area learners and for children in distressed areas or refugee camps. In addition, culturally appropriate content in the curriculum should be sought in terms of values and provide relevant skills for solving quotidian problems that face children and their communities, and endemic problems in the society at large. Learners have a right to education that is uninterrupted and free from disruptive behaviors, safe classrooms, and learner–centered teaching approaches in well managed classrooms and schools. Clearly, learners have a right to qualified teachers.

Inconsistency in the Areas of 'Teaching and Learning Resources'

Inconsistency in the areas of '*teaching and learning resources (environment)*' persist when learners are denied adequate resources and facilities such as a safe school infrastructure, playgrounds, toilets, furniture (desks, tables, chairs), reading materials (textbooks, writing utensils and exercise notebooks), laboratory apparatus (chemicals and other teaching and learning materials essential for science education—physics, chemistry and biology). When students are denied these facilities, they face barriers to learning. The classroom is no longer a place for learning and of joy. Instead, learners are relegated to poor academic performance, failure, poor education and ultimately receive inadequate preparation to meet societal challenges. In addition, children deserve learning environments which are healthy, protective, and gender sensitive, and an environment that guarantees safety and security.

Inconsistencies in Educational Outcomes

Inconsistency in the *areas educational outcomes* persist when results of educational outcomes are not known and adjustments to "good" education are ignored. Systematic continuous evaluation and annual evaluations of schools and teachers' performance contribute significantly to educational outcomes. Students' academic performance and overall school progress provide input to the necessary feedback so vital in decision making and budget allocations. Inconsistencies in educational outcomes persist when imbalances are not corrected in the educational system, so that every child receives high quality education regardless of gender, geographical background, ethnic group or ability to pay for education. Inconsistencies persist when policymakers, teaching staff and executives are not held accountable for the decisions they make, decisions that adversely affect children's lives.

For example, children are vulnerable or face hardships when education officers decide to locate a school far from the communities with children in most need. Children in these communities are denied the joy of schooling because they have to walk long distances to get to school, or must face obstacles like lack of transportation, wild animals, inclement weather, including floods, hurricanes, sandstorms, and the like. Education excellence should be the mantra for every school in the nation because every child deserved quality education and excellent academic results. Duty-bearers (those responsible for human rights realization and stakeholders in Southern Africa) cannot rest until they are satisfied with the education provided in terms of basic knowledge, social skills, rudimentary economic wellbeing—access to food, shelter and clothing—self-reliance and reasonable academic achievement, in both the short and long term timeframe. National educational goals are not meant for the few elite children from rich families and those who live in big cities, but rather for all children. *Human Rights Approach*es do not tolerate 'bare' minimum outcomes. The implementation of continuous (formative and summative) evaluations are necessary to monitor educational progress and educational outcomes.

In sum, educational outcomes are a function of many factors, including educational policies; relevance of curricula; teaching and learning materials, evaluation of learning outcomes, availability of qualified teachers, continuous professional development for teachers, quality of teaching and learning process, mastery of medium of instruction, including mother-tongue instruction in early grades; presence of all necessary school infrastructure; and integration of technology in teaching and the learning process. Children's rights to education must be understood to encompass *all children* and the effort to meet the needs of all children who pass through the education system, for these children to acquire the desired outcomes that enable them to face challenges for personal and national development, as well as the challenges of the 21st century and global competition. Inequities in gender, in access to educational inputs, and regional disparities account for the educational inconsistencies that need attention because persistence in this path denies many children their right to education and the preparedness necessary to be a full participant in the society when they reach adulthood. Clearly, science and technology are considered among the requirements of wealth creation and improvement of overall quality of life in the nation and local community.

WHY DO INCONSISTENCIES PERSIST IN CHILDREN´S RIGHTS TO HERITAGE EDUCATION

Heritage education, on the other hand, is not as obvious as the areas of "good" education outlined previously. For the purposes of this study, cultural heritage refers to local and traditional history of places, heroes and heroines, cultural practices—including rituals, foods, costumes or attire, plants, medicinal remedies, and areas of cultural, religious, and ceremonial significance (Smith & Jackson, 2006). Every child has a right to his/her cultural heritage, including religious and spiritual inclinations that form the prior knowledge or conception of their world of science before formal schooling and which "may be at variance with the meanings of scientific concepts learnt at school" (Semali, 2013, p. 17).

It is crucial for teachers to consider children's rights to cultural heritage and to identify such conceptions during their interactions with the learner to address them appropriately with the intention to improve comprehension and scientific concept(s) acquisition. Heritage education empowers students from local or pastoralist communities, remote areas, and people who have chosen to live indigenously, to make them visible, and thus, giving back to them a learning experience that necessitates showing appreciation about shared knowledge in their community (UN, 1998).

I argued previously that education is an indigenous value and throughout history, parents, grandparents, elders and the extended African families valued education as part of a child's heritage (Semali, 2014). Prior to the era of colonial imperialism in Tanzania, for example, family members held education—traditional education, language education, initiation education, and rituals education, to be essential for the sustenance and wellbeing of African children and their cultural heritage (Semali & Asino, 2013). However, the colonial project interrupted such indigenous values, and instead, children were left to compete against each other without safeguards or protections. This book addresses the needed conversation about human rights in education as outlined in the declaration of the rights of the child in 1959 (UNICEF, 1959).

It is important for schools and all "duty bearers" to respect and understand local customs and traditions, not only when they are carrying out relief work, or development aid, but also in understanding local values and connecting them to internationally recognized human rights. However, when promoting human rights standards, it is always important to point out that human rights are not culturally relative, but that certain universally accepted human rights are applicable to all human beings, no matter what the cultural setting (Jonsson, 2001). Some interventions require particular sensitivity to local customs. For example, in dealing with initiation rites involving various tribal communities, it is important to be aware of how families feel when traditional indigenous practices are perceived by foreigners, (particularly religious groups) as primitive, outmoded or uncivilized practices.

Another glaring example of human rights violation is the denial of use of local languages for formal academic instruction. In effect, education in a local language should be regarded as a human right (Babaci-Wilhite, 2014, p. 15). When students in early grades up to secondary school are denied the freedom to learn in their mother tongue or first language, inconsistencies persist in children's rights to good education and heritage education. German children learn in German; Chinese children learn in Chinese, Italian children learn in Italian—why can't Tanzanian children learn in Kiswahili; or Zambian children in Bemba or Chewa? When research on the use of first language or mother tongue for instruction is ignored, and when the benefits of research are disregarded, inconsistencies persist in children's rights in the areas of teaching and learning.

It is commonly accepted that the ability to speak local languages should give socio-economic benefits and strengthen cultural heritage and identity. It means giving consideration to local realities and direct education and its intellectual efforts and curriculum towards the achievement of freedoms that are consistent with education as a human right (Babaci-Wilhite & Geo-JaJa, 2011). Native languages are part of children's heritage, and for this reason, instruction in mother-tongue languages play a critical role in cognitive learning and in the development of logic, reason, critical thinking and new knowledge (Geo-JaJa, 2009).

Therefore, local languages should be seen as an integral part of culture and a child's heritage, thus should be designated as a human right in the education sector (Skutnabb-Kangas, 2000). Denying children their heritage education in the name of western or universal / global education is a violation of their human rights. Denying children the use of local languages and mother-tongue, learning their local history, using images and metaphors that come from their home culture in the early grades is a gross violation of children's rights to quality education.

Furthermore, the intent of decolonizing films' images as a form of cultural agency is an important stance to disengage the unequal relationship among polities, and uncover biased meanings, lopsided interpretations, and representations that distort African images with colonial legacies, archeology, and ideologies. For teachers to continue to expose children to biased educational content, found in the media and textbooks, is a violation of the child's rights to good education. In the literature, for example, bias is manifested in texts when authors present particular values as if they were universal; or denigrate one race over another, distort or replace local history in lieu of foreign beliefs.

Bias can be conveyed in the global media about pastoralists and other indigenous peoples through ways in which the selection of stories, sequence, and slant in newscasts; the placement or omission of stories in newspapers; who is interviewed and left out in radio or television talk shows and news programs; the advertisements on webpages, television, magazines, radio shows targeted at specific audiences; the lyrics of commercial jingles and popular music, and the images displayed with them in broadcast commercials and music videos; the goals, procedures, and the rules of video games (Semali, 2002). These examples show the extent to which child's rights to heritage education is denied them, clearly, a violation of human rights.

It is also apparent that apart from the unfavorable educational policies that particularly marginalize educationally pastoralist communities and their children, participation in schools is marginal. The argument used to explain pastoralists' marginal participation in school is directly connected with the outworking of the school system itself, which promotes the values and practices of the dominant culture and fails to accommodate the needs of marginalized populations (see Ezeomah, 1992; Semali, 1992).

For instance, ethnographic studies from nations with a high concentration of pastoralists point out that the long distances to schools, teachers' lack of interest to engage pastoralist communities, and the communities' alienation in the schooling process (biased curriculum content, irrelevant textbooks, and excluding teachers from pastoralist communities from hiring, etc.) are major obstacles influencing against school participation among pastorals. Similarly, the lack of learning systems that are more closely linked to the survival needs of pastorals, and lack of educational opportunities have been identified as formidable obstacles to their school participation (see Archibald et al., 1995; Dyer, 2000; Ezeomah, 1992; Semali, 1994).

This investigation of cultural heritage helps to explain the extent to which inconsistencies persist in *heritage education* which is embedded in many spheres of knowledge including the visual representations of cultural artifacts, stories or customary traditions—the objects and the spaces artifacts and visuals occupy in the relationships between the specimens and the architecture used to contain the objects, and the fact that children everywhere in Southern Africa are denied this heritage education as part of the educational process in schools.

WHY DO INCONSISTENCIES PERSIST IN CHILDREN'S RIGHTS TO STEM EDUCATION?

Inconsistencies persist in *children's rights to STEM education* when learners of science, technology, engineering and mathematics are denied the opportunities in science classrooms to use concepts, explanations, and interpretations that learners derive from personal and cultural knowledge, particularly when they cannot use such knowledge to improve or change their socio-economic and political leverage or use science as a screen to view and interpret knowledge experiences they encounter at school.

Barnhardt and Kawagley (1999, 2008) suggested that every indigenous culture has its own ways of living in nature and has an orientation to learning that is metaphorically represented in its art forms, communities' idiosyncrasies, languages, and varying associations between its culture and its relationship to its natural environment. Further, Ogawa (1995) referred to such contextually-bound science, found in a given culture, as its "*indigenous science*" (p. 585). Thus, indigenous science relates to the science knowledge of long-time residents, usually oral culture, and the scientific knowledge of all peoples, who as participants in a culture, encounter the effects of a particular worldview and the relativist interests of their "home communities" (Semali & Mehta, 2012). This indigenous science forms the learners' prior knowledge when they enter STEM classrooms. To deny children the opportunities to access and use this prior knowledge is a violation of children's rights to "good" education. To deny children culturally relevant science education is to deny them the opportunity to make learning meaningful to the learner by denying the children to draw from their socio-cultural environments like most children do in North America, Europe and other Northern countries.

Even though the study of STEM is widespread, educators' realities include contending with evolving classroom practices to meet 21st-century needs (Sanders & Binderup, 2000). We have indicated (see Semali & Mehta, 2012) that the divergences between implementation of STEM education in Tanzania, for example, leaves many students, after 4–6 years of secondary schooling, without sufficient skills or aspirations for pursuing employment in industry or manufacturing. And ignoring these barriers to universal education and the barriers' influence in thwarting the implementation of STEM is a violation of children's rights. To ensure their rights, teachers of science must accept their teaching responsibilities and respect learners' prior knowledge.

Even though progress has been made in implementing the EFA global goals for universal education, obstacles to STEM education persist. These obstacles are coupled with other barriers, such as: low and inadequate access to education (particularly for minority ethnic groups and children from pastoral communities or remote areas); irrelevant curricula (mostly influenced by colonial legacies and the insistence of using English for instruction at a time when Kiswahili is the lingua franca and widespread throughout the country); poor learning outcomes (mainly resulting from low quality teaching. inadequate textbooks and lack access to the Internet); lack of financing (schools and teachers rely on government bursaries for salaries, maintaining school buildings, and textbooks, exercise books, lab equipment, desks, toilets, etc.) (Semali, 2014).

In particular, the need and urgency for pastoral and nomadic education has been argued from political, environmental, economic, and health points of view (see Ezeomah, 1992; Semali, 1994). Politically, pastoralists have been viewed as a powerless and marginal segment of society. They are also viewed as critical because of the environmental pressures, and ecological disasters that come with overgrazing, and overstocking because of their pastoral activities. Economically, pastorals have been categorized as one of the poorest segments in their respective countries (For example, the Maasai of East Africa, the Fulani of Northern Nigeria, and the Nabian

San, to name a few groups in this category). Similarly, the direct contribution of pastoralists to food production in almost every society on the planet cannot be overemphasized. And because in recent years science has identified that diseases that affect animals can be directly transferred to humans, the lack of education among pastoral communities continues to be a concern to all nations, developed or underdeveloped.

These barriers coupled with low teachers' salaries, dilapidated buildings, impoverished libraries, poor teaching, and learning environments, collectively, portend a bleak picture of children's rights to STEM education.

CONCLUSION

The main task of this chapter is to question the reality of inconsistencies in education in Africa. The goal is to develop a sense of 'human rights' in education; a climate of schools that is sensitive to children's rights and a sense of gender equity and justice in the distribution of educational resources. The arguments supporting children's rights in "good education" aim to ensure quality education that produces results: high achievement in school and good preparation for adulthood. It is education that 'gives nourishment and self-respect' while it garners confidence in teachers and learners, as well as 'increase human capabilities, functions, and opportunities in societies' (Babaci-Wilhite, 2012, p. 19). Notably, "good education" is an entitlement for all school children as "rights holders" regardless of race, ethnicity, disability or socio-economic class, that is conceptualized within the common understanding that ensures "indivisibility, equality, participation and inclusion" (UNDP, 2006, cited in Babaci-Wilhite, 2012, p. 19).

Some scholars believe quality education must account for or match materials and physical facilities since primary schools form the foundation for secondary and tertiary education (Chonjo, 1996; Snyder, 2009). Reports show that the educational reform to result in universal education, gender parity and healthy children's programs is not happening. How then can we resolve such conundrum? African educationists insist that these challenges are historical—the indelible legacy of colonialism is the reason for the irreparable rupture to African education that introduced discontinuities and contradictions into the history of Africa, science curriculum and inequalities based on race, class and gender (Fafunwa, 1967). Discontinuities and contradictions pervade school systems everywhere within which the science curricula, content, and language of instruction emphasize factoids to be learned by rote. Students must memorize factoids applicable neither to their prior knowledge, indigenous knowledge and heritage culture nor to their lives outside the classroom. This chapter shed light on these inconsistencies and further research is ongoing to explore ways to overcome the contradictions that pervade educational policies and practice.

REFERENCES

Annan, K. (1999, April). *Secretary-general's statement to the commission on human rights, 55th Session.* New York, NY: United Nations.

Archibald, J., Hampton, E., & Newton, E. (1995). *Organization of educational services in sparsely populated regions of Canada.* Ottawa, Canada: Department of Indian Affairs and Northern Development.

Babaci-Wilhite, Z. (2012). A human rights-based approach to Zanzibar's language in education policy. Special issue on rights-based approach and globalisation in education. *World Studies in Education, 2*(13), 17–33. Retrieved from doi.org/10.7459/wse/13.2.03

Babaci-Wilhite, Z. (Ed.). (2014). *Giving space to African voices: Rights in local languages and local curriculum.* Rotterdam, NL: Sense Publishers.

Babaci Wilhite, Z., & Geo-JaJa, A. (2011). A critique and rethink of modern education in Africa's development in the 21st century (PED). *Journal of the School of Education, 30,* 133–159.

Barnhardt, R., & Kawagley, O. (1999). Education indigenous to place: Western science meets indigenous reality. In G. Smith & D. Williams (Eds.), *Ecological education in action* (pp. 117–140). Albany, NY: State University of New York Press

Barnhardt, R., & Kawagley, O. (2008). Indigenous knowledge systems and education. In G. Fenstermacher, D. Coulter, & J. Wiens (Eds.), *Why do we educate? Renewing the conversation* (pp. 223–241). New York, NY: NSSE.

Chonjo, P. (1994). The quality of education in Tanzanian primary schools: An assessment of physical facilities and teaching learning materials. *Utafiti (New Series), 1*(1), 36–46.

Dyer, C. (2000) "Education for All" and the rabies of Ketch, West India. *International Journal of Educational Research, 33,* 241–251.

Ezeomah, C. (1992, September). *Functional curriculum development: A means of retaining Nomadic Fulbe cultural identity.* Paper presented at the UNESCO Conference on Education, Paris France.

Fafunwa, B. (1967). *New perspectives in African education.* Lagos, Nigeria: Macmillan.

Fägerlind, I., & Saha, L. J. (1983). *Education and national development: A comparative perspective.* Oxford: Pergamon Press.

Freire, P. (1993). *Pedagogy of the oppressed, 1970* (Reprint). New York, NY: Continuum.

Geo-JaJa, M. A. (2009). Can globalization in Nigeria's Niger delta be harmonized for integration and development? In B. Brock-Utne & G. Garbo (Eds.), *Language and power: Implications of language for peace and development.* East Lansing, MI: Michigan State University Press.

Jonsson, U. (2001). *Human rights approach to development programing.* New York, NY: UNESCO.

Nkonongwa, P. (2012). *Quality education in Tanzanian context.* Paper presented at the African Federation of Heads of Schools. Conference, held at Mlimani City, Dar es Salaam, Tanzania.

Ogawa, M. (1995). Science education in a multi-science perspective. *Science Education, 79,* 583–593.

Rostow, W. W. (1971). *The stages of economic growth.* Cambridge, England: The University Press.

Sanders, M. E., & Binderup, K. (2000). *Integrating technology education across the curriculum* [Monograph]. Reston, VA: International Technology Education Association.

Semali, L. (1994, December). *The social and political context of literacy education for pastoral societies: The case of the Maasai of Tanzania.* Paper presented at the 43rd Annual Meeting of the National Reading Conference, Charleston, SC.

Semali, L. (2002, December/January). Crossing the information highway: The web of meanings and bias in global media. *Reading Online, 6*(5). Retrieved from http://www.readingonline.org/newliteracies/lit_index.asp?HREF=semali3/index.html

Semali, L. (2007). Challenges of rebuilding education in crisis: Access to universal primary education in Africa. In D. Baker & A. Wiseman (Eds.), *Education for all: Global promises, national challenges* (pp. 395–425). New York, NY: Elsevier.

Semali, L. (2008). Indigenous education: Culturalism, colonialism and politics of knowing. In D. Mulenga (Ed.), *Postcolonialism and education. challenging traditions and disrupting boundaries.* Toronto, Ontario: Palgrave Macmillan.

Semali, L. (2013). The iSPACES framework for rethinking a culturally responsive secondary science curriculum in Tanzania. *Journal of Contemporary Issues in Education, 8*(2), 32–46.
Semali, L. (2014). Rethinking quality education in Tanzania. In Z. Babaci-Wilhite (Ed.), *Giving space to African voices: Rights in local languages and local curriculum* (pp. 107–129). Rotterdam, NL: Sense Publishers.
Semali, L., & Mehta, K. (2012). Science education in Tanzania: Challenges and policy responses. *International Journal of Educational Research, 53*(2012), 225–239.
Sen, A. (2000). *Development as freedom.* Oxford, England: Oxford University Press.
Shostack, G. L. (1984). Designing services that deliver. *Harvard Business Review, 62*(1), 133–139.
Smith, C., & Jackson, G. (2006). Decolonizing indigenous archaeology: Developments from down under. *American Indian Quarterly, 30*(3/4), 311–349.
Snyder, C. W. (2009). *USAID/Tanzania education strategy for improving the quality of education, FY 2009–2013.* Washington, DC: Global Evaluation and Monitoring.
UN. (1948). *Universal declaration of human rights, 1948.* Cited in Babachi-Wilhite, Z. (2014). *Giving space to African voices: Rights in local languages and local curriculum* (p. 14). Rotterdam, NL: Sense Publishers.
UN. (1987). The Limburg principles on the implementation of the international covenant on economic, social and cultural rights (E/CN.4/1987/17). *Human Rights Quarterly, 9,* 122–135.
UN. (1998). The Maastricht guidelines on violations of economic, social and cultural rights. *Human Rights Quarterly, 20*(1998), 691–705.
UNESCO. (2005). *EFA global monitoring report 2005. Education for all: The quality imperative*. Paris, France: UNESCO.
UNESCO. (2010). *Dakar framework for action, education for all: Meeting our collective commitments*. Paris, France: UNESCO.
UNESCO. (2012). *Monitoring education quality: Indicators and data analysis*. Paris, France: UNESCO.
UNICEF. (1998). *Guidelines for human rights-based programming approach.* CF/EXD/1998-04, 21 April
UNICEF. (1999). *Programme cooperation for children and women from a human rights perspective*. E/ICEF/1999/11, June 1999.
UNICEF. (2009). *The quality of teaching and learning in Tanzanian primary schools: A baseline study*. Dar es Salaam, Tanzania: UNICEF.
UNDP. (2005). Human development report 2005. In *International cooperation at a cross-roads: Aid, trade and security in an unequal world.* New York, NY: UNDP.

Ladislaus M. Semali
College of Education
Pennsylvania State University
USA

ZEHLIA BABACI-WILHITE, MACLEANS A. GEO-JAJA
AND STEVE AZAIKI

15. HUMAN RIGHTS IN DEVELOPMENT AID FOR STEM EDUCATION IN NIGERIAN LANGUAGES

INTRODUCTION

This chapter assesses the weaknesses in the teaching and learning of Science, Technology, Engineering and Mathematics (STEM) subjects in Africa, and proposes new development approaches to STEM teaching and learning. Implicit to these approaches is an acknowledgement that African knowledge and African languages are critical to the effectiveness of STEM teaching and learning. This will be demonstrated using Nigeria as a case study. By defining access to effective STEM teaching and learning as a right in education, alternative development roadmaps can contribute to both sustainable development and to the satisfaction of the Paris Declaration of Aid Effectiveness. We argue that current problems of aid recipients have their origins in the lack of contextualization in development aid frameworks, which we see as mechanisms of dependency.

Many bilateral and multilateral donors are increasingly concerned with human rights. While some countries have adopted the rights-based approach to development aid, others seem not to see the benefits of ensuring the recipient's ownership of aid education programs. The premise of this chapter is that the right-capability based approach with a focus on linguistic rights will secure best educational practice and insure that education aid systematically ameliorates poverty's multi-dimensionalities. In highlighting roadblocks to human rights of transformational change ownership, in this chapter the authors recommend that indigenous knowledge and linguistic rights in STEM should be the driving force in development aid and should be defined as a human right in education for sustainable development in Africa.

In this chapter we will analyze the limited extent to which aid to education has concerned itself with the transformational and institutional dynamics of education. In this vein we will draw attention to current evidence on the "added value" that a concern with linguistic rights brings to development aid. This chapter draws attention to the reality that today in many countries in Africa, the choice of language-in-education policy disregards the socio-economic and rights to self-development hindered by the implementation of a non-local, non-indigenous language as the Language of Instruction (LoI) in STEM subjects. Despite the United Nations' Human Rights Declarations (UNHRD), education rights are still works in progress in developing countries (Babaci-Wilhite et al., 2012).

Z. Babaci-Wilhite (Ed.), Human Rights in Language and STEM Education, 255–269.

Africa does not have a robust strategic plan on STEM policies, nor do they have a framework to implement these subjects, yet the United States invests billions in STEM education knowing that 70% of African jobs–whether they are domestic or international–will require core scientific and technological skills. African leaders are concerned but do not invest in STEM education because implementation is expensive, despite the fact that the Information and Communication Technology (ICT) industry and investments are growing.

The exploitation of Africa´s natural resources requires that Africa to invest in STEM research, in order to train scientists, tech entrepreneurs, engineers, innovators to develop ways to resolve resource extraction and pollution problems locally, relying on local solutions instead of depending on the US and China. The outsourcing of educational infrastructure is a major issue, therefore it is time to create the right skills for Africans in order that Africa can develop on its own terms. African governments are signing infrastructure contracts with other countries outside of the continent instead of providing a future workforce trained in STEM subjects to create employment for African youths. Currently, most of the STEM jobs in Africa are performed by or outsourced to multinationals from the US, China and/or India. The construction teams are growing each year, as Africans are not trained to do the work, which is a serious issue that governments need to take into consideration.

Africa lacks well trained STEM teachers to provide the skills needed to build and innovate in several types of professions, such as engineers, scientists, designers, and accountants to name only a few. Investment in STEM will have a positive effect on employment as well as support the economic development of the continent. The focus needs to be moved from projects that teach English, such as the US supported Global Innovation Fund, to projects that support STEM teaching and learning. Training students in English is not providing the necessary skills to match the needs of Africa. The learning and teaching of STEM subjects in African languages will increase innovation and contribution to the African economy. Africa cannot compete globally without a strong development in STEM. The new focus on STEM subjects in African languages will address both literacy and science. It will address the linguistic rights crucial for STEM learning process as addressed in the next section, using Nigeria, Africa's largest growing economy as a case study.

LINGUISTIC RIGHTS IN STEM SUBJECTS IN NIGERIA

Nigeria has three major languages Hausa, Igbo and Yoruba. These languages have capacities for assimilating words, terms and expressions from other languages and other cultures (Olarenwaju, 2008). Olarenwaju argues that MT (Mother Tongue) or the language of the immediate community is considered a very important medium for achieving literacy and numeracy, but it is only used for the first 3 years of primary school.

In Nigeria, the importance of Nigerian languages to the protection, preservation and promotion of cultures and the enhancement of learning a major language for purposes of promoting national unity and integration is enshrined in the 1989 Constitution of the Republic of Nigeria. However this constitutional mandate has been difficult to implement due to the multiplicity of diverse local languages. For cultural identity and educational justification the government has settled on the use of 'immediate environment' local language in the first three years of schooling. Furthermore, decree 16 of 1985 on Education (National Minimum Standards and Establishments of Institutions) gave legal power to the enforcement of teaching in local languages. Any Nigerian languages in Nigeria can be used as a LoI.

Teaching in a foreign language adds to the difficulties for students in engaging easily in learning processes. While abstract knowledge is used in science education, comprehension will be improved when science teaching contextualizes knowledge in local examples using the local language. Recent studies on STEM subjects have demonstrated how learning is improved through the use of Nigerian languages.

A Case Study of Teaching Science in Yoruba Language

A recent research project has been investigating the use of Yoruba to improve the learning of science conducted in four schools in Ondo West Local Government Area using Yoruba language and English languages. The team researchers Osungbemiro et al. (2013) argue that teaching science and technology in the MT for example Yoruba requires that a large body of scientific literature be developed in the MT using the appropriate scientific words, terms and phrases. They point out that teaching biology is more than translating concepts, terms and principles such as "pectoral girdle" or "odontoid process" to Yoruba. Equivalent concepts need to be developed from the Nigerian cultural context.

> Teaching biology in Yoruba language provides a forum for enhancing a deeper understanding and knowledge of the topic. More importantly is the fact that interest of the student is aroused when some terms like cranium, scapula, cervical, thoraxic etc. hitherto expressed in English are now translated with their functions into mother tongue and this leads to more concentration on the part of students. Many scientific concepts, terms principles which do not have equivalents in Yoruba language can be absorbed through transliteration in spellings such words using Yoruba alphabets and of course, incorporating the total nature of the language. (2013, p. 5)

According to Osungbemiro et al. (2013) the use of Yoruba as LoI contributed to the higher achievement of the experimental group. They conclude that the difficulties in teaching and learning science was taught and learned with relative ease using Yoruba language as LoI. The authors of the study conclude that despite the Nigerian government's commitment to education, the quality of education in Nigerian schools has been declining rapidly, thereby drawing the concern of successive governments.

Furthermore, the authors argued that if we are interested in inculcating the spirit of science in our students, if we do not want to be left behind in the global world for scientific and technological development, then it is imperative for all concerned science teachers, science educators, educational administrators, curriculum developers, authors and policy makers-to join forces in order to teach and learn science in a language that gives minimum difficulty to the students (Olarewaju, 1991) thereby making the desired results a reality (Osungbemiro et al., 2013, p. 8).

Language is crucial to the learning process inside and outside of schools. MT instruction is the tool of learning and therefore what tool could be more easy to use than the local language as a language in all forms for education. Language influences the thought process of the learner and his understanding of his environment. Consequently in the learning of science, deliberate efforts should be made to enable students to learn science in their MT.

A Case Study of Teaching Mathematics in Igbo Language

Despite the importance of mathematics as a key subject and compulsory for both pupils and students in primary and secondary schools, there is ample evidence of continued low interest in the subject by Nigerian pupils/students, which Igbojinwaekwu and Njeji (2012) attributed to teachers´s non-utilization of appropriate teaching techniques. According to Akinsola and Popoola (2004) quoted by Dorgu and Igbojinwaekwu (2014) many teachers in Nigerian schools use only techniques they know, even if such techniques are not relevant to the concept under discussion. They cited Opara (2004) who advised that teachers should evolve strategies that involve the learner's active participation and generate interest. Furthermore, Igbojinwaekwu and Nneji (2012) attributed the high failure rate of students in Senior School Certificate Examination to the structure of questions on the examination and to the fact that no study has reported the effect of using MT as LoI on academic achievement of pupils in primary school mathematics. Therefore Dorgu and Igbojingwaekwu investigated the role of using Igbo language to teach mathematics in public primary schools in Oshimili North Local Government Area of Delta State and found that teaching mathematics using Igbo language as a LoI resulted in higher academic student achievement than among students who were taught mathematics using English language as LoI.

The team research Dorgu and Igbojinwaekwu (2014) studied a sample comprised of 1,600 primary school pupils randomly assigned to experimental and control groups. The sample consisted of pupils whose local language was Igbo. Six teachers, each from six selected schools taught the pupils. Three teachers, of Igbo origin, underwent a training programme for 3 weeks on how to teach mathematics, using Igbo language as LoI, while the other 3 of non-Igbo origin, underwent a training program for 3 weeks on how to teach mathematics, using English LoI. This was done to enable the teachers to blend their pedagogy and content knowledge with the LoI to be used in the classrooms.

Also, uniform lesson notes, on the concepts of numbers and counting, were prepared for the teachers. This was to ensure that the 6 teachers teach exactly the same thing, using the same steps, method, behavioral objectives and study questions or evaluative questions. The 6 teachers used in this study had the same year of experience in terms of service year and were all Nigerian Certificate in Education (NCE) holders in Mathematics/Physics. According to Dorgu and Igbojinwaekwu (2014) the primary 3 pupils who were taught Mathematics using Igbo language understood the concepts very clearly and it was easy for the pupils since the language was not alien to the pupils (2014, p. 11). Furthermore they emphasize the importance of the three weeks crash training programme given to teachers who used Igbo language as LoI to teach pupils mathematics and recommended to focus on first the training of the teachers since teachers were not adequately trained in the use of MT along with the pedagogy and content knowledge to effectively teach the pupils/students (Dorgu & Igbojinwaekwu, 2014). They concluded that using the local language as LoI was associated with higher academic achievement than using English as LoI and it facilitated a better understanding of the subject compared with pupils taught in foreign languages. This study is crucial in revealing that teachers in Nigeria do not have programmes in place that offer training courses in the use of local LoI to enhance their instructional methods, making the policy of using Nigerian languages as LoI difficult to implement (Dorgu & Igbojinwaekwu, 2014). This absence of training courses neglects the results of studies from the 1970s that demonstrated the benefits of teaching and learning in a local language. Nigeria experimented with the benefits of using Nigerian languages in the 1970s with the Six-Year Primary Project (SYPP), a project that was based at the University of Ife with two experimental classes and one control class in a rural school. The experimental classes used Yoruba as LoI throughout the six years of primary education. The results of this project confirmed that those who have their total primary education in MT proved more resourceful and better academically prepared. "The SYPP children have demonstrated greater manipulative ability in their relationship to their colleagues, they also tend to demonstrate a great sense of maturity, tolerance and other affective qualities that make them integrate easily and readily with those they come in contact with" (Fafunwa et al., 1989, p. 141). In terms of pure academic attestation, results show consistently that the experimental group performed highest on tests in all subjects including English.

Furthermore Olarenwaju (2008) states that the process of teaching students to think in a foreign language does not help students to be creative; on the contrary it reduces them to "robots" who merely memorize the notes given to them by their teachers and reproduce the knowledge when required without demonstrating appreciable degree of understanding of the scientific and technological information and processes under consideration. He furthermore noted that if students merely memorize facts, principles and generalizations, which are only to be regurgitated during examination, they will not be in a position to use the knowledge acquired since it has not been internalized. He concludes that the lack of internalization of scientific knowledge, process and skills by Nigerian students seems to have been largely responsible for Nigeria's inability to make a major breakthrough in scientific and technological development.

In terms of academic learning skills as well, he notes that students taught to read in their MT acquire such skills more quickly. Fafunwa (1990) points out the following:

1. Imposed LoI is an important factor militating against the dissemination of knowledge and skills, and therefore directly impact the rapid social and economic well-being of the majority of people in Africa.
2. There tends to be a correlation between faster assimilation and the use of a foreign language as the official language of a given country in Africa.
3. No society in the world has developed in a sustained and democratic fashion on the basis of a borrowed or colonial language. These are functions of MT negation in education and also an outcome of cultural assimilation which is structured in the context of the use of imposed languages.

The Federal Government of Nigeria should encourage the use of native and foreign languages as LoI by placing more premium on graduates who are versed in the effective use of an additional native language to the official *lingua franca*.

In this vein, the literature have shown that when people feel that they are outsiders or when the majority of students experience linguistic alienation from the education system, social problems and conflicts intensify, disturbing the contribution of education to cultural identity, nation-building and sovereignty, all of which are at the core of the right to own development (Geo-JaJa, 2006; Geo-JaJa & Azaiki, 2010). In sum the colonization of the mind that takes place when science subjects are taught in foreign languages becomes the major constraint to human rights and sustainable development.

THE HUMAN RIGHT FRAMEWORK TO IMPROVE DEVELOPMENT AID AND TEACHING OF STEM SUBJECTS

Aid donors has served as a roadblock to the inter-linkages between development and human rights (Tomasevski, 2006), as well as that of voicelessness and educational poverty. Moreover, while the linguistic rights in education – like all human rights – are universal and inalienable, several conventions have led to their enshrinement in various constitutions as rights to the people, thereby placing binding responsibility on ratifying States (Babaci-Wilhite, 2015a).

Quality education is an essential foundation for nation building and its theoretical core means that the principled framework for a just distribution of educational opportunities should be build on the '4As' framework developed by Tomasevski (2006). This education roadmap specifies that education reform should be derived from the following:

Availability: education should be free and government funded with adequate infrastructure and teachers

Accessibility: systems should not discriminate and positive steps should be taken to reach the most marginalized
Acceptability: the content of education should be relevant, culturally appropriate and of high quality
Adaptability: education should respond to changing needs of society and to different contexts.

This framework ensures the fact that education is first and foremost the vehicle through which communities reproduce themselves, rather than a tradable commodity. The facts suggest that only a revaluation that is clearly derived from international human rights treaties or conventions, could lead to development and human rights mutually reinforcing each other for self-development and rights in quality education.

In many ways, as an essential feature of self-development, the content of education as practiced should be relevant, culturally appropriate and of quality defined in human rights terms. Education engagement in human rights makes collective action feasible and productive, responds to changing needs of society and to different contexts to create a more structured and extended quality of life. With this new-shared understanding, all tiers of society and government have an opportunity to rethink and pursue focused rights respecting education (Tomasevki, 2006; Babaci-Wilhite et al., 2012a). This is not simply a function of having access to schools but a schooling system that does not discriminate, with positive steps taken to reach the most marginalized. This is schooling with the right to social and indigenous context in which all people have the opportunity to improve their economic, political, social and cultural rights in their respective desired well-being. This conceptualization of schooling, which relates principles of justice as fairness and the capability approach (Sen, 1990) to education that empowers, is significantly different from education as conceived in conventional aid based on neoliberal approaches to improved education.

Education is not a static commodity and in this new ideological fervor, education that holds its own inherent value is at the center of individual and community asset building, which expands the real freedom, that nation's value for both social and economic sustainability (UNESCO, 2002, p. 32).

Essentially, rights in education refer to the ability of a nation to convert and sustain self-determined education without external reform. The importance of redefining education rights is that education should not be considered in isolation from its greater context. This means that including that which is indigenous is essential for inclusion, since just access alone is not sufficient to the exercise of agency, because of structural power relationships, which undermine the marginalized position. Education that assigns high priority to human rights in both aid allocation and the choice of aid modalities will delink aid as a driver and maintainer of economic inequality. These factors that allow individuals and nations to flourish take into account valuable capabilities (Babaci-Wilhite et al., 2013), enabling people and structuring the opportunities to live valuable lives.

HUMAN RIGHTS IN EDUCATIONAL AID EFFECTIVENESS

Countries of Africa have come a long way from education that served colonial needs to promoting education with little or no impact at all on society's reproduction. The insensitivity to the local or the discarding of local initiatives by neoliberal reforms has direct negative impacts on rights in education. The above roadmap of aid donors' which ignore the contribution of human rights in education for human development, implies that human capital formation, or the creation of a "softly structured society," fail to acknowledge that market politics has not been an effective anti-poverty and capacity-building roadmap for sustainable development (Geo-JaJa & Mangum, 2003; World Bank, 2008). This is why the authors argue that human rights must be integrated into aid for development and a right-capability approach should be the core in defining aid development. The essential truth is that neoliberal approach to education aid that has its origin in colonization that does not incorporate cultural and social rights in curriculum is often in conflict with local ownership. This roadmap, caught up in the conflict dynamics of globalization, privatization of the knowledge sector and the liberalization of schooling intensifies disparities in opportunities, exacerbates income polarization, and has pushed communities into underdeveloped (Silova, 2009). Indeed, unless a strong synergy emerges between local communities, and structured aid from both bilateral and multilateral institutions, little will be gained in terms of the functionality of education through aid.

Schooling today is not an instrument to promote and strengthen empowering citizens and parents to critical thinking, but it is used to reinforce dependency, poverty and discrimination that push societies towards instability. In sum, donor aid support is not linked to quality education, which is an end in itself and has tremendous benefits to individuals and society. This also reflects the fact that higher quality schooling requires not only education about human rights but education for human rights – citizens should be empowered to take action to ensure their human rights and the human rights of others in their community are respected, protected and fulfilled. Although there has been progress made towards this end, the overarching message from the 2011 EFA Global Monitoring Report is that targets in all totality will be missed by a wide margin (UNESCO, 2010). Donors tend to not allow space for localized education – education rooted in African culture and settings. At the same time, learners are not sealed off from global knowledge (Babaci-Wilhite, 2012). In tandem, such education achieves a synthesis of the learners value first and that of universal values second, in order to create context in the critical mass of knowledge (Crossley, 2010). These salient facts attest to the increasing problem of reaching children who remain excluded from education due to the complex nature of policy associated with neoliberal capitalism. Above all, linking education with marketization and privatization, as characterized by commodification of knowledge, and globalizing learning, has left learners with educational opportunities that are of poor quality, as schools function as part of an ideological donor apparatus designed for donor reproduction and market penetration.

The Paris Declaration of Aid Effectiveness of 2005 emphasized the importance of donor aid's morality. It also specified the goals of strengthening quality education, viewed as the core for enlarging opportunities, enhancing participatory freedom and meeting human development challenges. These are matters of equity and morality that should be central in any reform agenda but are widely missing in orthodox aid roadmaps (Sen, 1999, p. 110; Geo-JaJa & Yang, 2003). For Sen (1999), States should ensure rights in education that include elements of availability, accessibility, acceptability and adaptability as a common denominator of education reform in all its manifestation and at all education levels.

The distinction we make of rights to education and rights in education is important for elevating education aid its intended moral high ground. The criticisms of education frameworks apply more readily to the narrow concept of market fundamentalism than to the broader rights-capability, which is able to explain deprivations and indicators of distributive inequities.

Within a wider conversation, rights in education accounts for such intrinsic values as cultural, political and social backgrounds of localities by curricularizing students' and societies needs and concerns, as well as re-orienting nations. What we elude to here, is that culture and social rights in education aid is a source for combating capability deprivation as it underscores the basic philosophy of self-development and learning for virtuous circle.

AID EFFECTIVENESS FOR QUALITY EDUCATION

The quality of aid or its effectiveness in education, and not the issues of quantity mitigate quality education or the promotion of sustainable development. Consequently, the motive of aid should not just be to accelerate adaptation towards global North-like market economies; its main purpose should be to promote, protect and spread a wide range of adequate standards of well-being. In Sen's thinking, aid that concentrates on just economic growth alone, results in systematic variations between donor's authoritarianism and recipient's desired liberation. For more moral debate on the performability of development aid we refer the reader to work of Alesina and Dollar (2000); and Berthémlemy (2006), who strongly stress the effective of aid in efficiency terms. On the contrary Easterly et al. (2003) and Geo-JaJa and Zajda (2005) stressed the significant failure of aid to achieve education intrinsic missions. Indeed, these roadblocks of education aid compound the broader sense of inclusive education, which is a crucial foundation to break out of critical social and economic insecurities. Rather than aid fostering real development or prompting a resurgence of focus on education in localized form, it has only tended to a globalize focus resulting to different dimensions of human deprivation (Tandon, 2008).

The combined observations and interviews enabled to achieve a detailed contextualization of the failure of neoliberal classroom interaction. Klees (2010, 2013) in line with Geo-JaJa (2009, 2013) argues that the World Bank roadmap has been an educational disaster, harmful to children around the world and that focus on private sector and market-based approaches to education has been an instrument to serve the global job market (Babaci-Wilhite, 2015b). These observations' of classroom activities and narrow roadmaps in school reform crucial for quality learning but missing is supportive of our study results that aid roadmap in particular attention to which language is used in classrooms does affect learning outcome, participation and comprehension, and last but not the least quality classrooms. These observations gave a holistic overview of the learning problems contextualized in rights in education.

In the schematic history of aid, the literature seems to be consistent with the ideology (Market-based adjustment of rolling back the state), focus (Macroeconomic reform), type of aid delivered (aid with conditionalities), and dominant institutions (rise of NGOs) starting from the 1980s. The NGOs that have come to stand in for bilateral and multilateral aid donors also in focusing on efficiency to the exclusion of other considerations have only yielded severe mismatched knowledge and development (see World Bank, 2003). Examples abound from our study to support this assertion, as scaled-up aid has denied or underestimated the intrinsic values of Nigerian languages, local knowledge in Nigeria and universal principles of human rights (Babaci-Wilhite & Geo-JaJa, 2014).

Market centered reform in privatizing the knowledge sector and the liberalization of schooling intensify disparities in opportunities, exacerbates income polarization, and have pushed communities into underdeveloped, third world, and developing socio-economic crisis within rapidly changing global wealth. Indeed, unless a strong synergy emerges between local communities, and structured aid from both bilateral and multilateral institutions, little gain will be made in terms of the viability or functionality of education through aid. Reforms in Africa are being undertaken on the basis of an unrealistic agenda that is incorporating Western curriculum and using Western languages. The reasons for this have to do with misplaced associations of development with modernization, where emulation of Western development and Western educational systems are regarded as the way forward for Africa.

Scientifically speaking, this does not form a basis for capability-based educational development, nor does it bring social justice and quality in education. It is time to recognize the wealth of African knowledge and to promote its languages and knowledge in education. This would make a significant contribution to African development on its own terms and for the benefit of the majority of Africans. There is no doubt that Africa is in need of education reforms to improve quality education, however, in the light of research on African educational systems and on the experiences of the others research in Nigeria and other countries, the substitution of a local with a foreign LoI cannot contribute to improve the teaching and learning of STEM subjects.

There is a substantial body of research Fafunwa (1990), Alidou (2003), Desai (2004), Bamgbose (2005), Brock-Utne (2012), which shows that students learn more quickly and effectively when taught in a familiar language than when first taught in a foreign language. For parents demand for quality outcomes, rather than demanding for larger freedom and critical thinking, facilitates education's use as a tool to reinforce dependency, poverty and discrimination that push societies towards instability (Spreen & Vally, 2006). While recognizing that right to education is still associated with limited access to education for children in Tanzania, the use of English as a LoI has contributed to unbalanced economic growth due to the low quality of education. Such use of dehumanizing language is exemplified by the subjugation and extermination of Africans during colonialism, and also of a significant population of African Americans during the era of slavery.

This chapter notes that if there is to be anything like equality of opportunity, it is impossible to justify providing language-enhancing facilities for some and not for others or in education curriculum reform in Africa. Last but not the least, if education is to be universal and compulsory, equity requires that it should be free, and common sense demands that it should be delivered in local language. The latter is central to any discussion of development in a community and the promotion of indigenous knowledge at all levels of education, long enough for learners to secure human rights consciousness-raising. According to Freire elements of cultural and language identity in schooling are rooted in a colonial past, that deepen disparities of opportunity through language polarization led to dehumanization (Freire, 1970).

The authors believe that such irrational post-colonialist aid roadmaps have set out to destroy indigenous language, as well as efforts to reclaim dignity and humanization through revitalizing indigenous communities and indigenous languages, which are centered on traditional teachings regarding the significance of interconnectedness across generations. As scholars, we argue for a reassessment of the current curriculum program in Nigeria (Babaci-Wilhite, 2015b). As there is no systematic evidence supporting the idea that curriculum using English to teach STEM subjects has brought better achievement in schools or has resulted to quality education, providing more effective aid and localized language curriculum is an important part of the equation to sustaining broad macro-education stability and creating learning spaces for critical schooling improvement and that the goal of universality of education that does not short circuit sustainable human development nor seriously threaten the heterogeneity of learners' rights in education.

CONCLUSION

In this chapter we have argued that quality education should be rooted in cultural reaffirmation including local knowledge and language that are imperatives for the control of economic, social and political rights. Our results question the current, blind-faith reliance on alien languages as LoI. Equally important, the political motives of aid givers must be questioned. Our findings encourage us also to ask why some donors such as the United States spend large amounts of 'foreign aid' money on the promotion of English in developing countries instead of using it for funding basic literacy acquisition in local languages and generating quality educational materials in native languages. The fact still remains that only when the socio-economic authorities of recipients are restored can education aid move schooling towards the promotion of social equity and sustainable development.

Reflections and debate on aid support must be contextualized in the broader context of integrating international rights into education. This chapter calls for a new roadmap of aid that is effective and, regarding education in STEM subjects, one that embodies the dynamics of 'power shift' from donor to the receiver in Africa with a contextualized teaching and learning.

Based on the current weaknesses in development aid to education, the authors strongly believe that the larger freedoms of citizens and nations sovereignty are not promoted. Aid needs adjusting, rethinking and localization. There is a need for African nations to take jurisdiction over their education with focus on unique classroom and local realities and follow the kind of development framework that is endogenous for self-determination.

These are some of the roadblocks and challenges that squarely face donor efforts in an effort to facilitate the provision of human rights, efficiency and efficacy in education. Our study demonstrates that for education to be an inherent right and a right for self-determination, education reform must break with the past mechanisms of emancipation of communities from donor political, economic and cultural penetration. These we find are mechanisms for dependency and a continuation of aid exploitation and the civilizing mission in Africa. Indeed, current education reform is a tool for social reproduction and has not promoted local language as a vehicle to eliminate capability deprivation in Africa. The pre-eminent reform, by using English in curricula-structured education, alienates learners and stifles self-development. More problematically, it promotes economic and political institutions that are alien to society. Furthermore, it undervalues indigenous languages, undermines any co-existence of rights, and short-circuits all ideational processes.

REFERENCES

African Charter on Human and People's Right. (1981). Organization of African unity, African charter on human and peoples' rights ("Banjul Charter"), 27 June 1981, CAB/LEG/67/3 rev. 5, 21 I.L.M. 58 (1982). Retrieved September 26, 2012, from http://www.unhcr.org/refworld/docid/3ae6b3630.html

Akinsola, M. K., & Popoola, A. A. (2004). A comparative study of the effectiveness of two strategies of solving mathematics problems on the academic achievement of secondary school student. *Abacus: Journal of the Mathematical Association of Nigeria* (JMAN), *29*(1), 67–68.

Alesina, A., & Dollar, D. (2000). Who gives foreign aid to whom and why? *Journal of Economic Growth*, *5*(1), 33–63.

Alidou, H. (2003). Medium of instruction in post-colonial Africa. In J. W. Tollefson & A. Tsui (Eds.), *Medium of instruction policies: Which agenda? Whose agenda?* (pp. 195–214). Mahwah, NJ: Lawrence Erlbaum Publishers.

Alkire, S. (2002). *Valuing freedoms: Sen's capability approach and poverty reduction.* Oxford, England: Oxford University Press.

Alston, P., & Robinson, M. (Eds.). (2005). Introduction: The challenges of ensuring the mutuality of human rights and development endeavors. In *Human rights and development: Towards mutual reinforcement.* Oxford, England: Oxford University Press.

Babaci-Wilhite, Z. (2012). A right based approach to Zanzibar's language-in-education policy. Special issue on right based approach and globalization in education. *World Studies in Education, 13*(2), 17–33. (Australia: James Nicholas Publishers Pty. Ltd.)

Babaci-Wilhite, Z. (2013). An analysis of debates on the use of a global or local language in education: Tanzania and Malaysia. In D. B. Napier & S. Majhanovich (Eds.), *Education, dominance and identity* (Vol. 3, pp. 121–133). Rotterdam, The Netherlands: Sense Publishers.

Babaci-Wilhite, Z. (2014, July 14). *New perspective on language & literacy in Africa and Asia.* A lead paper presented at the 17th Annual National Conference of the Association for Promoting Nigerian Languages and Cultures, Owerri, Nigeria.

Babaci-Wilhite, Z. (2015a). *Local languages of instruction as a human right in education: Cases from Africa* (Vol. 36, pp. 1–154). Rotterdam, The Netherlands: Sense Publishers.

Babaci-Wilhite, Z. (2015b). *Language, Development Aid and Human Rights in Education: Curriculum Policies in Africa and Asia.* UK: Palgrave Macmillan.

Babaci-Wilhite, Z., & Geo-JaJa, M. A. (2011). A critique and rethink of modern education in Africa's development in the 21st century. Papers in Education and Development (PED). *Journal of the School of Education, 30,* 133–154. (*University of Dar es Salaam, Tanzania.*)

Babaci-Wilhite, Z., & Geo-JaJa, M. A. (2014). Localization of instruction as a right in education: Tanzania and Nigeria language-in-education´s policies. In Z. Babaci-Wilhite (Ed.), *Giving space to African voices: Right in local languages and local curriculum* (Vol. 33, pp. 3–19). Rotterdam, The Netherlands: Sense Publishers.

Babaci-Wilhite, Z., Geo-JaJa, M. A., & Lou, S. (2012a). Education and language: A human right for sustainable development in Africa. *International Review of Education, 58*(5), 619–647. doi:10.1007/s11159-012-9311-7

Babaci-Wilhite, Z., Geo-JaJa, M. A., & Lou, S. (2012b). Human rights in development experience in Africa: The foreign aid and policy Nexus in OECD and China Aid. *World Studies in Education, 13*(1), 79–90. (Australia: James Nicholas Publishers Pty. Ltd.)

Babaci-Wilhite, Z., Geo-JaJa, M. A., & Mwajuma, V. (2013). Nordic aid and the education sector in Africa: The case of Tanzania. In C. A. Brown (Ed.), *Globalization, international education policy, and local policy formation* (Vol. 5, pp. 81–106). Dordrecht, The Netherlands: Springer Publishing.

Bamgbose, A. (2005). Mother-tongue education: Lessons from the Yoruba experience. In B. Brock-Utne & R. K. Hopsen (Eds.), *Language of instruction for African emancipation: Focus on post -colonial contexts and considerations* (pp. 231–257). Cape Town, South Africa: CASAS and Mkuki na Nyota Publishers.

Berthémlemy, J. C. (2006). Bilateral donors' interest vs. recipients' development motives in aid allocation: Do all donors behave the same? *Review of Development Economics, 10*(2), 179–194.

Brock-Utne, B. (2000). *Whose education for all? The recolonization of the African mind.* New York, NY: Falmer Press.

Brock-Utne, B. (2012). Language policy and science: Could some African countries learn from Asian countries? *International Review of Education, 58*(4), 481–503.

Crossley, M. (2010). Context matters in educational research and international development: Learning from the small states experience. *Prospects, 40*, 421–429. doi:10.1007/s11125-010-9172-4

Desai, Z. (2004). Starting a research project: Some lessons to be learnt. In B. Brock-Utne, Z. Desai, & M. A. Qorro (Eds.), *Researching the language of instruction in Tanzania and South Africa* (pp. 117–131). Cape Town, South Africa: African Minds.

Dorgu, T. E., & Igbojinwaekwu, P. C. (2014). *Ibgo as language of instruction and academic achievement of pupils in primary school mathematics.* Faculty of Education, Niger Delta University, Wilberfore Island, Bayelsa State, Nigeria.

Easterly, W., Levine, R., & Roodman, D. (2003). Low investment is not the constraint on African development. *Economic Development and Cultural Change, 51*(3), 547–571.

Fafunwa, B. A. (1990). Using national languages in education: A challenge to African educators. In Unesco-Unicef (Ed.), *African thoughts on the prospects of education for all* (pp. 97–110). [Selections from papers commissioned for the Regional Consultation on Education for All, Dakar 27–30 November 1989.] Dakar, Senegal: Breda.

Fafunwa, B. A., Macauley, J., & Soyinka, J. (Eds.). (1989). *Education in mother tongue: The Ife primary education research project.* Ibadan, Nigeria: University Press Limited.

Freire, P. (1970). *Pedagogy of the oppressed.* London, England: Penguin Books.

Geo-JaJa, M. A. (2006). Educational decentralization, public spending, and social justice in Nigeria. *International Review of Education, 52*(1–2), 129–153.

Geo-JaJa, M. A. (2009). *Can globalization in Nigeria's Niger delta be humanized for integration and development? Language and power. Implications of language for peace and development.* East Lansing, MI: Michigan State University Press.

Geo-JaJa, M. A. (2013). Education localization for optimizing globalization's opportunities. In S. Majhanovich & M. A. Geo-JaJa (Eds.), *Economics, aid and education: Implications for development.* Rotterdam, The Netherlands: Sense Publishers.

Geo-JaJa, M. A., & Azaiki, S. (2010). Development and education challenges in the Niger Delta in Nigeria. In M. A. Geo-JaJa & S. Majhanovich (Eds.), *Education, language, and economics: Growing national and global dilemmas* (pp. 53–69). Rotterdam, The Netherlands: Sense Publishers.

Geo-JaJa, M. A., & Mangum, G. (2003). Economic adjustment, education and human resource development in Africa: The case of Nigeria. *International Review of Education, 49*(3–4), 293–318.

Geo-JaJa, M. A., & Yang, X. (2003). Rethinking globalization in Africa. *Chimera, 1*(1), 19–28.

Geo-JaJa, M. A., & Zajda, J. (2005). Rethinking globalization and the future of education in Africa. In J. Zajda (Ed.), *International handbook of globalization, education and policy research.* Amsterdam, The Netherlands: Kluwer Academic Publishers.

Igbojinwaekwu, P. C., & Nneji, N. L. (2012). Restructured mathematics questions and academic achievement of students in senior school mathematics. *African Education Indices, 4*(1), 14–23.

Klees, S. (2010). Aid, development and education. *Current Issues in Comparative Education, 13*(1), 3–28.

Klees, S., & Qargha, O. (2013). The economics of aid: Implications for education and development. In S. Majhanovich & M. A. Geo-JaJa (Eds.), *Economics, aid and education: Implications for development* (pp. 15–29). Rotterdam, The Netherlands: Sense Publishers.

Osungbemiro, N. R., Olaniyan, R. F., Sanni, R. O., & Olajuyigbe, A. O. (2013). *Use of indigenous languages as media of instruction in teaching of biology* (A Case Study of Some Selected Secondary Schools in Ondo West Local Government), Onda State, Nigeria. Paper presented at the Cambridge Business and Economics conference, Cambridge, England.

Olarenwaju, A. O. (2008). *Using Nigerian languages as media of instruction to enhance scientific and technological development: An action delayed.* Lagos, Nigeria: Fafunwa Foundation Internet Journal of Education. Retrieved June 16, 2011, from http://fafunwafoundation.tripod.com/fafunwafoundation/id10.html

Samoff, J. (2003, re-edited 2007). Institutionalizing international influence. In R. F. Arnove & C. A. Torres (Eds.), *Comparative education: The dialectic of the global and the local* (pp. 52–91). Lanham, MD: Rowman & Littlefield Publishers.

Sen, A. (1999). *Development as freedom.* Oxford, England: Oxford University Press.

Silova, I. (Ed.). (2009). *Private supplementary tutoring in Central Asia: New opportunities and burdens.* Paris, France: UNESCO Institute of International Educational Planning (IIEP).

Spreen, C. A., & Vally, S. (2006). Education rights, education policies and inequality in South Africa. *International Journal of Educational Development, 26*(4), 352–362.

Tandon, Y. (2008). *Ending aid dependence.* Geneva, Switzerland: Fahuma books, South Center.

Tomasevski, K. (2006, August). *The state of the right to education worldwide free or fee: 2006 Global Report.* Copenhagen

UN (United Nations). (1948). *Universal declaration of human rights.* Adopted by the General Assembly on 10 December 1948. Geneva: United Nations.

UNESCO (United Nations Educational, Scientific and Cultural Organization). (1997). Declaration on the Human Genome and Human Rights (11 November 1997) UNESCO is currently evaluating the impact of the Declaration worldwide.

UNESCO (United Nations Educational, Scientific and Cultural Organizaiton). (2002). *Report 2002. Education for All - Is the world on track?* Retrieved from http://www.unesco.org/new/en/education/themes/leading-the-international-agenda/efareport/reports/2002-efa-on-track/

UNESCO (United Nations Educational, Scientific and Cultural Organization). (2010). *EFA Global Monitoring Report 2010. Education for All: Reaching the Marginalized.* Paris, France: Oxford University Press.

UNHCHR (United Nations High Commissioner for Human Rights). (1989). *Convention on the rights of the child.* Geneva, Switzerland: United Nations.

World Bank. (2003). *Annual review of development effectiveness. The effectiveness of bank support for policy reform 2004.* Washington, DC: The World Bank.

World Bank. (2005). *Global monitoring report 2005.* New York, NY: Oxford University Press.

World Bank. (2008, December 16–18). *The evolving allocative efficiency of education aid: A reflection on changes in aid priorities to enhance aid effectiveness.* Paper prepared for the World Bank High-level Group Meeting on EFA, The World Bank, Washington.

Zehlia Babaci-Wilhite
Graduate School of Education
University of California, Berkeley
USA

Norwegian Center for Human Rights
University of Oslo
Norway

Macleans A. Geo-JaJa
David O. McKay School of Education
Brigham Young University
USA

Steve Azaiki
The International Society of Comparative Education
Science & Technology, Yenagoa
Nigeria

GREGORY JOHNSON

AFTERWORD: ETHNOMATHEMATICS AND THE GEOMETRY OF ART

There is
Within
Each symbol
A life
Word/ a
Message
Indelibly linked
To
It's
people/
culture/
The individual nature/
(D'Ambrosio
Knew
This/
Like a manner
Of speech/
Mathematic/
Linguistic/
Set in the
Consciousness/
The
Spirit poem
Of
A
Continent/
That is
Hidden in the nature/
The stroke of a pen/
The splash
Of
Geometric
Scrawl/
As/is

Z. Babaci-Wilhite (Ed.), Human Rights in Language and STEM Education, 271–272.

Locked
Within
The ethno positive
Mode
Of
Expression

ART

A single note/
Demystified
By
Child logic/
Repeated
Until
A
Number
Form
Becomes a key/
With
The potential of
Ideas
(Collaboration

To liberate
The
Generations

ABOUT THE AUTHORS

Steve Azaiki is Professor of Agronomy with a PhD from Ukrainian Agricultural University. Azaiki was the President World Environmental Foundation For Africa (WEFFA) from (1993–2003) and Coordinator of the National Think Tank Nigeria (2007 to date). He has held numerous appointments, including serving as Commissioner and Secretary to the Bayelsa State Government of Nigeria (2002–2006). He is member Governing Council, Niger Delta University, Wilberforce Island, Bayelsa State (2013 to date) and also the President of the International Society of Comparative Education, Science and Technology Nigeria and Chairman of the 2014 & 2015 ISCEST Conference. President of the Institute of Science and Technology, Yenagoa, Nigeria and Visiting Professor/Fellow to a number of Institutes including Institute of Petroleum Studies, University of Port Harcourt, Rivers State; International Personnel Management Academy, Kiev, Ukraine; Ukrainian Agricultural University (Department of Phytopathology) Kiev, Ukraine; Sovereign Trust Insurance Plc; Bayelsa State Development & Investment Cooperation (BDIC); Bayelsa State Agricultural Development Board. Azaiki has published widely and recently received the Award of honorary professor by his Alma mater, Ukrainian National University of Life and Environmental Sciences, Ukraine, in September 2015.
Email: azaiki1@gmail.com and azaiki1@yahoo.com

Zehlia Babaci-Wilhite is a Visiting Assistant Professor at the University of California- Berkeley and an Adjunct Professor at the University of San Francisco, California, USA. She has taught courses and workshops in the USA, Norway, Japan, India, France & Nigeria on issues related to language and culture, development & human rights. Her current research interests include Language and Science Literacy as a Human Right in Education, Development Aid, African Higher Education and Technology Applications in Education. She is the author of many published articles in tier one academic journals, book chapters and has previously published three books on *Language, Development Aid and Human Rights in Education: Curriculum Policies in Africa and Asia* (2015, Palgrave Macmillan), *Local Languages as a Human Right in Education, Comparative Cases from Africa* in the series: Comparative and International Education: A diversity of Voices (2015, Sense Publishers), and *Giving Space to African Voices: Right in local languages and local curriculum* in the series: *Comparative and International Education: A diversity of Voices* (2014, Sense Publishers). She is fluent in French, English, Norwegian, Japanese and Berber with knowledge of Arabic, Portuguese, German, Spanish, Swahili and Igbo.
Email: z.b.wilhite@berkeley.edu and z.b.wilhite@nchr.uio.no

Alison K. Billman is a researcher and principal investigator at the Lawrence Hall of Science at the University of California, Berkeley. Her professional and research

interests focus on designing efficacious contexts for *reading to learn* in primary grade classrooms. She is particularly interested in developing integrated curriculum contexts and informational texts that support kindergarten and first grade students as they read to learn. At the Lawrence Hall of Science she she leads research focused on extending an integrated literacy and inquiry science curriculum to kindergarten and first grade and research focused on the design of informational texts for beginning readers. She received a Master's of Elementary Education degree from Edinboro State University of Pennsylvania and a PhD in Educational Psychology from Michigan State University.
Email: alison_billman@berkeley.edu

Inga Bostad is currently Director of the Norwegian Centre for Human Rights, University of Oslo (UiO) and associated Professor of Philosophy. Her research interests are philosophical pedagogy, disability studies and human rights. She has previously served as Vice-Chancellor at UiO (2006–2013), chaired the National Commission on General education and Liberal Arts in Higher Education, and the Bostad Committee – a Government appointed Committee for new Objective clause for primary and secondary education and Kindergarden in Norway. Bostad is also member of the Executive Committee of AHRI (Association of Human Rights Institutes), the council of the Global Advisory Board, Human Dignity and Humiliation Studies, and a board member of Roskilde University and the European Wergeland Centre. She chairs the Board of University of Oslo Human Rights Prize Committee.
Email: inga.bostad@nchr.uio.no

Birgit Brock-Utne is affiliated to the University of Oslo, Norway, as a Professor in Education and Development. In 1997 she built up a master's degree course of comparative and international education, of which she was a director for ten years. Before that she was a Professor of Education at the University of Dar es Salaam, Tanzania, for four years. She was the Norwegian coordinator of the Language of Instruction in Tanzania and South Africa project (see www.loitasa.org). She is now in the leadership group of the NORAD supported project TRANSLED. She has been a Visiting Professor at several universities in the US, at the University of Otago in New Zealand, the University of Hiroshima in Japan, at the European University of Peace in Austria and Universitat Jaume Primero in Spain and also works as a consultant around the globe. She has her own consultancy company, EDCON (Education and Development Consulting) (see: http://www.edcon.no). She has worked for NORAD, MFA, SIDA, DANIDA, UNICEF, UNESCO, DEZA, DSE, the Namibian Ministry of Education, Save the Children and the World Bank. In 2011 she was the President of BAICE (the British Association of Comparative and International Education). She has written, co-authored and coedited many books, chapters in books and articles within the field of peace studies, education and development and languages in Africa, among these: *Whose Education for All? The Recolonization of the African*

Mind (Falmer Press 2000, republished in 2006 in Korea). She speaks Norwegian, German, English, Kiswahili and French.
Email: birgit.brock-utne@iped.uio.no

Salaeya Butt has many years of experience of conducting research on various education policy issues such as equitable access to education, medium of instruction and assessment practices. She has been responsible for the complete project cycle from writing proposals; grant management, research design, data collection and analysis to writing and editing of the final report for publication with expertise in both qualitative and quantitative research methods. Butt holds a Master's in International Development Management from University of Westminster, London.
Email: salaeya.butt@cqe.net.pk and butt.salaeha@gmail.com

Ubiratan D'Ambrosio is an Emeritus Professor of Mathematics, State University of Campinas/UNICAMP, São Paulo, Brazil (retired in 1994). Received the degree of Doctor (Mathematics) from University of São Paulo/USP, Brazil, in 1963, with a thesis on Geometric Measure Theory and the Calculus of Variations. Currently, Professor of Mathematics Education and History of Mathematics at UNIAN/Universidade Anhanguera de São Paulo, Brazil. He is member of the *Academia de Ciências do Estado de São Paulo/ACIESP* (1975), Elected Fellow of the AAAS/American Association for the Advancement of Science with the citation "For imaginative and effective leadership in Latin American Mathematics Education and in efforts towards international cooperation."(1983), Awarded the Kenneth O.May Medal of History of Mathematics, granted by the International Commission of History of Mathematics/ICHM, affiliated of the IUHPS and IMU (2001), Awarded the Felix Klein Medal of Mathematics Education, granted by the International Commission of Mathematics Instruction/ICMI, affiliated of the IMU (2005), Elected Member of the *International Academy of the History of Science* (2014) and Received the International Mathematics Education Achievement Award pelo *Teachers College Program in Mathematics* da *Columbia University*, New York (2014).
Email: ubi@usp.br

Fatima Dar earned her PhD from the School of Education and Cognitive Science, Asia e University, Malaysia. She also has an MA in Curriculum and Teaching from Teachers College, Columbia University, USA and an MA in English from Government College University, Lahore. She has served as a Teacher Educator, consultant and faculty with various organizations including Ali Institute of Education, Lahore, Lahore Grammar School, Society for the Advancement of Higher Education (SAHE), FAST, Aga Khan University, Karachi, Beaconhouse National University, Lahore and IoBM, Karachi. She has also worked with ESRA-USAID as a consultant for the Professional Development component on Action Research. She has also worked on a number of teacher training, curriculum and materials development projects all over Pakistan. She has also been a visiting tutor at the Institute of Education,

University of London and University of Warwick and was awarded the Charles Wallace Trust visiting fellowship for the year 2000 to the Institute of Education, University of London She has also presented papers on aspects of teacher education and English Education both nationally and internationally. Her publications, national and international, also discuss the current issues and aspects of English language education and teacher education. At present she works as an Assistant Professor, Department of Education at SZABIST, Karachi.
Email: fatimadar@yahoo.com

Aled Dilwyn Fisher is a Doctoral Research Fellow at the Faculty of Law, University of Oslo, as part of the European Joint Doctorate in Law and Development (EDOLAD). His research interests are ecology, human rights and development, with his PhD focusing on India. He has previously worked at the Norwegian Centre for Human Rights, and has published articles on climate change and human rights, indigenous peoples' rights and other rights-related issues. In addition to his academic work, he is an active trade unionist in the Norwegian Civil Service Union and involved in various environmental campaigns. He is originally from Cardiff, Wales, and has a bachelor's degree in history and international relations (London School of Economics and Political Science) and a master's degree in human rights (Norwegian Centre for Human Rights).
Email: a.d.fisher@jus.uio.no

Macleans A. Geo-JaJa is Professor of Economics and Education in the Department of Educational Leadership, at Brigham Young University. As an experienced educator and researcher, he combines the scholar's critical approach and the experience of a researcher in making educational development and enlarging opportunities for sustainable development. He studies and teaches about economics of education, poverty and system-wide change in education and development in the world of development practitioners. As a Fulbright Senior Specialist Fellow, Geo-JaJa serves as technical and program adviser to universities in different regions of the world. He is the author of numerous articles published in tier one journals and book chapters on human rights in development aid, structural adjustment and decentralization in education, and human development in globalization. He serves on the Advisory Board of the International Review of Education Journal of UNESCO and numerous other top tier Journals. He is a Guest and visiting Research Fellow at Zhejiang Normal University in China. He has thought in many universities around the world. He is a consultant to numerous international organizations – The World Bank, DFID, UNDP, USAID, and others. Geo-JaJa recently was appointed a member of the Nigerian National Think Tank.
Email: geo-jaja@byu.edu

Megan Goss is a senior curriculum developer with the Learning Design Group at the Lawrence Hall of Science, University of California, Berkeley. She designs

instructional sequences for the Learning Design Group's integrated science and literacy curriculum. In addition she leads several teams of designers and participates in research endeavors through the University, and presents at conferences. A former elementary teacher, Goss also provides professional development opportunities for teachers nationwide.
Email: mgoss@berkeley.edu

Page Hersey, Ed.D., has been an educator in the San Francisco Bay Area for over 15 years as a classroom teacher, literacy specialist, and teacher educator. She holds an MA in Language and Literacy from San Francisco State University and a doctorate in Multicultural Education from the University of San Francisco. She places Bay Area teachers and students at the center of her research, highlighting voices from classrooms that connect local and global issues. Currently Hersey divides her time between working in schools as a literacy and professional development coach and a professor of education teaching courses in literacy and social studies education. Hersey is a passionate human rights educator and integrates issues of human rights and social justice into her work with teachers and students. She is proud to serve on the steering committee of Human Rights Educators-USA, a national organization dedicated to expanding and strengthening human rights education. As a teacher consultant with the Bay Area Writing Project she conducts professional development workshops with local teachers and had the privilege to work as a writing teacher in Seoul, South Korea for three summers. She has presented her research at the American Educational Research Association's annual conferences and conducted numerous workshops and seminars at home and abroad.
Email: pagehersey@gmail.com

Lanette Jimerson, Ph.D., is the program director of the Multicultural Urban Secondary English (MUSE) Credential and Masters Program at the University of California, Berkeley. Her research focuses on teacher development and writing instruction with a particular emphasis on the contexts that support teachers in locating spaces of agency to counteract the impact of traditional writing instruction. Prior to becoming the program director of MUSE, Jimerson worked as a program manager for the Stanford Center for Assessment, Learning and Equity. In this role was a key member on the team that designed the Smarter Balanced Assessment Consortium (SBAC) Writing Performance Tasks for students. She also was charged with developing an induction level teacher performance assessment for the state of Ohio. In addition to her university research, as a practitioner, Jimerson is a teacher-consultant and former co-director of the Bay Area Writing Project. She was a co-author for the National Writing Project's "Writing Assignment Framework and Overview" and the chapter "Developing Domains for Multimodal Writing Assessment: The Language of Evaluation, the Language of Instruction." At the community level Lanette serves on the board of ForWords Literacy Lab, a non-profit that supports the literacy development of English Language Learners. Her

research and work in the community has been recognized through numerous awards including the 2003 BWOPA Educator of the Year, 2014 California State University, East Bay Young Alumna of the Year and the 2015 Educator of Excellence Honoree.
Email: ljimerson@berkeley.edu

Gregory Johnson is an Artist/Researcher from Berkeley, California (USA), who received a Degree in Art History at the College of Wooster, Wooster, Ohio (USA). Johnson continued his study at the California College of Arts and Crafts (CCAC) in Oakland, California; at the Vancouver Art Institute, Vancouver, British Columbia (Canada), and at Madison Commercial Art School at Madison, Wisconsin (USA). He additionally worked for COVANCE Research Labs as a Study Coordinator while working on several Art projects. Among his Art projects, Johnson contributed to a mural project for the National Democratic Party Government in BC (1980s), a grant mural project for city arts in Madison and Dane county arts counsel with youth (1994), and an Art on sight and sound for the Wisconsin counsel for the blind funded by the Koehler Foundation in (2000). Currently Johnson works with the Geometry of Arts linked to Ethno-Mathematics as well as being affiliated as a Field Manager with the Grassroots Campaign Incorporated (GCI), an Advocate for the Southern Poverty Law Center (SPLC) located in Berkeley, California.
Email: grmanisme@yahoo.com

Ellen-Rose Kambel is the co-founder and director of the Rutu Foundation for Intercultural Multilingual Education. She holds a law degree and received a Doctorate in Social Sciences from the University of Leiden (the Netherlands). Originally from Suriname, she has worked closely with indigenous peoples, governments and donor agencies on building capacity on indigenous rights. Since 2011, she is the founder and director of the Rutu Foundation, an NGO aimed at supporting mother tongue education worldwide (www.rutufoundation.org). She has published several articles relating to language revival and mother tongue based bilingual education. Prior to Rutu, she worked for 15 years as an international human rights and development consultant with the ILO, UNHCR, Inter-American Development Bank and the Rainforest Foundation.
Email: ellen-rose@rutufoundation.org

Emmanuelle LePichon-Vorstman is assistant professor at the department of Modern Languages at Utrecht University (The Netherlands). She is particularly interested in the relation between the development of language skills and verbal communication in school context. At present, she leads a research project founded by the European Commission on the integration of newly arrived migrant children at elementary and secondary school in Europe (EDINA project). She has been involved in the European Comenius project TRAM, 'Transitions and multilingualism'. The goal of this project was to provide preschool and primary school teachers with skills that would allow them to better accompany children with

different ethnic backgrounds and mother tongues. In May 2014, she was invited to come to Paramaribo (Suriname) to train teachers there. She is the author of several scientific papers on plurilingualism and at present divides her time between linguistic research and teaching.
Email: E.M.M.lePichon@uu.nl

Jabari Mahiri is the inaugural holder of the William and Mary Jane Brinton Family Chair in Urban Teaching. He directs the TEACH Project (Technology, Equity, And Culture for High-performance schools), a research initiative that collaborates with urban and international schools and community partners on uses of new media for increasing student achievement and educational equity, and for improving teacher professional development. He is Faculty Director of the Multicultural Urban Secondary English Credential and Master's Degree Program and also Faculty Director of the Bay Area Writing Project. He is a member of the College Writing Advisory Board as UC Berkeley, a Senior Scholar for the National Urban Alliance for Effective Education, and an elected member of the American Educational Research Association governing council. He received the Chancellor's Award for Advancing Institutional Excellence, the Chancellor's Community Service Award, the AERA Outstanding Mentorship Award, the Leon Henkin Equity and Inclusion Award, the Berkeley Graduate Assembly Faculty Mentor Award, and the SMASH Prep Mentorship Award from the Level Playing Field Institute. Mahiri is author or editor of seven books including *Digital Tools in Urban Schools: Mediating a Remix of Learning* (2011), *The First Year of Teaching: Classroom Research to Improve Student Learning* (2014), and *Deconstructing Race: Micro-cultures Shifting Multiculturalism* (Forthcoming, 2016).
Email: jmahiri@berkeley.edu

Sam Mchombo received his PhD in Linguistics in 1978 from the University of London (SOAS). From 1972–1984 he pioneered studies of Chichewa and Linguistics in the University of Malawi (Chancellor College). He was the first Chair of the (current) Department of African Languages & Linguistics. In 1984 he held appointment as visiting Fulbright scholar in Linguistics at MIT (spring), at Stanford University (summer), then as post-doctoral fellow at MIT (fall). He has taught courses on African linguistics, linguistic theory, Chichewa, and Swahili at San José State University (1985-1988), University of Limpopo (South Africa) (spring 1997), University of London (spring 2003), and at Universidad de Sonora, in Mexico. Since 2009 he has held appointment as honorary professor in the African Studies Program at Hong Kong University, teaching courses during the summer in African linguistics and Swahili. He has delivered linguistics presentations in many countries including France, Germany, Mexico, Norway, South Africa, and published articles on sports, the media, religion and politics in Malawi, and linguistic theory. He is the author of *The Syntax of Chichewa*, Cambridge University Press (2004), editor of *Theoretical Aspects of Bantu Grammar,* CSLI, Stanford (1993), and was guest editor of the

journal *Linguistic Analysis* for a special issue on African linguistics (1999). Since 1988 he has been on the faculty of the University of California, Berkeley, where he is Associate Professor of African Languages and Linguistics in the Department of African American Studies.
Email: mchombo@berkeley.edu

Helen Min is a Curriculum Specialist on the literacy team for The Learning Design Group. She completed her PhD at the School of Education, University of California, Berkeley in the Language, Literacy, Society and Culture Program. Her research interests focused on immigrant students and their academic and social experience of learning English in public schools. Prior to graduate school, Min was a public school teacher.
Email: helenmin@berkeley.edu

Irfan Muzaffar's work has been expanded to embrace teacher education, education research, education policy analysis, programme design, and evaluation in a variety of cultural contexts that include Africa, South Asia, and the United States. Director at the Education and Social Research Collective (ESRC). As a teacher-educator, Irfan has conducted mathematics education courses at the Ali Institute of Education, Lahore and Michigan State University, USA. His current research interests centre on the history and philosophy of education reforms in general, and citizenship education in Muslim countries in particular. He regularly offers seminars on research methods and connections between research and policy. Muzaffar´s academic contributions on education policy issues related to public-private education debates have been widely published.
Email: imuzaffar@gmail.com

Basil Amarachi Okonkwo is a PhD candidate in philosophy at Nnamdi Azikiwe University, Awka, Anambra State, Nigeria. Okonkwo did his B.A and M.A in Philosophy at Imo State University, Owerri, Imo State, Nigeria. Members of Sons of Mary Mother of Mercy Congregation, he hails from Uwakonye Aniche Arondizuogu in Onuimo Local Government Area, Imo State, Nigeria. His works include: The Human Condition in Jean Paul Sartre Master's Thesis, 2008. He publication includes: Hermeneutics in Gadamer: the Quest for Understanding in Philosophy, Amamihe: Journal of Applied Philosophy Vol 7, 2009, the Relevance of Culture in Nation Building (Journal of Nigeria Languages and Culture Vol. 12, 2010, the Adumbration and Language use in Jerome Ikechukwu Okonkwo: The Phenomenological Method Explained in Issues in Languages and Human Development; A Festchrift for Jerome Ikechukwu Okonkwo). In 2014 he published The Importance of Promoting Mother Tongue in Nigeria. Russian American Education Forum: An online journal, Vol. 6, Issue: 3, 2014.
Email: basilonline2@gmail.com

P. David Pearson is a faculty member in the programs in Language and Literacy and Cognition and Development at the Graduate School of Education at the University of California, Berkeley, where he served as Dean from 2001-2010. With degrees from UC Berkeley (BA in history) and Minnesota (PhD in education), his research focuses on reading instruction and assessment practices, with special attention to disciplinary literacy. With colleagues at Lawrence Hall of Science, he conducts research and development work on integrating language and literacy into inquiry based science. He has received many awards for his nearly 50 years of research and service to reading research, most notable the William S. Gray Citation of Merit by the International Reading Association, the Oscar Causey Award from the Literacy Research Association, the Distinguished Contributions to Research in Education Award from the American Educational Research Association. In 2001, the Literacy Research Association established the P. David Pearson Scholarly Influence Award to honor research that impacts practice and policy. Professor Pearson received his B.A. in History from the University of California at Berkeley, after which he taught elementary school in California for several years, and went on to complete his Ph.D. in Reading Education at the University of Minnesota. He completed post-doctoral study at the University of Texas, Austin and Stanford University.
Email: ppearson@berkeley.edu

Abbas Rashid has been engaged in education policy analysis for many years and is currently the Executive Director of the Society for the Advancement of Education (SAHE), a Lahore-based institution which has been active in the field of education research and advocacy for over 30 years. He is also the convener of the Campaign for Quality Education (CQE). He has a particular interest in the areas of language and learning, privatization of education and education for citizenship, especially in the context of Pakistan. His work on these issues has been published at home and abroad. He also contributes frequently on matters relating to education quality and equity in the media.
Email: abbasrh@gmail.com

Ladislaus Semali is professor of Education at the Pennsylvania State University, specializing in adult literacy education, comparative and international education and non-Western educational epistemologies. Professor Semali has extensive experience in organizing internships and workshops for students in Africa, and is currently directing research on the crisis of interest in STEM science in East Africa. Dr. Semali's research and teaching interests include Higher Education, Comparative and International Education, Adult Education, communications media and the use of technology to enhance and develop curriculum. He teaches undergraduate and graduate courses in Language and Literacy Education; Media Literacy in the Classroom; Perspectives in Intercultural Communication in Multicultural settings; Cross-Cultural Research Methods in Education; Issues in Indigenous and Ecological Literacies; Perspectives in African Education; and courses in Comparative and

International Education. His work has been published in the *International Review of Education* and *Comparative Education Review*. He is author of *Literacy in Multimedia America* (Routledge/Falmer), *Postliteracy in the age of democracy* (Austin & Winfield) and editor of *What is Indigenous knowledge: Voices from the academy* with Joe Kincheloe (Garland).
Email: lms11@psu.edu

Jeremiah Sims is a PhD Candidate in Language, Literacy, and Culture at UC Berkeley's Graduate School of Education where he also earned a B.A. in Rhetoric with Honors and an M.A. in Education. During his doctoral studies, he received the Chancellor's Fellowship. He has been a lecturer for a number of courses in the Education and Rhetoric Departments including co-instruction of the graduate course, Urban Education. Additionally, he is an adjunct Instructor for Berkeley City College and a faculty member on special assignment for the four campuses of the Peralta Community College District where he works to create a holistic transfer and matriculation program for African American and Latino males through the Office of Student Success and Equity. He is Founding Director for the MAN UP Program that is the focus of this chapter.
Email: jsims@berkeley.edu

Carrie Strohl draws on years of K-12 and university teaching experience to engage in educational research, curriculum development, and professional leadership in science and literacy education. Before pursuing her doctorate in Science Education at UC Davis, she was a member of the Learning Design Group at the Lawrence Hall of Science at UC Berkeley. Her research interests include classroom discourse, special education and science learning in school gardens.
Email: castrohl@ucdavis.edu

Brad Washington is an Assistant Professor in the International and Multicultural Education Department in the School of Education at the University of San Francisco. Dr. Washington's areas of research focus on the global historical tensions of language shift, sociolinguistics, and linguicism in the 21st century. In partnerships with schools and organizations in the San Francisco Bay Area, Dr. Washington was granted funding for the purpose of promoting literacy and advocacy for newcomer communities in the United States. In 2015, Dr. Washington was elected to a 3-year term to become the coordinator of the Bay Area Teachers of English to Speakers of Other Languages chapter for the State of California. Prior to his current post, Dr. Washington spent 6 years working with the University of California Berkeley to secure Title VI federal funding from the U.S. Department of Education for the purpose of promoting study abroad and multilingual education initiatives for post baccalaureate students in the University of California System. His current work spans linguistic educational development in North America and Asia.
Email: washingtonb@usfca.edu

CPSIA information can be obtained
at www.ICGtesting.com
Printed in the USA
LVOW01s0046280317
528654LV00002B/12/P

9 789463 004039